BEYOND SOUTHERN SKIES

BEYOND SOUTHERN SKIES

RADIO ASTRONOMY AND THE PARKES TELESCOPE

Peter Robertson

CAMBRIDGE
UNIVERSITY PRESS

For Stuart

Published by the Press Syndicate of the University of Cambridge
The Pitt Building, Trumpington Street, Cambridge CB2 1RP, UK
40 West 20th Street, New York, NY 10011-4211, USA
10 Stamford Road, Oakleigh, Victoria 3166, Australia

First published 1992

Printed in Melbourne by Brown Prior Anderson Pty Ltd

National Library of Australia cataloguing in publication data
Robertson, Peter.
Beyond southern skies, radio astronomy and the Parkes Telescope.
Bibliography.
Includes index.
ISBN 0 521 41408 3.
1. Parkes Radio Telescope – History. 2. Radio telescope. 3. Radio astronomy – New South Wales – Parkes – History. I. Title.
522.682099445

Library of Congress cataloguing in publication data
Robertson, Peter.
Beyond southern skies, radio astronomy and the Parkes Telescope.
Peter Robertson.
Includes bibliographical references and index.
ISBN 0 521 41408 3.
1. Radio telescopes – Australia – New South Wales. 2. Radio astronomy – Australia – New South Wales. I. Title.
QB479.2.R63 1992 92-3077
522'.682'099445 – dc20 CIP

A catalogue record for this book is available from the British Library
ISBN 0 521 41408 3 hard back

Contents

Foreword

by Sir Bernard Lovell

The core of this book concerns one of the great instruments of modern science—the radio telescope at Parkes in New South Wales. This instrument began probing the heavens in 1961 and for more than thirty years has been the envy of the postwar corpus of radio astronomers. Indeed, like the subject itself, the Parkes telescope is a legacy of the war.

Historians may well contend that radio astronomy began with the discovery by Karl Jansky in the early 1930s of radio waves reaching the earth from regions of the universe beyond the solar system. But neither Jansky, nor Grote Reber the other pre-war pioneer, created this new science. That remarkable feature of the contemporary scene—the opening of a new window for the observation of the heavens—is a postwar creation of young people released from their wartime tasks. The science emerged simultaneously in England and Australia in almost identical circumstances. In England the key discoveries were made by those released from the Telecommunications Research Establishment or still within the Army Operational Research Group, and in Australia the foundations of the new science over a broad front were laid by the young wartime scientists of the CSIRO.

By the 1950s the new techniques for studying the universe became the science of radio astronomy, recognised to be of such importance that rapid developments occurred in many countries. This book is a salutary reminder of the humble beginnings of this revolutionary discipline and, in particular, of the small group in CSIRO who, in the late 1940s, made discoveries of such importance that the classical astronomers were forced to recognise that the universe was of a complexity beyond any understanding that could be derived from the observations by optical telescopes.

It was a fortunate and accidental legacy of a wartime dispute in the Telecommunications Research Establishment that in those critical years E. G. (Taffy) Bowen was in Australia as the head of the Radiophysics Laboratory of CSIRO. Bowen was one of the small group of pioneers who in the few years before 1939 developed the coastal chain of radar stations that played the vital role in the Battle of Britain. He then achieved the astonishing feat of producing a miniaturised version for the RAF night fighters. In September of 1940 he journeyed to America as a member of the Tizard mission to introduce the cavity magnetron to American scientists. The Parkes telescope is a direct result of Bowen's pioneering spirit and of his wide circle of colleagues—for, as will be evident to any reader of this book, the engineering problems are one aspect only of the construction of a major scientific instrument.

Those who derive assessments of scientists from lists of their publications will find little evidence of Bowen's significance in the development of postwar astronomy. It is, therefore, of especial value that this book places so clearly on record that the Parkes telescope is his creation. A host of scientific reputations have been made by those using this telescope but the significance of the instrument goes far beyond those researches. In the diplomatic battle that occurred in England throughout the late 1950s and early 1960s as to whether the UK should join with the Europeans in the formation of the European Southern Observatory or with the Australians in the construction of a large optical telescope in Australia, the eminence of Australia in radio astronomy was a crucial factor. The scientific judgment of the importance of creating a large optical facility to parallel the work of the Parkes telescope could not be questioned. The Anglo-Australian agreement of 1967 led to the 150 inch Anglo-Australian telescope at Siding Spring and to the 48 inch UK Schmidt on the same site. These telescopes, in parallel with the radio astronomical studies at Parkes and by other groups in Australia, have transformed our knowledge of the southern sky. They make an epic success story of twentieth century astronomy.

Indeed, there were 'other groups' in Australian radio astronomy. Until 1960 the CSIRO group was dominant, yet expansion and success creates its own problems and divisions of opinion. Not everyone agreed with Bowen's plan for the Parkes telescope and even those who agreed in principle were unhappy with the plan to build the largest possible. The splinter groups, particularly those of Bernard Mills and Chris Christiansen who moved to the University of Sydney, were of great importance to Australian and world astronomy. The sad case of Joe Pawsey who so dominated the early years and decided to move to the United States before his lamented death is a less happy story, as is that of the relation of the Australian radio astronomers with Martin Ryle and his group in Cambridge. In the free exchange and international collaborative scientific world of the 1990s it is hard to imagine that isolationism of the early years in radio astronomy. But the isolationism of Cambridge and the internal difficulties of the CSIRO group are parts of history that have had such a significant effect on the radio astronomical world of today and it is good that they are related here with such accuracy and sympathy.

Accuracy is a welcome aspect of this book. In an age when scientists are only too ready to claim to be first, and when a historical perspective is so often lacking, this book is a great corrective. Whilst rightly recognising and giving credit to the great achievements of the Australian group, the international perspective is maintained. The interpretation of the techniques and results of modern science in a form which is both readable and accurate is a most difficult art. *Beyond Southern Skies* is one of the few books that pass these most rigorous tests.

Sir Bernard Lovell

Jodrell Bank, Cheshire

Preface

My first glimpse of the Parkes telescope occurred during a holiday to Queensland with my father. The inland route from Melbourne involves long distances between small towns, with few natural features that stand out in the landscape. Perhaps that is why the sight of the telescope from the highway created such a lasting impression. It was not so much the size of the structure, more the contrasting, incongruous image the telescope creates as it first appears among the rolling plains.

Several years later my interest in the history of science prompted me to learn more about the telescope. To my surprise I found that the story of Parkes had not been written down in any detail. It seemed odd in view of the great public interest in the telescope. Indeed, for many people the telescope is the very symbol of Australian science.

After deciding to write this book my plan was to divide it into two parts, the first dealing with the planning and construction of the telescope, the second surveying its contributions to astronomy. Inevitably my research drew me further back in time to the origin and early development of radio astronomy in Australia. The structure of this book reflects these three distinct, though closely related topics.

Part 1 looks at the beginnings of radio astronomy, starting with the American pioneers of the 1930s and then the postwar expansion of this fledgling science in countries such as Australia, Britain and the Netherlands. The Australian group at the CSIRO Radiophysics Laboratory in Sydney rapidly grew to be the largest, the most generously funded, and the most diversified in the range of new problems it tackled. Numerous articles and several books have already been published on the early history of radio astronomy; my aim here is to put the Australian contribution into a new and more coherent perspective. The four chapters which comprise Part 1 are dedicated to Joe Pawsey, the founding father of Australian radio astronomy.

Part 2 tells the story of the funding, design and construction of the telescope during the 1950s. It was Australia's first successful venture into the world of big science, one which was to have major repercussions for the radio astronomy group. While the telescope's completion in 1961 marked a new beginning, it also brought to a close what might be called the golden era of Australian radio astronomy. The material in these chapters has been based on the correspondence files and documents held in the CSIRO Archives in Sydney and Canberra. Part 2 is dedicated to Taffy Bowen, the builder, whose drive and determination made the telescope a reality.

The achievements of the astronomers at Parkes are reviewed in Part 3. Over one thousand research papers have been produced from

Parkes during the past thirty years, so I make no claim that my account is a comprehensive or balanced overview of such an extraordinary variety of topics. Instead my aim has been to highlight a number of general themes which, at least in my view, have been the most interesting and fruitful. Part 3 is dedicated to John Bolton, the first director at Parkes, who did more than any other to make the telescope such an outstanding success.

In the preparation of this book I have benefited significantly by my association with the Department of History and Philosophy of Science at the University of Melbourne. Although I have written this book with both astronomers and historians of science in mind, my ambition has been to make the Parkes story accessible to a much wider circle of readers. For this reason I have attempted to keep the amount of technical material in the text to a minimum. References and other source material have been relegated to notes at the end of the text. Occasionally, a theme of special interest, but which has no immediate bearing on the story, is developed in more detail in a separate boxed item, referred to as a Tangent. Time will tell whether I have bridged the gap between the expectations of the academically oriented reader on one side and the general reader on the other. In Denmark, they have a saying which points out the fate of someone who tries to sit between two stools.

Throughout this project I have had the good fortune to be helped by a large number of people, many of whom have played a prominent role in the development of radio astronomy. My sincere thanks go to those who took part in interviews, who responded to my questionnaires and who commented on an earlier draft: Taffy Bowen, Ron Bracewell, Ron Burman, Jim Caswell, Chris Christiansen, Dave Cooke, Brian Cooper, George Day, Frank Gardner, Ben Gascoigne, Linden Gillbank, Robert Hanbury Brown, Cyril Hazard, Rod Home, Dick Hunstead, Ken Kellermann, Max Komesaroff, Sir Bernard Lovell, Dick Manchester, Don Mathewson, Bernard Mills, Doug Milne, Ray Norris, Jack Piddington, V. Radhakrishnan, Grote Reber, Jim Roberts, Brian Robinson, John Shimmins, Bruce Slee, Rick Twardy, Kevin Westfold, Sir Fred White, John Whiteoak, Paul Wild, Alan Wright and Don Yabsley.

In particular I am especially grateful to three of my reviewers for their extensive comments, constructive advice and generous encouragement. Woody Sullivan made many helpful suggestions which, combined with his own publications on the history of radio astronomy, have been both invaluable and inspirational. Harry Minnett patiently guided me through the complex technical issues and sequence of events involved in the design, construction and upgrading of the telescope. John Bolton also made a major contribution by providing me with a wealth of information not readily found in official records or in the literature. Even though I have benefited a great deal from so much expert advice, the responsibility for any remaining errors of fact or interpretation is entirely my own.

At various stages I have been fortunate to have the support of a number of my colleagues in CSIRO. Among them I would like to pay tribute to Peter Alston and Susan Reed for tracking down numerous and often obscure references; Robert Leydon for his skilful editing of the manuscript; John Masterson, Franz Spranger and Vonnie Thompson for their assistance with the photographs and illustrations; Michael Moran and Colin Smith at the CSIRO Archives in Canberra; Mary Morris and Chris Toshack for their keyboarding skills; and John Thawley and Paul Reekie for their advice and encouragement.

I owe a special thanks to Kim Greene who keyed the manuscript and then cheerfully put up with a seemingly endless series of revisions. Above all, I am indebted to Sally Atkinson who helped me find my way among the extensive collection of material held in the Radiophysics Archive.

Last but not least, *Beyond Southern Skies* would not have been possible without the loving support of Diane, Katie and Laura.

Peter Robertson

Melbourne

PART ONE

JOE PAWSEY

THE FOUNDER

CHAPTER ONE

In the Beginning

> Twinkle twinkle radio star
> We can't see you: you're too far
> Now we hear from outer space
> How you run your cosmic race
>
> Lord De L'Isle
> at the inauguration

The official opening of the Parkes telescope on Tuesday, 31 October 1961 marked a special day for science in Australia. An idea first conceived ten years earlier, the telescope had taken four years to design and a further two years to construct. From the outset the project had been coordinated on an international level. A substantial part of the finance had been provided by American philanthropic foundations, the design had been carried out by British engineers, the construction by German steelmakers and the overall planning of the project directed by CSIRO's Radiophysics Laboratory.

The site for the telescope, a dry and dusty plain in central New South Wales, provided a sharp contrast to the offices and boardrooms in New York, London and Sydney where plans for the telescope had taken shape. A short distance south of the site, the small town of Parkes looked forward to the opening of the telescope with great excitement. Not only had one of the marvels of modern science been erected in its own backyard, but Australia's Governor-General, Lord De L'Isle, was to perform the opening and combine it with a visit to Parkes. Some locals said this was the most important event in the life of the town since the discovery of gold in the area eighty years before.

At precisely 11 on the morning of 31 October a piercing sound from the siren at the Parkes Fire Station announced to all that the Governor-General's Convair had touched down at the nearby airport. The town was well turned out for the occasion. Most businesses had closed their shutters for the day. Flags and decorations flapped in a strong wind, cars crowded the main street, and people edged forward onto the roadway trying to catch sight of the approaching Governor-General's party. The official motorcade arrived amidst the cheers and stopped outside Tattersalls Hotel where the mayor of Parkes stepped forward to greet

Lord De L'Isle. After a wreath-laying ceremony at the war memorial, the official party made its way to a civic reception at the town's sports oval. De L'Isle thanked the gathering of several thousand schoolchildren and adults for their warm welcome:

> 'We are happy to be here on such an auspicious occasion to take part in an event of world importance in the scientific field—the inauguration of the most modern and powerful radio telescope in the world. This instrument will attract the attention of scientists the world over and its importance in helping to unravel the secrets of space cannot be overestimated.' [1]

The procession of official cars then drove out of town and through the rolling wheatlands to the site of the telescope, set in the shallow Goobang Valley twenty kilometres to the north. At isolated points along the way small groups standing on the roadside cheered the passing motorcade. Almost 500 official guests had already gathered at the site. Some had made the long drive from Sydney, Canberra or Melbourne, but most had arrived on chartered flights. For several hours the small Parkes airstrip had seemed like a major air terminal with aircraft arriving at 15 minute intervals. The guest list read like a 'who's who' of Australian science, and also included a large number of politicians, senior public servants and representatives of the companies that had designed and built the telescope. Many of those unable to attend sent telegrams of congratulations instead. One came from Bernard Lovell, in charge of the Jodrell Bank radio telescope in England, the first of the very large dishes to be built: 'All at Jodrell Bank greet our Australian colleagues on this great occasion. We are delighted that a big dish has been built in the southern hemisphere and we look forward to close collaboration in the future.' [2]

Before the ceremony commenced the visitors inspected the telescope, clambering along gangways and out on the catwalks directly beneath the surface of the dish. A photographic display specially set up in the tower building recorded the construction of the telescope and showed some of the wide variety of other radio telescopes in operation around the world. By the time the vice-regal party had arrived, the guests who had already lunched in marquees pitched near the telescope sat waiting in front of the official dais. The weather conditions could hardly have been worse. The temperature climbed into the mid-thirties and gale-force winds gusting as high as 80 km per hour lashed the site, whipping up thick clouds of red dust and drowning out the sound from the public address system. Ironically, one reason for selecting the Parkes site had been its excellent record of low wind speeds!

Press, radio and television reporters surrounded the dais as the chairman of CSIRO, Frederick White, welcomed the guests and introduced the Governor-General. De L'Isle began by reading a message of congratulations from Prince Philip, the Duke of Edinburgh:

> 'The development of radio astronomy has been one of the great events of modern science, and the scientists of CSIRO have played a leading part in this important work. Their studies of the sun and the exploration of extra-terrestrial space to the remotest parts of the universe have gained the admiration of astronomers everywhere and have brought great credit to Australia.
>
> Today the radio astronomers of Australia are to be rewarded by the inauguration of a magnificent radio telescope which will enable them to penetrate still further the secrets of the cosmos. This occasion marks the culmination of much planning and hard work and I send you my best wishes for the success of your investigations with this new instrument.'[3]

The theme of De L'Isle's speech was the way in which the giant radio telescope had come about as the result of a combination in Australia of several distinct, but favourable circumstances. The next speaker was Donald Cameron, the science minister in charge of CSIRO. Cameron commenced by saying that the inauguration was a special event for CSIRO because it marked the fulfilment of a large project in which a great deal of thought and planning of the highest quality had been invested. The main topic of the speech was one that has often dominated the politics of Australian science. Whereas a major part of CSIRO's work lies in the promotion of applied scientific research, it is also of great importance to support pure research in areas such as astronomy which may not yield any immediate economic benefit to Australian society: 'Many discoveries when first made seemed to be of little immediate significance and yet out of them have come things which have profoundly affected human life and events. What I think we can say is happening here is a measure of the growth and expansion of Australian science, and in a way too of the growth and development of the nation.'

Cameron then introduced his predecessor as science minister, Sir Richard Casey. Casey had helped launch the telescope project and played a central role in securing funds from both private and government sources. His speech presented a brief history of radio astronomy, a new science which had developed at breakneck speed during the postwar years. The plan at this stage of the proceedings had been for Casey to call on the Governor-General to press a button which would slowly tilt the telescope's 500 tonne dish from the upright position down to its lowest point facing the assembled guests. Unfortunately, the strong gusty winds far exceeded the design safety limit. Driving the telescope would have involved a high risk of serious damage to the gear and bearing mechanisms. In a last minute arrangement, Norm Broten, a visiting Canadian radio astronomer, together with Cliff Smith, a foreman at the site, made their way up to the small aerial cabin perched high above the centre of the dish and unfurled a large Australian flag.

The wind tore at the flag, holding it in an almost perfectly extended position and bending the makeshift flagpole over into a dangerous arch. Broten and Smith gingerly made their way back to ground level, both shaken by the climb.

To close the ceremony an outline of the future plans for the telescope was given by E. G. 'Taffy' Bowen, chief of the CSIRO Radiophysics Laboratory. More than any other factor, it had been Bowen's energy and enthusiasm that had made the telescope a reality. It is worthwhile reproducing his words in full:

> 'Not often in the history of science in Australia are we as highly privileged as we are today in being present at the commissioning of this magnificent scientific instrument—an instrument which is destined to extend our vision and understanding of the mysteries of outer space.
>
> Since the dawn of history, man has gazed at the stars and wondered at the mystery and immensity of the universe. He did this first with his naked eye and then with crude lenses no better than a burning glass. Over the years his understanding has improved and his instruments have become more powerful, culminating in the great 100 and 200 inch optical telescopes of the northern hemisphere.
>
> Only in the past twenty or thirty years have radio telescopes appeared over our horizons, and today, less than ten years after its conception, we have this great 210 foot instrument—unique in the southern hemisphere and with a performance unlikely to be matched for many years to come.
>
> Radio telescopes exist in a bewildering array of types, but this one is regarded as the basic type: it may be steered to any part of the sky, it may be used over a wide range of frequencies and for an infinite variety of astronomical researches. The instrument we have commissioned today is the prime example of its kind in the world at the present time, and for a generation it is likely to remain a key instrument in our search for the ultimate truth about the universe.
>
> What shall we find and what purpose will it serve? If we knew what we would find, we would not be looking. Over the years there have been many theories and much speculation about the nature and origin of the universe, but there is very little evidence of a factual nature. There are, in fact, many more questions than there are answers. We do not know whether the universe is bounded, or whether it reaches to infinity in every direction; whether it has a beginning and an end, or whether it will have a perpetual existence; whether there are intelligent beings on other planets. The questions go on and on and it is our profound belief that this instrument and instruments like it are the ones which will reveal the ultimate truth about our universe.
>
> If we do not at present know the answer to these questions, we certainly know where we should look. In the first place the telescope will be used to study in much greater detail those parts of the universe already known to us in a crude way—the planets of our solar system, our local galaxy and the nearer external galaxies. But the really exciting task will be to explore the

most distant reaches of the universe, far beyond the range of any previous instruments, radio or optical, and at distances which have never yet been probed by mankind. The signals from these regions will have travelled five thousand million years or more before reaching us—signals which began their journey before this earth of ours even existed. In this way we shall be looking back into the history of the universe itself and so may discover how it had its origin and how it all began.

The existence of this instrument has attracted the interest of our fellow astronomers the world over. When the first urgent tasks of the next few months have been performed, it will be thrown open to astronomers from every part of the world with worthwhile problems to attack. We pledge ourselves to work in close collaboration with our fellow scientists from any country and will make this observatory a rallying point for equipment and information, for techniques and intellects which, together, will lead to further great advances in the science of astronomy.

It has truly been said that the search for truth is one of the noblest aims of mankind and there is nothing which adds to the glory of the human race or lends it such dignity as the urge to bring the vast complexity of the universe within the range of human understanding.

It is now my pleasant duty to thank Your Excellency for so graciously undertaking the inauguration of this great telescope, and to thank our other distinguished speakers for the part which they have played.'

A feature common to each of the speeches delivered at the opening ceremony was unbridled optimism. As the most technically advanced instrument of its kind in the world, the giant radio telescope at Parkes would open up new areas of astronomical research and enable Australian astronomers to stay at the forefront of radio astronomy, an exciting new field of science they had done so much to pioneer.

To understand how the idea for the Parkes telescope came about, and how it came to be built, we need to go back to the earliest days of radio astronomy. In Australia, radio astronomy grew out of the wartime research and development of radar by CSIRO's Radiophysics Laboratory. At the end of the war, with secret radar work for the military no longer required, the Radiophysics Laboratory cast about for new peacetime areas of research to pursue. With a large and highly skilled staff and an ample store of suitable equipment, the Laboratory made an accelerated start in the new science of radio astronomy.

The story of the Parkes telescope really begins with the wartime activities of the Radiophysics Laboratory. But before we take up that story in Chapter 2, it is interesting to take a brief look at the origins of radio astronomy itself, in particular the contributions by the pioneers Karl Jansky and Grote Reber. Although the research by these two American radio engineers formed the first two landmarks in the history of radio astronomy, their work in the 1930s was largely overlooked. It was not until immediately after the Second World War that radio astronomy rapidly grew into one of the most exciting areas of modern science.

A Mysterious Cosmic Hiss

> There is plenty to speculate about, isn't there? I've got to get busy and write another paper right away before somebody else interprets the results in my other paper in the same way and steals the thunder from my own data.
>
> Karl Jansky in a letter to his father, 1932

> According to legend, someone has a bright idea called a theory. Apparatus is assembled to test this theory. If the result is positive, a discovery is made. Such an idealistic scenario does happen once in a blue moon. Usually, the drama consists of an observant and diligent person noting some anomaly in an ongoing set of measurements. The anomaly may be trivial and of no importance to the intended measurements. However, if the observer has curiosity and resourcefulness, the unexpected anomaly may turn out to be of far greater interest than the intended measurements.
>
> Grote Reber on Jansky's discovery

Somewhere between patchy and non-existent is probably how one would describe the coverage of science by the Australian news media over the years. Even so, the inauguration of the Parkes telescope did reasonably well. On a local level, the Parkes *Champion Post* understandably made the most of the occasion with a four-page spread which reported the text of most of the speeches and gave a detailed account of the day's events. The news diarist found added drama to report as the Governor-General's party left the telescope site at the end of the day. Speeding towards the airport the rear tyre on one of the motorcycle escorts blew out at 100 km per hour. The patrolman managed to avoid serious injury bringing the bike to a halt only by some very skilful riding.[4] On a national level, several newspapers featured the inauguration with stories and photos. In the *Sydney Morning Herald* only one item received more coverage. The main story brought news that the Soviet Union had exploded a 30 megaton hydrogen bomb, only a day after a similar report on the explosion of a 50 megaton device.

For the public at large these two stories perhaps summed up all that was good, all that was bad about science. On the one hand, the H-bomb was the most powerful expression of the destructive forces science could unleash, forces that threatened the future of humanity; on the other, the Parkes telescope symbolised the opposite, the search for fundamental truths about the universe, truths that would enrich and uplift the spirit of all humanity. The train of events which led to these two newspaper stories on the first morning of November 1961 had their origins in two scientific discoveries made almost thirty years earlier. In February 1932 James Chadwick at the Cavendish Laboratory in England announced the discovery of the neutron, a fundamental particle that provided the key to understanding the structure and behaviour of the atomic nucleus. The importance of the discovery was recognised immediately and earned Chadwick the Nobel Prize for physics in 1935. The new field of nuclear physics developed rapidly leading to the discovery of nuclear fission, the construction of the first A-bomb and then to the far more destructive H-bomb. One month before Chadwick's announcement of the neutron, from the other side of the Atlantic in New Jersey, a young radio engineer had also announced the discovery of a new phenomenon—the detection of radio waves from space. Although it took much longer to realise its significance, Jansky's discovery also marked the birth of a major new science.

Unwittingly Jansky succeeded where others before him had failed. The first attempt to detect radio waves from space now goes back over one hundred years. In 1887 laboratory experiments by the German physicist Heinrich Hertz conclusively demonstrated the existence of radio waves, which had been predicted earlier by James Maxwell in his theory of electromagnetic radiation. These experiments by Hertz showed that radio waves are simply another type of electromagnetic radiation, differing from the ordinary light visible to our eyes only in that their characteristic wavelengths are very much longer. Besides opening up a range of potential applications in communications, the discovery of radio waves raised an obvious question—are there any natural sources of radio emission on the earth, or from the stars above?

The first attempt to detect radio waves from the sun was made shortly afterwards in 1890 by the brilliant American inventor Thomas Edison. By then Edison had already achieved worldwide fame from some of the hundreds of ideas he eventually patented in a lifetime of invention. Edison planned to suspend a long loop of telephone wire around the perimeter of a large deposit of iron-ore in a New Jersey field. Edison reasoned that if radio waves are produced from violent electromagnetic disturbances on the sun, their arrival on earth might magnetise the iron-ore and set up a small flow of electric current in the telephone wire. Old records show that the wooden poles for the telephone wire were delivered to the field, but there is no evidence that this huge radio receiver was ever erected. With the benefit of hindsight, we know that this attempt to detect radio waves from the sun could not have succeeded. First, even this large iron-ore detector would not have been sufficiently sensitive (devices to amplify weak electric currents had not yet been invented) and, second, radiation of wavelengths long enough to be picked up by this device cannot penetrate the earth's ionosphere, a reflecting layer unknown at the time consisting of electrons in the atmosphere at heights above 60 km.

The English physicist Oliver Lodge made the next attempt in an experiment carried out in 1894. Lodge constructed equipment in his Liverpool laboratory to receive radio waves at wavelengths short enough to penetrate the ionosphere but, similar to Edison's attempt, the sensitivity of the equipment proved completely inadequate. Even if the sensitivity had been improved, the radio interference from electrical machinery in industrial Liverpool was already strong enough in the 1890s to force Lodge to abandon the search. A third, but again unsuccessful attempt was made in 1901 by a young French scientist Charles Nordmann, who decided to move well away from industrial sources of radio interference by setting up an aerial on a glacier high in the French Alps. Nordmann guessed correctly that radio emission from the sun, if it did exist, might vary with the solar cycle and, as a result, that the emission might originate from the violent disturbances associated with sunspots. Unfortunately Nordmann got cold feet, perhaps literally, and gave up the search after only a few days. With more perseverance he might well have succeeded as 1901 turned out to be a year of minimum sunspot activity.[5]

In the same year as Nordmann's alpine search for radio emission from the sun, Guglielmo Marconi carried out his historic demonstration of transmitting radio signals clear across the Atlantic. Radio waves, like any form of electromagnetic radiation, travel in straight lines, yet somehow Marconi's result seemed to defy the natural curvature of the earth. This led Arthur Kennelly (Edison's assistant at the time of the unsuccessful search in 1890) and Oliver Heaviside in England to predict the existence of a layer high in the atmosphere which reflected radio waves. It was not, however, until the mid 1920s that experiments by Edward Appleton in England and others in the United States convincingly demonstrated the existence of the ionosphere. Intensive investigations then followed on the basic properties of this ionised layer so that its effect on long-distance radio communication could be better understood and predicted. At the same time the 1920s saw a research and development boom in radio technology at numerous universities and in some of the growing industrial giants in the United States and Europe. Oddly enough, even with this rapid progress in understanding the physical and engineering aspects of radio, no one made a systematic attempt to take up where Edison, Lodge and Nordmann had left off at the turn of the century.

Grote Reber's remarks on Jansky's discovery, quoted above, point out that often a new field of science starts as a result of good fortune rather than theoretical insight, serendipity rather than design. So it was with radio astronomy. Karl Guthe Jansky grew up in a small university town in Wisconsin and developed a strong interest in radio science, encouraged by his father who was a professor of electrical engineering. After graduating in physics Jansky joined the Bell Telephone Laboratories in 1928 and was assigned the task of investigating any source of atmospheric static which might interfere with a new system of ship-to-shore and trans-Atlantic radio communication then under development at Bell. At Bell's field station near Holmdel, New Jersey, Jansky constructed an aerial array 30 metres long and 4 metres high from timber and brass pipes. The array was mounted on four wheels taken from an old model T Ford, with a small motor and chain drive which could turn the array through one revolution every twenty minutes. The contraption earned the name 'merry-go-round'.

Jansky was able to distinguish three distinct types of radio static. The first arose from the intermittent crashes of local thunderstorms and the second was a weaker, steadier static due to the combined effect of many storms far off in the atmosphere. The third type was composed of a very weak and steady hiss of unknown origin. As he wrote in his monthly work report for January 1932: '... a very steady continuous interference—the term 'static' doesn't quite fit it.' [6] Initially Jansky thought that this weak hiss would be traced to some source of industrial interference, but then he noticed that the maximum strength of the signal came from a direction which moved around the sky each day and seemed to correspond roughly with the position of the sun. Confident that he had succeeded where others thirty years earlier had failed, Jansky presented a paper on his discovery in April 1932 to a conference of radio engineers in Washington in which he concluded that 'the source of this disturbance was somehow associated with the sun.' *

The observations continued throughout 1932 and Jansky found that with the passing of the months, contrary to what he believed, the direction of the static began to drift further and further away from the position of the sun. Obviously the sun could not be the source of the radio noise. The daily period of variation of the noise in fact turned out to be 23 hours and 56 minutes, four minutes less than the daily period of the sun. This is known as the sidereal day—the period of the earth's rotation with respect to the stars—and so Jansky could conclude that the source of the noise must lie beyond the sun, and beyond the solar system as well. In another paper, published early in 1933 in the British journal *Nature*, Jansky could now cautiously state that 'the source of this noise is located in some region that is stationary with respect to the stars.' [8] Further, the strongest source of this 'cosmic static' seemed to come from the centre of our galaxy, located in the direction of the constellation Sagittarius.

When time permitted Jansky continued observations for another year or so, publishing his final paper on the subject in 1935. The distribution of the cosmic static across the sky was shown to be disc-like and to approximately coincide with the distribution of stars visible to us along the Milky Way. Because he had been unable to detect radio waves from the sun itself, Jansky ruled out the far more distant stars as the source of this cosmic radiation. Instead, he speculated that the source of this new radiation arose from charged particles in rapid motion through the vast stretches of interstellar space, an idea that is basically correct.

* Jansky published an account of his discovery in the *Proceedings of the Institution of Radio Engineers* in 1932, the first in a burgeoning bibliography of papers on radio astronomy which now numbers well in excess of 100,000 publications (see note 7).

Karl Jansky and the 'merry-go-round' aerial used to discover radio waves from space.

The publicity department at Bell Labs made sure that Jansky's discovery received wide exposure. In May 1933 the *New York Times* carried a front-page report with the banner 'New radio waves traced to center of Milky Way', and similar reports appeared in other newspapers and magazines across the United States. Two weeks later a radio station featured a special evening program in which radio astronomy went on air for the first time. Cosmic static picked up by Jansky's aerial in New Jersey was relayed to the New York station, and the 'hiss of the universe' then broadcast to a national audience.

Despite Jansky's research papers and the national media coverage of the discovery, not a great deal of notice was taken where it mattered most—the astronomical community. Several astronomers got in touch with Jansky, a research paper was written on a theory of the cosmic static's origin, but basically no one seemed to know what to do next. For radio engineers, Jansky's discovery was peripheral and of interest only as far as the effect cosmic static might have on long-distance radio communication. For astronomers, it was of course an intriguing discovery but the radio techniques required to confirm and extend Jansky's work were completely beyond their professional ken. At this stage, perhaps an enterprising physicist could have bridged the gap between the fields of astronomy and radio engineering but, by and large, most physicists were much too preoccupied with the startling pace of discovery in their own field. The year 1932 has been called the *annus mirabilis* of physics. In addition to the neutron mentioned earlier, discoveries were made of the first particle of antimatter (the positron), the hydrogen isotope of mass 2 (the deuteron), and the first types of linear and cyclotron particle accelerators for probing subatomic matter came into operation. All eyes focused on the new glamour field of nuclear physics.[9]

Jansky did not pursue his discovery any further. From the point of view of Bell Labs, Jansky had completed his original objective in building the merry-go-round, finding the intensity and directional properties of static noise, and cataloguing the effects the noise might have on radio communication. There is evidence that Jansky planned to build a much larger aerial specifically designed to study the cosmic static, but the proposal was shelved—too expensive and too far outside the primary business of the Laboratories. Jansky continued his work on sources of radio interference, a research program he carried out until he died of a stroke at the age of 44.

Jansky's discovery opened up an entirely new approach to astronomy. Until 1932 almost our entire knowledge of the sun, the planets, the stars and the universe beyond had been gained because our eyes are sensitive to a small part of the electromagnetic spectrum—the optical region extending from the red through to the violet. Radiation from astronomical objects in this part of the spectrum penetrates the earth's atmosphere without appreciable absorption, and so this part of the spectrum provides a 'window' through which we can look out into the universe. At wavelengths shorter than violet or longer than red light, electromagnetic radiation is absorbed by water vapour and other molecules in the earth's atmosphere. From the laws of physics, astronomers had known that stars give out most of their energy in the visible part of the spectrum, and so this inability to study radiation from space beyond the red or the violet had never been considered too serious a limitation. With the continual improvement in the size and power of telescopes over many years, astronomers had been able to build up a remarkably detailed picture of the countless millions of stars and star systems and of the many astrophysical processes at work in space.

The new window opened up by Jansky provided an opportunity to observe the universe in a completely new 'light'. The radio window is about a thousand times wider than its visible or optical counterpart, covering a range of wavelengths from roughly a centimetre in length up to several tens of metres. The lower cut-off at wavelengths shorter than one centimetre is caused by water vapour in the atmosphere, while the upper cut-off at wavelengths longer than about twenty metres is caused by reflection or absorption of incoming radio waves by the ionosphere—as mentioned earlier, the layer of charged particles in the atmosphere that had defeated the first attempts to detect cosmic radio waves at the turn of the century.

With the failure of any of the astronomical observatories to follow up Jansky's discovery, the new science of radio astronomy might have fallen into limbo had it not been for the extraordinary initiative of Grote Reber, an engineer with a passion for all aspects of radio science. At the age of 15 Reber had built his own transmitter–receiver and, from his hometown in Wheaton, Illinois, communicated with fellow radio hams in over sixty countries. By the end of his student years, Reber had achieved his ambition of 'working all continents' and began looking for a fresh challenge. After reading Jansky's research papers and realising their significance, Reber discussed the possibility of continuing these observations with a number of astronomers at the Yerkes Observatory in Chicago. As he recalled years later, none showed any particular interest:

> 'The problem was really philosophical. They couldn't think of a way Mother Nature produced these radio waves—straightforward thermal emission from the stars didn't seem to work. Some claimed Jansky's work was a total misinterpretation of the data and, at worst, a complete hoax. Part of the problem for me as an engineer was trying to communicate across a professional gulf. I probably knew more of astronomy than they did of radio engineering. It became clear that here was a first-class opportunity for me to do something new and different. By then I was working for a Chicago radio firm and had access to the best equipment. So, in the summer of 1937, I decided to take some time out and follow up the discovery myself.' [10]

Jansky had exploited his discovery to the technical limits of his receiver and merry-go-round aerial, so Reber realised further progress would require the construction of new equipment specifically designed to observe the cosmic static. The 'radio telescope' would need not only to determine in sharper detail how the intensity of cosmic radiation is distributed over the sky, but also how the intensity varies with wavelength. The best solution seemed to be a large parabolic reflector or dish which could be accurately pointed across the sky. Incoming radio waves of any wavelength would be reflected from the dish's surface and focused to a single point—much like a car headlight working in reverse. Different radio wavelengths could be investigated simply by changing the receiver placed at the focus of the dish. Hertz had made use of this principle in his experiments leading to the discovery of radio waves in 1887, and parabolic dishes several metres in diameter were also being secretly developed in the 1930s by the early radar investigators.

Reber's plans for a large dish received an initial setback when quotes from various construction firms all proved 'astronomical'. Undeterred, he decided to build the radio telescope himself, in the backyard of his parents' home in Wheaton. The dish was surfaced in sheet metal, 31 feet in diameter, and mounted on a movable wooden support structure. Apart from steel joints cut and drilled by a local blacksmith, Reber made each component, completing the entire construction by himself over a period of four months. By the standards of 1937, the cost of $1300 was a large sum for a temporarily unemployed engineer in pursuit of an untried idea:

> 'The dish usually emitted snapping and banging sounds every morning and evening due to unequal expansion in the reflector's metal skin. When parked in the vertical position great volumes of water poured through the centre hole during a rain storm. This caused rumours among the locals that the contraption was for controlling the weather! The telescope must have also looked spectacular from the sky. Air navigation rules were practically non-existent at this time and often small planes buzzed back and forth overhead. One day a plane about a thousand feet up developed engine trouble and had to make an emergency landing in a nearby paddock. This led to another rumour that the telescope was a new secret weapon for putting the ignition systems of planes out of action. That would have been far better than radar!'

Reber's first attempts to detect radio waves from prominent objects such as the sun, moon, planets and several bright stars at a wavelength of 9 cm produced no response. Another receiver working at 33 cm also proved unsuccessful. Over a year passed and Reber feared that his wild idea was doomed to failure. He persevered and again built a receiver working at a wavelength just short of 2 metres. Early in the spring of 1939, he was at last rewarded by receiving unmistakable signals coming from the plane of the Milky Way. Full of enthusiasm, Reber then launched into an observing schedule that would have tested any astronomer. At night he carried out measurements from midnight to 6

Grote Reber and his backyard radio telescope in Wheaton, Illinois. As a tribute to Reber's pioneering efforts, a replica of the telescope was later erected at the US National Radio Astronomy Observatory at Green Bank, West Virginia.

am when local radio interference, chiefly from car ignition systems, was at a minimum, and then drove 50 km to a new job with the Chicago radio company. After the evening meal he would sleep to midnight, and then resume the observations.

In his first paper, published in the same radio engineering journal used by Jansky, Reber reported that the source of the cosmic static lies predominantly along the Milky Way, thereby confirming Jansky's earlier discovery.[11] To make his results known where it counted, Reber sent a similar paper to the *Astrophysical Journal*, referred to in the United States as the astronomer's bible. The editor, Otto Struve, based at the nearby Yerkes Observatory was so perplexed by the new data he put the paper aside for several months. Eventually, after favourable reports from others who had inspected the backyard telescope, Struve decided to publish the paper, reassured that possibly a radio engineer did have a contribution to make to astronomy.

Unlike Jansky's aerial, Reber's telescope could be pointed fairly accurately which meant that he was able to make detailed measurements of the variation in radio strength across the sky. Plotting the contours of strength gave a celestial radio map—similar in appearance to the more familiar maps which show the elevation of the earth's terrain. In addition to the main radio strength which peaked at the centre of our galaxy, other subsidiary peaks were prominent in the constellations of Cygnus and Cassiopeia. Reber could not, however, detect any appreciable strength from nearby stars. This seemed to support Jansky's hunch that the radio emission did not originate from the stars themselves, but probably from energetic electrons and protons moving freely in interstellar space. Later, in 1944, Reber managed to detect signals from the sun but, as we see in Chapter 3, the existence of solar radio waves had already been reported in classified military documents by wartime radar operators in both England and the United States.

While Jansky's discovery laid the first foundations for the new science of radio astronomy, Grote Reber became its first real practitioner. For a lone-hand working in his spare time, his achievement was exceptional. The Wheaton radio maps of the sky were not improved upon until the work of Australian radio astronomers early in the 1950s. Reber can also claim credit for having built the first true radio telescope, his parabolic dish beginning a line of development of similar instruments which led directly to the Parkes telescope.

CHAPTER TWO

Physicists Take to the Air

> The old dream of wireless communication through space has now been realised in an entirely different manner than many had expected. The cosmic shortwaves bring us neither the stockmarket nor jazz from distant worlds. With soft noises they rather tell the physicist of the endless love play between electrons and protons.
>
> Albrecht Unsöld, 1946

> Radio astronomy was not born with a silver spoon in its mouth. Its parents were workers. One parent was the radio-telephone, the other was radar.
>
> Chris Christiansen, 1952

Grote Reber's survey maps of the radio sky in the early 1940s had a much greater impact on astronomers than Jansky's earlier work, but by then the significance of these results was swamped by events on a grander scale. Scientists and engineers alike put aside the pursuit of such esoteric lines of research and joined in the war effort. Some enlisted as regulars in the armed forces, but there were many who were mobilised into special research teams and directed to the task of improving the weaponry of war. The radio techniques employed by Jansky and Reber to explore the cosmic static now found a far more pressing application—that of radar. The military recognised early on the strategic and tactical importance of radar in warfare. Teams of scientists and engineers were formed, first in England, and then in countries such as Australia, Canada and the United States, not to mention Germany and Japan, and given the urgent task of improving radar technology. Under the pressures of war, rapid advances were made in the research and development of new aerials, receivers, transmitters and a host of new electronic devices and techniques. Largely as a result of this extraordinary progress in radar technology, several countries immediately after the war—in particular Australia—were ideally placed to develop the new science of radio astronomy. By 1945 the equipment used by the pioneers Jansky and Reber had receded into the horse and buggy days of radio astronomy.

The word radar was coined by the Americans late in the war to describe what for many years had been known instead as radio direction finding. As early as 1922 Marconi and others had pointed out that it might be possible to locate ships at sea by means of reflected radio waves. In 1931 engineers at the British Post Office developed the idea and built a workable system for detecting nearby ships. At about the same time several radio operators reported radio interference apparently caused by planes flying in the vicinity, and this led to speculation that this effect might be used as a way of detecting aircraft. In January 1935 Robert Watson Watt wrote a confidential report which argued that this effect should be developed into a major part of Britain's system of air defence. It was a bold proposal. Detecting slow moving steel battleships in coastal waters was one thing. To detect fast approaching aircraft early enough to mobilise fighter squadrons was another. Rough calculations showed that the radio energy reflected from a typical wooden-framed aircraft would be only a million, million, millionth part of the radio energy transmitted. Against understandable scepticism in official circles, an air ministry committee chaired by Henry Tizard decided to provide financial backing to test the idea. A small research station was secretly established near the small village of Bawdsey on the English Channel. An encouraging start was made in June 1935 when echoes from a stream of transmitted radio pulses were detected after bouncing off an aircraft at a distance of 30 km. Three months later the detection range of the equipment had been extended to 150 km. After intensive development this radar equipment formed the basis of the air warning system which proved crucial to eventual success in the Battle of Britain.

One of the first arrivals at the Bawdsey research station was Edward Gordon Bowen. Bowen grew up in Swansea, studied physics at the University of Wales and then carried out research for his doctorate at King's College in London under the supervision of Edward Appleton. Bowen's talents came to the notice of Watson Watt, who recruited him as one of the original team of four stationed at Bawdsey. Following the successful development of a coastal chain of radar stations, Tizard at the air ministry then pressed for a radar set sufficiently light and compact to be mounted in aircraft. Bowen was appointed leader of the airborne radar group. A recent history of early radar observed:

> 'The Bawdsey spirit found its epitome in an enthusiastic young Welshman, Taffy Bowen, who had been working on a doctorate in physics before joining the team and whom Watson Watt liked to portray as a Celtic brigand. Bowen now took responsibility for squeezing radar gear from the size of a room into the size of a two-foot-square box that could be jammed into an airplane. Always smiling and often nearly electrocuting himself, the 24-year-old Bowen set about single-handedly creating the first airborne radar equipment.' [1]

After hundreds of flying hours, the first airborne radar system was made operational by the end of 1937. Bowen's group also developed the first radar screen, known at the time as the plan-position indicator. As we see later, two of Bowen's associates, Bernard Lovell and Robert Hanbury Brown, helped to give Britain its postwar start in radio astronomy.

At the outbreak of the European war in September 1939, the radar equipment in use along the east coast of England operated at long radio wavelengths in the range 1 to 40 metres. A major goal in radar research became to develop sets that could operate at much shorter wavelengths. The shorter the wavelength, the greater the precision (resolution) in detecting ships or aircraft and, particularly for airborne radar, the equipment could be miniaturised because the size of the antenna receiving the reflected signals varies according to the wavelength. No devices were available at this stage which could operate at short wavelengths and still generate enough signal strength to power a radar set. The critical breakthrough occurred early in 1940 by a group at the University of Birmingham under the direction of the Australian physicist Mark Oliphant. John Randall and Henry Boot invented a new device known as a resonant cavity magnetron which could deliver more power at 10 centimetres than the bulkier transmitters operating at metre wavelengths. The exceptional performance characteristics of the magnetron led to major improvements in every type of radar then under development—ground-based, shipboard and airborne. The magnetron gave Britain a lead in radar technology that lasted until the end of the war.

The invention of this small metal device raised an important diplomatic issue for the British. Should the secret be shared with the Americans and, if so, when? The United States had not yet entered the war, but the industrial might of Britain's main ally could well determine its eventual outcome. Earlier in 1940, before the magnetron, Britain had sent a scientific delegation to the United States with the task of learning what progress the Americans had made in radar, while at the same time not revealing any of the British secrets. The mission achieved very little. In response, Henry Tizard as scientific adviser to the Air Ministry in Britain argued that not only the magnetron but all recent scientific developments of military value should be shared with the Americans, without asking for anything in return. At the top of the list was an item of even greater importance than the magnetron. Two refugee physicists from Germany, Otto Frisch and Rudolf Peierls, had recently written a top secret report outlining how the fission of the uranium nucleus might trigger a chain reaction of devastating explosive power. Whitehall accepted Tizard's argument and in September 1940 he led a mission to the United States consisting of military and civilian advisers. The task of delivering the magnetron and demonstrating its outstanding characteristics was entrusted to Taffy Bowen. Taking one of the first magnetrons produced by General Electric in Wembley, Bowen flew the Atlantic carrying in his briefcase what one historian has called the most valuable item of cargo ever brought to the shores of the United States.[2]

Above The magnetron taken to North America by the Tizard mission in September 1940, now on display in the Museum of the National Research Council, Ottawa. *below* Taffy Bowen (left) and the physicists Lee DuBridge and Isidor Rabi at the MIT Radiation Laboratory. Both DuBridge and Rabi helped to promote the growth of radio astronomy in the United States in the early 1950s.

Bowen remained in Washington after the rest of the Tizard mission had returned home. His brief was to facilitate the exchange of scientific and technical information set in train by the visit, and then to advise the Americans on the establishment of a major radar laboratory at the Massachusetts Institute of Technology. The MIT Rad Lab, as it became known, became the powerhouse of American radar research and development for the rest of the war. At the height of its activities the Lab employed one thousand scientists and engineers who developed a wide variety of magnetrons and electronic devices, incorporating them into over one hundred new types of radar systems. This was a mobilisation of American scientific resources second only to the development of the atomic bomb by the Manhattan Project.

The Japanese attack on the US naval fleet at Pearl Harbor in December 1941 broadened the European war into global conflict. In terms of radar development, Australia had already forged ahead of the Americans. Over a year before the Tizard mission to the United States, Britain had decided to share its knowledge of radar with several countries in the British Commonwealth by inviting each country to send its best qualified physicist to learn of the secret research. Australia chose David F. Martyn for the clandestine mission. Martyn had already gained an international reputation for his theoretical studies on the upper atmosphere through such work as his explanation of the 'Luxembourg effect', a phenomenon where the broadcast from one radio station sometimes carries the background sound from another station, even though this second station is transmitting at a completely different frequency. During 1939 Martyn spent five months touring British radar establishments and returned home with a lead-lined trunk containing hundreds of classified reports, several of the latest radio components, as well as numerous blueprints and technical specifications for constructing radar equipment. Arriving in Sydney, Martyn reported to two of the key figures in Australian science. One was David Rivett, chief executive officer of the Council for Scientific and Industrial Research (the forerunner of CSIRO) since its inception in 1926. The other was John Madsen, professor of electrical engineering at the University of Sydney. The three men drew up a set of recommendations for the establishment of a national laboratory for radar research, and then travelled to Canberra to present their proposal to Richard Casey, the minister in charge of CSIR. Following a brief meeting between Casey and the prime minister Robert Menzies, the recommendations were accepted without change. A sum of £50,000 was allocated for building costs and a further £30,000 for laboratory and workshop equipment. The name chosen—the Radiophysics Laboratory—was a fairly innocuous one designed to draw as little attention as possible to the laboratory's real purpose.

With government backing for the radar project secured, John Madsen set sail for England in December 1939 to hold discussions with Tizard, Watson Watt and other British experts. Madsen wanted to clarify the role the new laboratory would play in relation to the British radar effort. The outcome was that the laboratory would function basically as an outpost of the main centres in Britain with a research program arranged, after consultation with London, to avoid duplication where possible of the development work. The role envisaged for the laboratory was essentially to act as the main radar depot in the Pacific theatre and to provide support for military forces in the region, such as the strategically important base in Singapore. Because of the vulnerability of supply routes to Britain the laboratory was to build up a store of essential equipment and technical information. Later, in 1940, the laboratory's charter was changed to give it more autonomy. At its own initiative, the laboratory could carry out research on the basic principles of radar, on new types of radar systems, and on how best to adapt British radar sets to suit the different wartime conditions in the Pacific.

The site chosen for the new laboratory was in the grounds of the University of Sydney. CSIR had already begun construction of a National Standards Laboratory on the site and the first stage had been completed late in 1939. The new radar laboratory would form an additional wing of the main laboratory which would have the advantage of making its secret work a little less conspicuous. Under the urgency of war the construction pushed ahead rapidly and by March 1940, only three months after the foundations had been laid, the building was near enough to completion for staff to move in. David Martyn was appointed chief of the new laboratory and by the end of 1940 about forty scientists, engineers and support staff had been recruited.

Most of the scientists came from a small section in CSIR known as the Radio Research Board. The RRB had been formed in 1927 at John Madsen's suggestion and had been the first step by CSIR to venture away from almost a total concentration on research aimed at benefiting the agricultural and mining industries. Although not having the same direct bearing on the economic welfare of the nation, radio would obviously have an important future in a large and sparsely populated country such as Australia, both for internal communications and for contacts overseas. The RRB initially appointed four full-time scientists from England (one was Martyn) to carry out research into radio communications relevant to Australian conditions. However, the major activity of the Board turned out to be to promote the growth of radio science by providing funds to support postgraduate research students. The two main beneficiaries were Madsen's own department of electrical engineering at the University of Sydney and the school of physics at the University of Melbourne under the direction of Thomas Laby. The research produced by the full-time staff of the Board and the young university physicists and radio engineers ranked highly in comparison with similar centres in Britain and the United States. By the end of the 1930s the RRB had nurtured a small group of talented scientists well schooled in both the theory and practice of radio science. To borrow a modern term, the RRB established Australia's first 'centre of excellence' in the physical sciences. On the outbreak of war, no other section of CSIR was better prepared to play an active role in the national war effort.[3]

Among the first to join the new Radiophysics Laboratory were Frank Kerr, recruited from Laby's school in Melbourne, and Brian Cooper from Madsen's department in Sydney. Most of the recruits were recent graduates, but one of the more experienced scientists was Jack Piddington who had studied under Appleton at Cambridge and had gained firsthand experience of some of the early developments in radar. In 1938 Piddington had constructed the first radar set produced in Australia, and with this experience he was appointed leader of the radar design section. Another senior appointment was Joseph Lade Pawsey, recruited to lead the basic research group at Radiophysics. Born at Ararat in Victoria in 1908, Pawsey studied physics under Laby at Melbourne and then began a thesis at the Cavendish Laboratory in Cambridge on the phenomenon of broadcast fading, caused by the irregular reflection of radio waves by the ionosphere. The head of the Cavendish, Ernest Rutherford, described the qualities of the research student who would become the founding father of radio astronomy in Australia:

> 'Pawsey has attacked his experimental problems with energy and enthusiasm and has made good progress. He has proved a very competent experimenter with good judgement in the interpretation of his results. Mr Ratcliffe can speak with more authority than I can of the value of his original work but I understand he has shown excellent promise as an investigator and made very satisfactory progress with his work. I should judge Pawsey would prove a very useful investigator on radio problems and I am sure he would work well with his colleagues.' [4]

Upon completing his doctorate in 1934, Pawsey joined Electronic and Musical Industries (EMI), a firm that had just won a BBC contract to develop a national television system. At the BBC station at Alexandra Palace, Pawsey was chiefly responsible for the design of transmission lines and aerials necessary for television's broad bandwidth.

Initially, the Radiophysics Laboratory concentrated on constructing radar sets to suit the needs of the three Australian armed services. In 1940 it seemed that the most likely attack on Australia would come from the sea. In response, the first major project tackled was the development on behalf of the Army of a shore defence radar system to be used by coastal artillery for locating enemy ships. A prototype model was erected in May 1940 at an army testing ground on the clifftops of Dover Heights, south of the entrance to Sydney Harbour. This site was later to become the Laboratory's first main radio astronomy station. The trials were promising and led to a major innovation by Joe Pawsey and Harry Minnett, who developed an ingenious switching device so that only one tower and one aerial were needed for both transmission and reception of the radar signal. In England at this time, radar systems had their transmitting and receiving equipment isolated on separate towers which added substantially to the cost of the system.

After Pearl Harbor it became obvious that air attack would be the greatest danger to Australia. Jack Piddington and his group worked at top speed to improvise an air-warning system based on the shore defence developed for the detection of ships. In its first week of trials at Dover

One of the early experimental radar units developed for shore defence at the Dover Heights field station. This unit was used for the detection of ships along the Australian coast, but then a new version was hurriedly developed for air warning following the Japanese attack on Pearl Harbor.

Heights the new system detected aircraft at distances more than 100 km out to sea. The prototype was pressed into production and the new radar hurriedly installed along the east coast of Australia. Arrangements were also made to set up a radar system in Darwin on the north coast of Australia, but arguments broke out about which of the three armed services should bear the cost and the installation was delayed. In mid February 1942, without warning, the Japanese staged a devastating bombing raid. Piddington and his team immediately flew to Darwin and assisted in getting the system operational in time to detect the approach of a further Japanese bombing raid three days later. The attack was intercepted and scattered off the coast by Allied aircraft.[5]

Over twenty major radar projects were begun during the course of the war, including range finding equipment for anti-aircraft guns, air-to-surface radar for the location of ships by aircraft, and a small flashlight-size detector used by commandos landing in enemy territory to locate Japanese radar installations. Projects carried out on behalf of the RAAF eventually came to occupy the whole of the Lab's program. Most of the work concentrated on the development of the light-weight air-warning (LW/AW) radar set. The initial prototype, hastily built in December 1941, underwent extensive improvement and culminated in the Mark V version completed for the RAAF in 1945. The LW/AW radar set was the

The light-weight air-warning (LW/AW) radar developed by Radiophysics especially for tropical conditions in the Pacific. The unit could be brought through the surf by landing barge or airlifted to forward areas and then rapidly assembled. It played an important part in the offensive island-hopping campaign later in the war.

Lab's finest achievement during the war years. Official Australian war historian D. P. Mellor wrote:

> 'No matter how efficient a radar set was technically, it was useless if it was so cumbersome that it could not be got to the place where it was required. On the other hand, the situation was not much better if a set was made so light that, though it could be got to the place where it was required rapidly enough, it was liable to break down after a short time. For use in New Guinea and other tropical regions something between these two extremes was needed; something that could be transported by air and sea, and if necessary by hand through the jungle, without damage, and that could be assembled and put in operation within a few hours of its arrival at the site. Equipment evolved to meet these conditions, known as the Australian lightweight air-warning set, came as near as any made by the Allied countries to being ideal in respect of reliability and portability.' [6]

The Radiophysics Lab expanded rapidly as the war progressed with the number of staff and the size of the budget increasing each year in leaps and bounds. There was also a change of leadership which would later prove crucial to the Lab's postwar future. Late in 1940 David Martyn planned to visit England to learn first-hand of the latest developments in radar. The CSIR Executive decided to appoint a temporary replacement for Martyn and extended the invitation to Frederick White, the professor of physics at Canterbury University College in Christchurch, New Zealand. With a background similar to that of several of the Radiophysics staff, White had completed a doctorate at Cambridge in 1932 and then been appointed lecturer in physics in Appleton's department at King's College, London. White returned to New Zealand in 1937 and set about a vigorous program of radio research in Christchurch.

As it turned out, Martyn cancelled his proposed trip to England and, instead, John Madsen travelled to both London and Washington to establish wartime arrangements for the exchange of technical information on radar. By the time of White's arrival in May 1941, relations between the Radiophysics Lab under Martyn's leadership and the armed services had become severely strained. Similar tensions had also arisen between the Lab and other government departments which had the important task of carrying out full-scale production of the radar prototypes developed in the Lab. In an attempt to ease the tension, White took over the administrative duties leaving Martyn in charge of research. However, by the end of 1941 and the outbreak of the Pacific war, relations between the Lab and these outside bodies had deteriorated further. David Rivett and the CSIR Executive were forced to act. In May 1942 Martyn was transferred to the post of director of an army operational research group. Four months later, White was officially installed as the new chief of the Radiophysics Lab.

Martyn felt humiliated by what he saw as a demotion and harboured a bitterness that was to adversely affect personal and institutional relation-

The Radiophysics Laboratory in the grounds of the University of Sydney. By the end of the war the staff had grown to over 300.

ships in Australian science for many years after the war. As a scientist, Martyn's exceptional talent had played a large part in placing Australia at the forefront of international radio research in the 1930s, but behind this brilliance lay a personality one writer has described as impulsive, intensely competitive and sensitive to criticism.[7] Traits such as these can often be an advantage to a scientist working in relative isolation, but in dealings with hard-headed service chiefs and the heads of government departments working under the pressures of war, they were bound to lead to difficulties.

Fred White soon proved himself a gifted manager with the qualities to steer a firm course through the complex and often contradictory responsibilities and priorities faced by the Lab. In mid 1943 he toured radar establishments in Britain and the United States. At the MIT Rad Lab he met with Taffy Bowen, the airborne radar specialist, and renewed a friendship that had begun at King's College when Bowen had been a research student and White a lecturer. In an inspired move White invited Bowen to become assistant chief of the Radiophysics Lab and to take charge of the research activities. Bowen agreed and arrived in Sydney in January 1944, pleased with what he found:

> 'It was already a flourishing concern with a staff of two hundred people; it had an excellent mix of highly qualified scientists and engineers and the permanent building included first-rate workshop facilities. For its size, it was one of the best equipped laboratories in which I had worked up to that time. The laboratory had good connections with industry and in spite of the enormous distances and the difficulties of wartime travel, it had kept closely in touch with radar research in Britain and America. Members of the staff made frequent visits to the Telecommunications Research Establishment (TRE) in England and the MIT Rad Lab and were familiar with the latest technology.' [8]

By the time of Bowen's arrival, the war in the Pacific had swung decisively in favour of the Allied forces. As the immediate danger to Australia receded, there was a gradual shift in emphasis in the Lab away from a total commitment to applied military projects towards longer term studies in basic research. The time had come to make plans for the postwar era. The Lab itself had grown rapidly during the war, from a total of 19 staff early in 1940 to 300 by war's end, 60 of whom were classified as research scientists. Over the same period the annual budget grew at a similar rate, from £12,000 to £154,000. By 1945 the Lab had outgrown most of the other older CSIR divisions, even though CSIR itself had grown rapidly during the war. The Lab's workshops and laboratory benches were fitted out with the most advanced equipment available. At least in the physical sciences, the staff formed the most talented and motivated group of scientists ever assembled in Australia, a group that had been right at the forefront of research and development of wartime radar.[9]

A critical decision made at the end of the war, largely at the insistence of David Rivett, was that CSIR would turn over all secret military work to other government agencies. CSIR was to concentrate solely on research which would bring direct economic benefit to Australia or add to the general store of scientific knowledge. Two options were open. First, the Lab could fulfil its military commitments and then steadily wind down its staff and activities to more modest peacetime proportions. This was the course followed at the two largest radar centres overseas, the TRE in England and the Rad Lab in the United States. At the end of the war hundreds of scientists in both countries made a mass exodus back to the universities and corporate research laboratories from where they had originally been recruited. A high proportion of the Radiophysics scientists had, however, been recruited as students straight from university and the possibility of pursuing a full-time research career in any of the Australian universities in the 1940s was limited, even more so in industry.[10] For most of the staff, the second option open to the Radiophysics Lab was the one they preferred: basically to keep the Lab intact even though this would mean a major redefinition and redirection of its research program.

Early in 1945 Bowen set about the challenging task of drawing up a list of possible peacetime research areas. In Bowen's view, the Lab would need to pass through a transition phase lasting about two years. This would consist of a type of postwar spring cleaning where the knowledge and techniques of radar would be handed over to the armed services and, where appropriate, to other areas of government and industry. At the same time the Lab would begin to strengthen its commitment to fundamental research and this would require recruiting new staff to replace the relatively small number of those who decided to go their own way after the war. Bowen drafted a report which divided the proposed research program into nine main areas, ranging from studies of radio propagation, radio methods of air and marine navigation, airborne surveying techniques, vacuum tube research, through to radar studies of cloud and rain formation. The first of these nine areas, under the heading 'Propagation of radio waves', was further subdivided into five categories covering topics such as ionospheric propagation and the scattering of radio waves in various layers of the atmosphere. In the second of these five subcategories, with the innocuous title of 'Radio noise', Bowen noted that there is a certain type of radio noise which is thought to originate in the stars or in interstellar space: 'Little is known of this noise and a comparatively simple series of observations on short wavelengths might lead to the discovery of new phenomena or to the introduction of new techniques.' [11] A humble sentence, but one that contained the seed of a major new science in Australia.

In July 1945 Bowen's future program for the Radiophysics Lab was presented to a meeting of the CSIR Council, the organisation's full governing body. The document was assured of a favourable reception.

Earlier in the year Fred White had been promoted to the CSIR Executive in recognition of the leadership skills he had displayed during the tough baptism of war. For a few months John Briton served as White's replacement before returning, as arranged previously, to the position of technical director of EMI in Sydney. Having been groomed for the position by White, Bowen's appointment as chief of Radiophysics early in 1946 completed the formalities. The top management of CSIR also underwent a major change at this time. Its Executive consisting of David Rivett, George Julius and Arnold Richardson had remained virtually unchanged since the formation of the national research body in 1926. In 1945 Julius retired as chairman and, with both Rivett and Richardson approaching retirement, a decision was made to appoint two new members to the Executive—Fred White to represent the physical sciences and a brilliant young veterinarian, Ian Clunies Ross, to represent the biological and agricultural sciences.

The enthusiastic endorsement by the new Executive of Bowen's postwar research program signalled a major reorientation in CSIR. The passing of the old guard coincided with increasing political pressure on CSIR to move away from its traditional emphasis on research directed to the agricultural and mining industries. The new emphasis was to be on areas of modern technology which underpinned Australia's emerging manufacturing industries, some of the newer fields of research pointed to in Bowen's report. The Executive's support of Bowen's program also carried with it an implicit understanding. As chief, Bowen would be free to direct the research program of Radiophysics with a minimum amount of interference by the Executive. This idea of individual responsibility in CSIR had been championed by Rivett himself who believed that the best science comes from choosing the best qualified people and then giving them a free hand. After leaving an outstanding career as a research chemist to join CSIR, Rivett maintained that CSIR's approach to research should be more in line with the academic freedom offered to university scientists, rather than the approach usually adopted in government laboratories where scientists were moulded to fit structured research programs. Rivett expressed this idea of self-rule in a letter to a friend in 1945:

> 'Unless we can keep CSIR free from all the straitjackets that are all too freely offered to it from all sides, we are not going to count very much in 20 years' time, even if we do succeed in the meantime in doing a job or two that wins favour from the press, populace and politicians. I fully believe ... that we shall fail in the end unless quite 50% of our effort is directed to finding out how the machine of Nature works, without a thought as to whether that knowledge may or not be useful in this decade, or next century, in showing farmers how to save sixpence or politicians how to increase revenue from taxation.' [12]

This promise of research freedom had in fact been an important precondition in Bowen's decision to accept the position of chief early in 1946.

Still at the relatively young age of 35, brimming with self-confidence, Bowen shared Rivett's ideals and his faith in the value of small, independent research teams. A proposal at this time to amalgamate the Radiophysics Laboratory with the National Standards Laboratory to form a monolithic Commonwealth Physical Laboratory was successfully opposed by Bowen for just these reasons.

With virtually carte-blanche to pursue any of the research fields listed in his report, Bowen's strategy was simple and flexible. Small groups would be turned loose on a wide variety of topics, but then a kind of Darwinian natural selection would prevail. Projects that showed early promise would be rewarded with the support of more staff and funds at the expense of those that failed to make headway. Promising starts were made in a number of fields, but by the early 1950s basically only two areas had come through in the contest of 'survival of the fittest'—radio astronomy and cloud physics. Radio astronomy proved to be the wild card in the pack, emerging in just a few years to become the largest and most successful of the Lab's activities. Although it had rated only a few lines in Bowen's research proposal in July 1945, the rapid growth of radio astronomy can be gauged by a summary of the Lab's research activities[13] issued in December 1947—one half of it was devoted to radio astronomy.* Before turning to the early work in radio astronomy in the next chapter, let us survey briefly the progress and fate of the other research work carried out at Radiophysics in the postwar years.

☆ ☆ ☆

The research program most closely related to the wartime work of the Lab was the application of radar to air navigation. During the war radar had been developed to guide aircraft to and from their home bases, and the main task in peacetime was to adapt this military equipment to civil aircraft. Numerous proposals for navigation systems were made throughout the world and among the most successful was the distance-measuring equipment (DME) developed by Brian Cooper at Radiophysics. First installed on commercial aircraft in 1950, the DME provided pilots with a direct and reliable reading of the distance to and from airports. In 1954 the DME was made compulsory on all passenger-carrying aircraft in Australia, about ten years before similar systems were adopted in the United States and elsewhere.

Further research was done at Radiophysics to improve the DME system, but by the mid 1950s air navigation had ceased to be a significant program. In 1971 the Lab re-entered the field and, under the direction of new chief Paul Wild, developed the highly successful Interscan landing system, now adopted as the international standard of the future.

* The term 'cosmic noise' was used in these early annual research summaries. In 1947 a group at Cornell University in the United States coined the name 'microwave astronomy', but neither Bowen nor Joe Pawsey warmed to the new term. Pawsey himself used the term 'radio astronomy' in a letter dated 14 January 1948, apparently the earliest use of the word (see note 14).

Research on electronic components and circuits also continued after the war and led to a device known as the resonant cavity X-ray tube, developed by Bernard Mills and capable of delivering over a million volts. One obvious application for such a device was in the construction of particle accelerators for nuclear physics, but basically Radiophysics was not in the same league when it came to the massive investment and development occurring elsewhere, particularly in the United States. As one example, in 1948 Stanford University announced plans to build a one-billion-volt electron accelerator at a cost of over one million dollars. In Australia, plans were also being developed at a number of centres to build large and expensive accelerators, including the Australian National University, Melbourne University and the Australian Atomic Energy Commission. The development of the million-volt X-ray tube was a notable achievement, but the possibility of using it to carry out basic research in nuclear physics was simply beyond the resources then available to Radiophysics.

This research on electronic devices and techniques at Radiophysics led to another outstanding development—the design and construction of one of the first electronic computers. The pace of scientific and technological change during the war had stimulated a tremendous demand for machines with computational powers greater than the mechanical or electric calculators then available. The spin-offs from wartime radar research—devices such as memory circuits, pulse counters, and generators delivering pulses as short as a microsecond in duration—provided the technology for the next step to be taken. Immediately after the war construction of the first all-electronic computers began at several centres in the United States and Britain. Not far behind, Trevor Pearcey and Maston Beard at Radiophysics began building Australia's first computer in 1947. Christened the CSIRAC, the computer came into operation in June 1951 at about the same time as two machines known as ILLIAC (University of Illinois) and EDSAC (Cambridge University). It proved just as good. Bristling with over 1500 radio valves and able to carry out 500 arithmetical calculations per second, the CSIRAC eased the workload of a roomful of staff (mainly women) who had done most of the laborious, repetitive 'number crunching' calculations using electric desktop calculators. For the next five years, the computer not only provided a service for the research programs at Radiophysics, but also for a number of outside organisations. One task was the design of power-generation systems for the Snowy Mountains Project, a vast hydro-electric scheme built in Australia in the 1950s.[15]

CSIRAC gave Australia a flying start in computer technology, but that was about the end of the story. One commentator later wrote:

> 'In 1955 a committee of two famous English mathematicians and an Australian was established to decide whether computer development should continue. It reported that Australia should concentrate on primary industry

Trevor Pearcey with CSIRAC, the first, and last, computer wholly designed and built in Australia. In 1956 CSIRAC was carted south on two trucks to the University of Melbourne. Later, it was retired to what is now the Museum of Victoria. CSIRAC is believed to be the only first-generation computer still in existence.

> and not waste resources attempting to compete with Britain and the USA. Towards the end of 1955 it was clear that, although attempts were made to interest Australian electronics manufacturers in the commercial exploitation of the CSIRO computer, they could not have competed in terms of cost and speed of production with American manufacturers. CSIRO had reluctantly to cease computer development.' [16]

Deference to overseas experts and the conservative nature of local industry added up to a missed opportunity. Currently, most of Australia's computer hardware, a billion dollar investment annually, is imported from overseas.

Aside from radio astronomy, research into cloud and rain physics became the only program to flourish at Radiophysics over the long term. During the war radar operators had been familiar with the echoes produced by cloud and rain; it had simply been regarded as an operational nuisance. Laboratory experiments had already revealed, at least in broad outline, how clouds form and also the important role played by condensation nuclei, but by 1945 surprisingly little was known about the fundamental causes of rain formation. The main barrier had been the difficulty in making direct measurements within clouds of the complex and interrelated physical quantities involved. This had to await the technological developments that occurred during the war—suitable aircraft to provide an airborne laboratory and radar to provide a way of 'seeing' deep within cloud systems.

The cloud and rain physics group at Radiophysics grew steadily after the war and by the early 1950s rivalled the radio astronomy group in staff numbers. With fewer expensive items such as large radio telescopes, however, the cloud and rain group consumed a comparatively smaller proportion of the Radiophysics budget. The aircraft and aircrew used in the experiments and trials were, for example, supplied free of charge by the RAAF. The research program eventually evolved into three closely related fields: cloud physics, the physics of rain formation, and the artificial stimulation of rain. The third field, rain-making, became the one most intensively studied and the one, even more than astronomy, that gave the Lab a high public profile. Great expectations were held for the success of this work. To a country noted alike for its droughts and floods, and still largely dependent for its prosperity on agricultural and pastoral industries, no natural phenomenon is more important. Since the time of the first settlers, farmers had alternated between elation and despair over the unpredictable patterns of rainfall. If the processes of rain formation could be understood well enough for even a small degree of control to be exercised, the consequences for primary industry and the national economy would be far-reaching.

Radiophysics had its first rain-making success in February 1947 when an aircraft seeded a large cloud inland from Sydney. The cloud showered up to 12 mm of rain over an area of almost 80 square kilometres, while none of the surrounding unseeded clouds produced rain. This was the first occasion anywhere in the world where a scientifically controlled experiment led to an appreciable artificial rainfall. After such an encouraging start, regular trials began on a variety of cloud types using a number of seeding agents such as frozen carbon dioxide (dry ice) and silver-iodide crystals. The trials were not without their risks. In November 1952 a Dakota aircraft disappeared in cloud off Cronulla, near Sydney, with the loss of the RAAF crew and two of the Lab's scientists.

By 1955 plans had been made for four intensive long-term experiments to be carried out over South Australia, the Snowy Mountains, an area west of Sydney, and a region near Armidale in New South Wales. For each trial a target area up to 8000 square kilometres was chosen, together with a neighbouring control area of the same size which would remain unseeded. At the end of the first year the results seemed almost too good to be true—increases of rainfall in the target areas by up to one third. However, the longer the trials progressed, the more baffling and the more disappointing the results became. The rainfall in the target areas slipped back to its unseeded level and in all four areas the occurrence of rainfall became more variable, more unpredictable than before the trials had commenced. Extensive though they were, the trials eventually proved inconclusive.

The same difficulty in finding hard statistical evidence in favour of rain-making also bedevilled attempts to understand some of the underlying causes of rain formation. An interesting case is a radical theory put forward by Taffy Bowen in 1953. While Joe Pawsey had taken on the leadership of the radio astronomy group, Bowen himself directed the research program of the cloud and rain physics group. After an extensive analysis of rainfall records going back over fifty years, Bowen found evidence that exceptionally heavy rainfalls occur over wide areas of the globe on specific days of the year. This pattern seemed to be repeated year after year. Bowen proposed the following explanation. Several days before these exceptional dates of heavy rainfall, the earth on its yearly orbit around the sun passes through dense clouds of tiny meteors, particles which are far too small to appear as the familiar 'shooting stars' entering the earth's atmosphere. This meteoritic 'dust' then filters down through the atmosphere to cloud level. Similar to dry ice or silver-iodide crystals, the dust provides a sudden large increase in the number of nucleation centres and leads to the occurrence of heavy rainfall in widely separated parts of the world.[17]

Bowen's theory caused quite a controversy. The existence of meteoritic dust clouds was not challenged, and several leading astronomers such as Fred Hoyle publicised the idea. Other meteorologists and physicists however attacked Bowen's statistical evidence, chief among them David Martyn. Nowadays, the theory has not been conclusively proven wrong, but it is certainly one that has fallen from favour.[18]

Overall, the research carried out by the Radiophysics group under Bowen's leadership made substantial progress during the 1950s. A leading American meteorologist wrote: 'It is probably correct to say that no other organisation contributed more to practical cloud physics during the period approximately from 1950 to 1961.' [19]

CHAPTER THREE

From Venus to Cygnus

Australia's history was linked with astronomy from the start—Captain Cook's first voyage was as much to observe the 1769 transit of Venus across the solar disc in Tahiti as it was to explore for *Terra Australis Incognita*, leading to the discovery of Australia's eastern coast the following year. But 175 years would pass before Australians became part of the first rank of world astronomical research. And when this happened, from cliff edges only a few miles removed from Cook's landing site at Botany Bay, it was in a most unlikely manner, for they did their astronomy not with glass lenses, but with rods of metal.

Woody Sullivan, 1988

We turn now to the origins of radio astronomy in Australia and the first observations which soon flourished into a major new science. To begin, we briefly discuss three factors that enabled the Radiophysics Lab to get off to a flying start in radio astronomy: staff, stuff and spirit.

By far the most important ingredient in the postwar success in radio astronomy was the *staff* at Radiophysics. As mentioned, only a small fraction of the staff left Radiophysics at the end of the war and the Lab was maintained at full strength by a batch of new recruits, most of them young men eager to carve out their reputations in the years ahead. A significant number of the staff who joined during the war years had benefited from months, and in some cases years, stationed at radar laboratories in Britain or the United States, an experience that instilled self-confidence and brought first-hand knowledge of how top scientific research is carried out elsewhere. On their return, most were determined to pursue research careers of their own and keen to see the Lab develop as a leading centre for research. Until then, the usual way for a young scientist in Australia to pursue a first-class research career was to go overseas, and then not think too much about returning except for the occasional holiday.

In view of the Lab's sudden plunge into radio astronomy it now seems odd that not one astronomer could be counted among the staff of 300 at the end of the war, nor for that matter anyone who had even done a university course in the subject. In retrospect this lack of formal training may have been more of a help than a hindrance. The new radio astronomers were in no danger of being bound to a set of existing preconceptions or 'paradigm'. Astronomy from standard textbooks could be picked up on the run. Similarly, the knowledge possessed initially by Radiophysics staff of the pioneering work by Karl Jansky and Grote Reber was also minimal. It had been generally known among radar operators that Jansky's cosmic hiss placed a limitation on the sensitivity of their equipment, but few had actually read Jansky's or Reber's research papers or were aware of their astronomical implications. A total commitment to the research and development of wartime radar had meant that subjects like astronomy—whether traditional or radio—were irrelevant to the task at hand.

A second factor that certainly helped Radiophysics make a start in radio astronomy was that it had all the right *stuff* at hand. Antennas, receivers, recorders, all the necessary pieces of equipment had been accumulated and all had undergone major improvement during the course of the war. While the same advances in radio technology would have come eventually in response to the postwar boom in civil aviation, radio broadcasting and television, the particularly rapid advance in wartime radio technology had yet to be exploited in a number of areas of scientific research. On this occasion the usual belief that basic scientific research leads to technological applications or spin-offs has to be reversed. Instead, by 1945, radio technology itself was looking for a new science, and it found it in radio astronomy.

The Radiophysics Lab also had an unexpected windfall. At the end of the war a section of the Pacific Fleet gathered in Sydney Harbour before returning to the United States. The Americans had considered it impractical to take back all of their vast collection of radar and communications equipment, so orders had been issued to destroy a large part of it. Bowen learnt of the plan and received permission to collect whatever Radiophysics needed. After several busy weeks loading trucks on the dockside, Radiophysics staff had salvaged enough equipment to fill two warehouses at Botany Bay. The remaining stocks of unwanted American equipment were loaded onto the deck of an aircraft carrier, taken a safe distance beyond Sydney Heads and bulldozed into the sea.

The third and remaining factor is one that differs from tangible assets of staff and stuff, one that can perhaps best be described as *spirit.* It came from two directions. One was the high morale built up after years of working under pressure, the confidence and cohesiveness that comes from victory through a team effort. The other was the enthusiasm and optimism which came with the prospect of years of peace ahead. Pioneering a new science gave free rein to this spirit. Taffy Bowen has written:

> 'In retrospect, it is clear that morale was another important factor in the Australian resurgence of research at the end of the war. Many scientists, especially those from nuclear physics, came out of the war with severe reservations about the part they had played. They were worried by the fact that their products had been used for unprecedented destruction, often at the expense of civilian populations, and they were horrified by it. Radar scientists, on the other hand, had no such qualms. Radar had been used predominantly to detect enemy aircraft and enemy submarines with the prime objective of destroying them. In other words, radar had been a means of destroying the engines of destruction and we were comforted by this fact. At the same time, we knew for certain that after the war there would be an enormous number of peacetime applications, like the navigation of ships and aircraft, to which radar could contribute and which were waiting to be exploited for the benefit of mankind.' [1]

Even though the the war had prevented scientists from following up the early work by Jansky and Reber, radio astronomy had not been completely dormant. In fact, the next major development came as a direct result of wartime radar research. Late in February 1942 radar stations throughout England had reported severe bursts of radio noise which had made normal operation impossible for several days. At this time two German warships had brazenly passed through the Straits of Dover without British radar warning of their approach. The War Office knew that the Germans had been developing methods of jamming radar and assigned J. S. Hey of the Army Operational Research Group the top priority task of investigating the problem. A short time later further reports of jamming were received from radar stations throughout Britain. After an analysis of the records, Hey realised that on this occasion the mysterious interference was not the result of enemy jamming, but came from a direction in the sky which seemed to coincide with the sun. Hey checked with the Royal Greenwich Observatory in London and learnt that during the days of maximum interference an exceptionally active sunspot had been in transit across the solar disc. In his report to the War Office, Hey concluded that despite earlier German successes in jamming British radar, this time the real culprit had been the sun. Hey also conjectured that the sunspot had been the source of the unwelcome interference.

Across the Atlantic at about the same time, a group led by George Southworth at the Bell Telephone Labs in Holmdel, New Jersey, made the same discovery. During trials of a new receiver operating at short wavelengths, Southworth detected a definite increase in the noise output from the receiver whenever the antenna pointed at the sun. Similar to Hey's work, the security regulations in force meant that these results received only limited circulation in secret reports, and publication of the discovery in the scientific literature had to await until the end of the war.[2] Hey and Southworth had succeeded where Thomas Edison and others at the turn of the century had failed. It later came to light, however, that

James S. Hey pioneered radio astronomy in Britain. As well as his wartime studies of the sun, Hey made two other important discoveries. One was that radar could be used to detect and track meteors more accurately than visual observations. This marked the birth of *radar* astronomy, a field quite distinct from *radio* astronomy. Hey was also the first to report fluctuations in the strength of the strong Cygnus source. This led the Australian group to the discovery and identification of the first radio sources. After the war, Hey founded a radio astronomy group at the Royal Radar Establishment in Worcestershire.

radio emission from the sun had made its presence felt as early as 1936, a time of maximum sunspot activity. A number of radio operators around the world had filed reports of an unusual hissing sound through their headsets, but no one guessed that the cause of the interference had come from the sun.

The existence of solar radio emission had been known in the Radiophysics Lab well before the end of the war. Both Bowen and Joe Pawsey had visited the Holmdel Laboratory and heard of Southworth's work, but there had been no immediate effort to confirm the discovery in Sydney. In 1944 Frank Kerr made some preliminary observations but then had to abandon the project to concentrate on more urgent wartime work. Similarly, in April 1944 Pawsey made an attempt to detect sources of extraterrestrial radio noise in an experiment that has to be admired for its simplicity. Assisted by Ruby Payne-Scott, Pawsey stuck a small parabolic dish out a laboratory window and pointed it around the sky, hoping to register an increase in the noise level above that generated by the receiver. The experiment failed, but contained within in it was the seed of an idea that eventually led to the Parkes telescope.

About a year later Pawsey's curiosity was again aroused by the receipt of two reports that supported Southworth's work. The first dealt with what had been termed the 'Norfolk Island effect', bursts of radio noise from the sun detected by New Zealand radar operators stationed on this small island in the Tasman Sea, north–east of Sydney. A report on these bursts and similar observations from the New Zealand mainland had been prepared by Elizabeth Alexander at the Radio Development Laboratory in Wellington (the Kiwi equivalent of Radiophysics).[3] Shortly afterwards, the receipt of a letter from J. S. Hey enclosing a copy of his secret 1942 report was enough to spur Pawsey into action. Pawsey chose Ruby Payne-Scott and Lindsay McCready to be his collaborators. McCready had been one of several engineers at Amalgamated Wireless Australasia (AWA) to join Radiophysics at the beginning of the war and had then risen to become Pawsey's second-in-command. Observations began in October 1945 using one of the existing RAAF radar antennas overlooking the sea at Collaroy, a northern Sydney suburb. Success came immediately. Enough data were soon gathered to confirm the overseas reports, but then an analysis of certain features in the signals, received at a wavelength of 1.5 metres, yielded a surprising result. Even with sunspot activity at a minimum, the radio emission seemed to indicate that some regions of the sun were at temperatures as high as one million degrees.

Initially the result was greeted with disbelief. Since the invention of the telescope early in the seventeenth century, no astronomical object had been studied in greater detail than the sun and much had been learnt about its physical characteristics. In the 1930s physicists had worked out the thermonuclear reactions that power the sun, steadily converting hydrogen into helium and releasing tremendous quantities of energy, in accordance with the famous Einstein mass–energy relation. The sun's innermost core was calculated to be at about 20 million degrees, with the temperature then steadily falling out from the centre, reaching a low of 6000 degrees on the visible surface (the photosphere). Astronomers had also carried out detailed studies of the chromosphere, the layer of 'atmosphere' extending to an altitude of roughly 20,000 km. Measurements of discrete emission lines from highly excited hydrogen and calcium atoms had indicated that the temperature of the chromosphere was higher than the photosphere, and also that it steadily increased with altitude. Above the chromosphere extending to an altitude of about a million kilometres is the region known as the corona. Studies in the early 1940s had provided evidence that the extremely rarefied gas in the corona might be at temperatures much higher than in the chromosphere, but the interruption of war meant most astronomers were unaware of this result.

The Radiophysics discovery of million-degree temperatures on the sun came to the attention of David Martyn, who had recently taken up a position at the Mt Stromlo Observatory near Canberra. Martyn realised immediately that these extremely high temperatures corresponded to radio emission high in the solar corona, and that the observations provided decisive support for the tentative evidence found earlier by astronomers. Martyn then set about developing a general theory to explain how the wavelength of the radio emission varies according to the

Ruby Payne-Scott, Alec Little and Chris Christiansen at the Potts Hill field station in 1949. Payne-Scott became the first woman to make a major contribution to radio astronomy, but only because she kept her marriage secret before leaving Radiophysics to raise a family in 1951. Australian public service rules at the time did not allow married women to hold permanent staff positions.

temperature in the solar atmosphere. Beginning with the photosphere at a temperature of 6000 degrees, which emits at visible wavelengths, the temperature climbs steadily to 20,000 degrees in the chromosphere, accompanied by radio emission at centimetre wavelengths (as discovered by Southworth in 1942). The temperature then reaches a peak of about one million degrees high in the corona with radio emission at metre wavelengths. Unknown to Martyn, two Russian physicists developed similar theories at the same time, but it was Martyn's work that became the most widely known.[4]

These observations with the radar antenna at Collaroy not only marked the beginning of radio astronomy in Australia, but also the first occasion radio astronomy had provided important information on a problem in traditional optical astronomy. Pawsey, Payne-Scott and McCready summarised their results in a letter to *Nature*, published in February 1946.[5] As an indication of how radio astronomy was still in its infancy, the letter contained only four references to earlier work—the papers by Jansky and Reber, and the wartime reports by J. S. Hey in England and Elizabeth Alexander in New Zealand. A short time later Pawsey set out in more detail the evidence for the million-degree corona, the paper appearing alongside one from Martyn summarising his theoretical study.[6] In the view of one radio astronomer: 'The simultaneous publication of these experimental and theoretical results marked the real beginning of radio astronomy.'[7]

These first papers dealt with the 'quiet' or undisturbed sun. However, like the wartime observers, Pawsey's group had noticed day-to-day variations in the intensity of emission, with the strongest signals occurring when sunspots were visible on the sun's disc. A closer investigation of this correlation seemed the best step to take next. The problem confronting Pawsey was to devise a way of achieving enough resolution to pinpoint the sunspots as the source of the strong fluctuating radio noise. The radar antennas available were quite useless for the task because of their broad 'cone of acceptance': in technical terms their beamwidth was about ten degrees or more. As the visible disc of the sun itself is only half a degree in angular extent, no existing antenna came close to resolving detail as small as sunspots. Pawsey found a way around the problem by adopting a technique known as sea interferometry, based on a trick radar operators had often used during the war to estimate the altitude of aircraft flying over the sea. The central idea in Pawsey's plan was to use an antenna high on a clifftop and facing due east. As the sun slowly rose at dawn, the antenna would pick up radio signals along two paths, one direct from the sun and the other reflected from the surface of the sea. The two signals received by the antenna would be superimposed to form an interference pattern, which could then yield accurate information on the size and position of the radio source. In principle, the technique of sea interferometry devised by Pawsey was no different from an effect in optics discovered in 1837 by the Irish scientist Humphrey Lloyd, who demonstrated how interference fringes could be formed by combining one beam of direct light with another reflected at shallow angles from a mirror.

The Dover Heights field station south of Sydney Heads provided a better vantage point than Collaroy and that is where Pawsey decided to begin the observations early in 1946. Fortune smiled after only a few days. One of the largest groups of sunspots ever recorded began its slow transit across the sun's disc, causing major disruption to radio communications throughout the world. The early-morning interference patterns recorded at Dover Heights established beyond doubt that the strong increase in the level of radio emission originated directly from the sunspot region. In their paper Pawsey, McCready and Payne-Scott reported that the intense emission was different in character to the steady type of radio emission from the chromosphere and corona. Some type of new and unknown mechanism for generating radio waves was at work in the violent electrical disturbances accompanying the sunspots.[8]

Confirmation of these results came four months later from radio astronomers at Cambridge University led by Martin Ryle. The English group had independently developed an interferometer based on the same principle as the sea interferometer, but one which employed two antennas spaced at a suitable distance and both pointing towards the sun. The signals from each antenna combined to form the same type of interference pattern as in the sea interferometer. Ryle and his group also

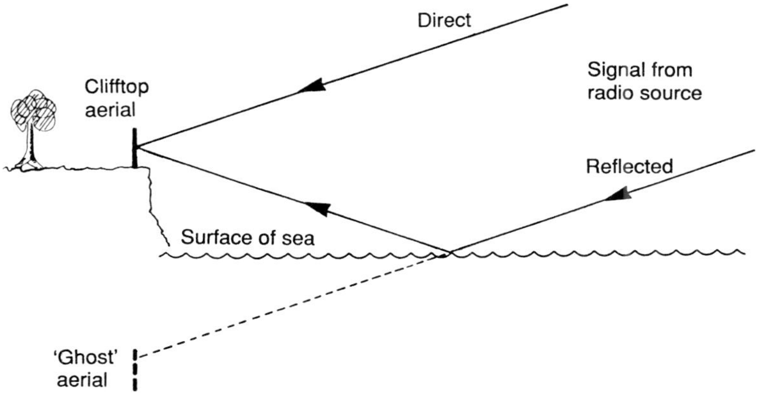

Radio astronomy began in Australia with the sea interferometer. The clifftop aerial combines the direct signal with the signal reflected from the sea to form an interference pattern. The reflected signal simulates an imaginary or 'ghost' aerial, spaced from the real aerial at a distance equal to twice the height of the cliff.

established that strong and variable emission originates from sunspot regions and, in fact, succeeded in getting their work into print ahead of Pawsey's group.[9] This sparked off what became a long and intense rivalry between the two groups (see Chapter 4).

Radio Waves over the High Cliffs of Dover

> The secret of science is to ask the right question, and it is the choice of problem more than anything else that marks the man of genius in the scientific world.
>
> Henry Tizard

The summer months following the end of the war had seen a brilliant start to radio astronomy in Australia. Pawsey and his group had pioneered radio interferometry, a technique of central importance to the future development of radio astronomy. Their observations had done much to clarify the nature of radio emission from the quiet sun and had also pinpointed sunspots as the source of violent radio outbursts. It was just the type of start Bowen had pledged to support under the principle 'nothing succeeds like success'. Resources at Radiophysics were mobilised immediately to pursue these discoveries in more depth. Some of the established staff transferred to the new 'radio noise' group (as it was then called) and they in turn were joined by some of the new faces.

One of these postwar recruits was John Gatenby Bolton, later to become the first director of the Parkes telescope. Bolton began his career as a research scientist more by accident than design. Born in Sheffield in 1922, he had gone to the local grammar school and then won a scholarship to study physics and mathematics at Trinity College, Cambridge. After graduating in 1942 Bolton joined the navy as a radar officer but was soon recruited into radar research, first in Scotland and then at the Telecommunications Research Establishment in Worcestershire. In 1944 he transferred to the Pacific and served as radio officer on the aircraft carrier *Unicorn.* By the end of the war Bolton had no strong desire to return to England. His student years already seemed remote and he knew it was difficult for a Yorkshireman to break into English academic life. Instead, Bolton decided to follow up a few contacts he had made in Sydney. Early in 1946, the day after his discharge from the navy, he visited the Radiophysics Laboratory. After a brief session together, Bowen took an immediate liking to Bolton and encouraged him to apply for a new position coming up at Radiophysics. This meeting marked the beginning of a collaboration that would play a major part in shaping the future direction of Australian radio astronomy.

Bolton only just got the position ahead of another applicant, John Paul Wild, who in turn was appointed to the next position to be advertised. Both men had remarkably similar backgrounds but, oddly enough, had not met at the time. Wild was also born in Sheffield, studied physics and mathematics at Cambridge, and then served as a radar officer on the battleship *King George V.* As we see later, their careers continued in parallel: Wild became the driving force in solar radio astronomy in Australia, and Bolton emerged as the outstanding figure in cosmic radio astronomy. Both men were even appointed joint vice-captains of the Radiophysics cricket team!

Most scientists only begin to make their mark in a field of science after serving a long apprenticeship, sometimes lasting many years; for Bolton, it only took a few months. In August 1946 Pawsey received a copy of a new paper by J. S. Hey and his group in England. Following on from his discovery of emission from the sun, Hey carried out a survey of radio strength over the sky and produced radio maps similar to those published earlier by Grote Reber. Hey noticed that whereas radio signals from any given direction were quite constant in strength, a region in the direction of the Cygnus constellation exhibited peculiar fluctuations, changing in intensity during times as short as a minute. On general physical grounds Hey concluded that this intense, variable emission must come from a relatively small region of space, possibly from some unknown star-like object.[10] Pawsey jumped at the opportunity. Within days he had confirmed the existence of the signals from Cygnus and also noted how the fluctuations were not unlike the type of 'bursts' observed in solar noise.

Pawsey then turned his attention back to the sun, assigning the recruit Bolton to a study of the polarisation properties of the sunspot radiation. Bolton began the construction of two new aerials for sea interferometry at Dover Heights in November 1946, assisted by Bruce Slee, another recruit who had spent the war years working on the radar defences of Darwin. Slee was no novice. Towards the end of 1945 he had often observed sea-interferometer fringes with the radar pointing west at sunset, and had correctly guessed that the sun was the cause. Upon reading a newspaper article describing the Radiophysics work at Collaroy, Slee decided to summarise his own observations in a letter. Pawsey was impressed by what he read and, shortly after, offered Slee a position.[11]

By the time the new aerials had been completed the sun had entered a quiet phase. After several days of fruitless observation it seemed pointless to continue. Bolton had become aware of the mysterious Cygnus signal and decided to start where Pawsey had left off. Bolton and Slee turned their aerials out to sea and began hunting for signals with interference patterns that might reveal the presence of radio sources in the sky similar to the strange fluctuations received from Cygnus. But, as Bolton recalled years later: 'Our efforts did not bring any success. After a week or so they were cut short by an unheralded visit to Dover Heights from Pawsey. He noticed straight away that our aerials were not set up to look at the sun. Needless to say, he was not amused and we were both ordered back to the Lab.' [12]

The insubordination was soon forgotten. Back at Radiophysics Bolton joined in the work building two new receivers for an expedition Pawsey planned to make to Brazil to observe a total eclipse of the sun. The passage of the moon's disc across the sun would allow the source of sunspot radiation to be located with pinpoint accuracy. At the last minute the expedition was judged too expensive and cancelled. Pawsey decided to let Bolton use the new equipment for whatever purpose he chose. Bolton decided first to take a close look at the intense outbursts of radio emission which accompany the appearance of solar flares, usually from the vicinity of sunspots. In February 1947 Bolton, together with a young electrical engineer, Gordon Stanley, set up three aerials at Dover Heights each operating at a different wavelength. Early in March an intense outburst was recorded by all three aerials but with a time delay of several minutes from one aerial to the next, in increasing order of wavelength. The same type of delay had already been noticed earlier in a series of observations by Ruby Payne-Scott and Don Yabsley at the inland Potts Hill station. The observations at both field stations showed that solar flares consist of a stream of charged particles, electrons and ions, ejected at very high velocities up through the solar atmosphere. As the charged particles reach the more rarefied upper parts, radio emission occurs at progressively longer wavelengths, thus accounting for the delay observed between the signals. Particularly violent solar flares often lead to magnetic storms on earth, caused by a shower of charged particles arriving a day or so later.[13]

By May 1947 the sun had again quietened to a phase of low activity and, growing tired of routine monitoring runs, Bolton and Stanley decided to return to the topic which Pawsey had brought to an abrupt

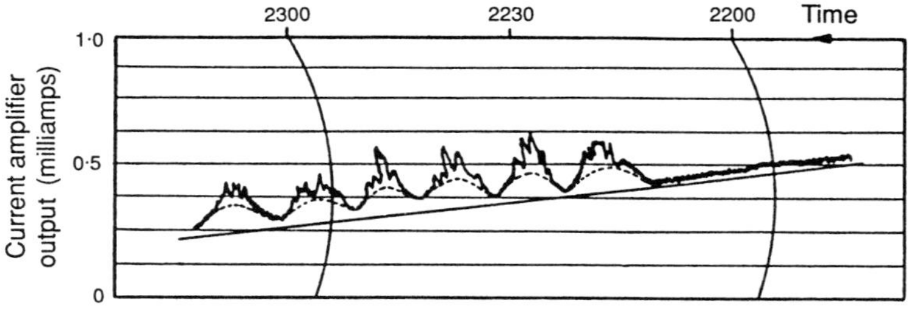

Signals from Cygnus A recorded on the evening of 19 June 1947 at Dover Heights. The strength of the signal decreases (towards the left) as the source rises higher above the horizon. The spiky structure in the signal is the 'twinkling' caused by the earth's ionosphere. The information of real interest is given by the wavy dashed line, which provides a measure of the angular size of the source.

halt six months earlier. This time the search was an outstanding success. Again using the sea interferometer at Dover Heights, Bolton and Stanley pointed the aerial to the north–east horizon and detected a strong source of emission as the Cygnus constellation rose just after ten in the evening, presumably the same source reported by Hey. Of key importance, the 'spiky' interference pattern showed conclusively that the angular size of the source was no more than eight minutes of arc in diameter (one minute of arc is one-sixtieth of a degree). Whereas Hey inferred that the radio source had to be small or point-like, here now was direct evidence.

The detection of the point-like Cygnus source proved to be of fundamental significance, a discovery that raised a host of intriguing questions: Were there other point-like or discrete sources distributed over the sky and, if so, how many? Was it possible that most of the 'cosmic static' originated from discrete sources, rather than large clouds of ionised interstellar gas as Jansky and Reber had believed? Furthermore, what was the nature of this discrete source? In their paper Bolton and Stanley reported that apart from two ordinary looking stars in the region, the position for the Cygnus source did not coincide with any known object listed in the standard star catalogues.[14] Could it be that the new 'radio star' in Cygnus was the first of a new class of objects unknown to astronomers?

Bolton and Stanley decided next to tackle the first of these questions. After improvements to the interferometer's receiver made by Stanley, a systematic search began across the eastern horizon. It soon became clear that the sky had a lot more to offer than just the Cygnus source. By March 1948, after three months of observations, Bolton felt confident enough to announce the detection of a further six new sources. The three best candidates were named Taurus A, Virgo A and Centaurus A. Similar to the Cygnus source, the name gives the constellation in which the source is found, with the letter A indicating that this is the strongest source in this constellation, the letter B the second strongest, and so on. This convenient nomenclature introduced by Bolton is still in use.

So far it had been possible to detect these new sources and to show that they were essentially point-like objects in the sky. The next and much more difficult issue to be confronted was to find out just what these objects were and this, first of all, meant overcoming a major problem. The sea interferometer gave a fairly accurate measure of the angular size of a radio source, but not its precise position in the sky. This is largely because the source is observed very low on the horizon and the incoming radio waves are bent by refraction as they pass through the earth's atmosphere, but by how much was unknown. If the position measured was out by only a fraction of a degree it would make any attempt at identification useless. Even in a small area, say half a degree square, a large optical telescope sees the sky peppered with astronomical objects.

Bolton's solution to the problem was to propose a second interferometer, this time facing westward out to sea, which would provide independent but complementary data to the Dover Heights aerial. Two sites were chosen in New Zealand, north of Auckland, the first on a hilltop on a narrow peninsula near the town of Leigh, at an elevation 300 metres above sea level. In June 1948 observations were made on Cygnus A as it rose and set along the northern horizon. Next Bolton and Stanley towed their mobile radio telescope to the second site near the resort town of Piha on the west coast and observed the other radio sources as they set on the horizon. In the meantime, Bruce Slee had remained at Dover Heights to make the observations of the same sources rising in the east.

The analysis of the New Zealand data back in Sydney proved long and laborious, but by the end of the year Bolton had worked out positions for four sources, each accurate to within 7 minutes of arc in at least one of the two celestial coordinates. Although the accuracy was not good enough to be absolutely sure, Bolton consulted a star atlas and hazarded a guess for three of the sources. Each later proved correct. The identity of Cygnus A remained a mystery, but Taurus A, Virgo A and Centaurus A all appeared to coincide with unusual nebulous objects reasonably well known to astronomers.

Taurus A corresponded to a supernova remnant known as the Crab Nebula, the remains of a star that had violently exploded in the year 1054. No account of this supernova can be found in any European chronicle surviving from this time, but astrologers in the court of the Chinese emperor kept a detailed record of this spectacular event. The supernova appeared suddenly and was said to develop spikes leaving it in all directions. Its reddish-white colour remained clearly visible even in bright daylight for three weeks and for months afterwards it dominated the night sky. The sudden appearance of the supernova had been the result of a star undergoing gravitational collapse and leading to an

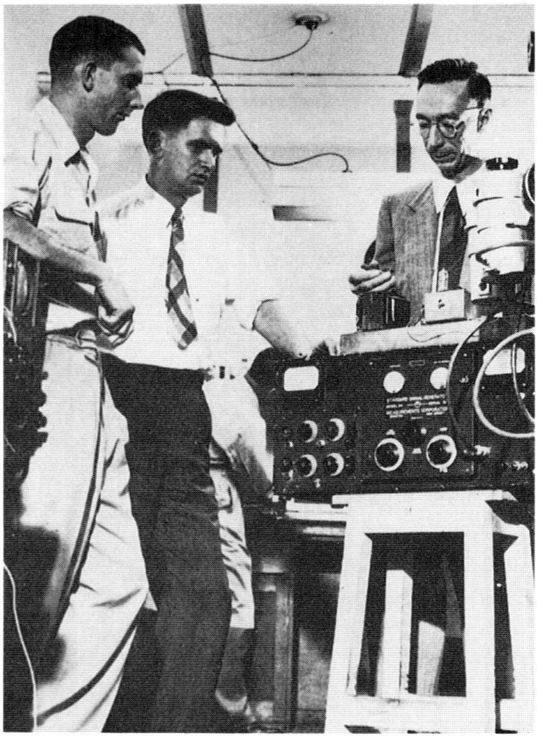

Right John Bolton, Gordon Stanley and Joe Pawsey at the time of the discovery of the first radio sources. *below* The mobile sea interferometer taken by Bolton and Stanley to New Zealand to measure accurate positions for the new sources. The four Yagi aerials forming this radio telescope were essentially the same as early TV aerials.

extraordinarily violent series of nuclear reactions. So much energy is released in explosions of this type that some supernovas, initially, outshine the combined output of several hundred millions of normal stars. Put another way, in a matter of seconds the supernova releases more energy than our sun will radiate in its entire lifetime. For the Crab supernova, nine hundred years after the initial explosion, the shell of gaseous material is still expanding at over one thousand kilometres per second and generating enough energy to make it one of the most powerful radio sources in the sky. As astronomers knew, the occurrence of a supernova remains a rare event. Even among the billions of stars that comprise our own galaxy, a supernova occurs only a few times every century.

The two other radio sources identified by Bolton provided an even bigger surprise. The Crab Nebula is in our galaxy at a distance of 6000 light-years and, relatively speaking, is a close neighbour. The other two sources on the other hand turned out to be galactic systems well beyond our Milky Way. Virgo A coincided with a giant elliptical galaxy, believed to be at a distance of 50 million light-years and classified under the label M87 (or NGC 4486), which has an unusual bright blue jet of material extending from its centre. The other, Centaurus A (NGC 5128), turned out to be a bright and peculiar galaxy with a dark dustband straddling its disc—its distance unknown at the time was later measured to be about 15 million light-years.

Bolton, Stanley and Slee wrote up their work on the identification of these three sources towards the end of 1948.[15] Before dispatching the paper, Bolton sent letters summarising the work to a number of leading astronomers, notably Jan Oort in Holland, Bengt Strømgren in Denmark and Rudolph Minkowski in California, receiving from each an enthusiastic reply. Although several key figures had been awakened to the potential of the new astronomy, according to Bolton, the response overall was disappointing:

> 'Few of them believed us initially. Before the identification paper on Taurus, Virgo and Centaurus I had published another one which listed the six radio sources and said that here was a whole new field of study. No one wrote to me to follow up this paper. Later I heard that the Commonwealth Astronomer in Canberra had remarked that even if these objects did exist they could be of no possible interest to astronomers.' [16]

It took some time before the significance of this work came to be widely appreciated. Indeed, a case could be made that the importance of this Australian work has never been given the recognition it deserves.

In the meantime the handful of radio sources first listed by Bolton's group began to grow. In 1948 Martin Ryle and Francis Graham-Smith adapted the Cambridge interferometer, used initially to study sunspot emission, and discovered a strong source in the constellation of Cassiopeia.[17] Later work showed the source to consist of fast moving

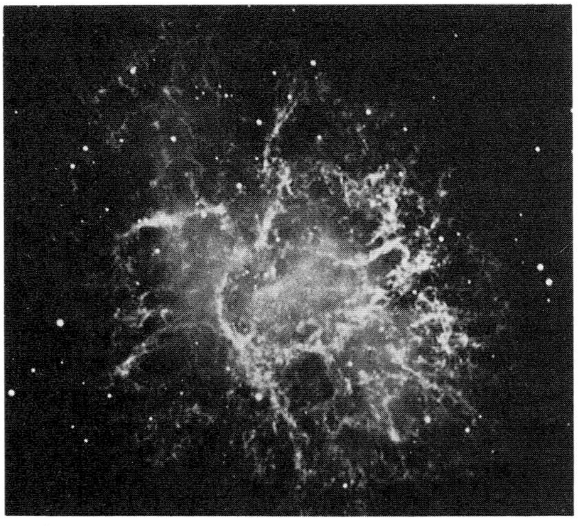

The first three radio sources to be identified with visible objects: *right* Taurus A (the Crab Nebula), *centre* Centaurus A (NGC 5128) and *below* Virgo A (M87).

filaments of material, apparently the remnant of a supernova that exploded sometime late in the seventeenth century. Unlike the observation of the Crab supernova by the Chinese, the Cassiopeia supernova had gone unnoticed by astronomers at the time, presumably because intervening clouds of interstellar dust had blocked out the catastrophic explosion. Another group led by Bernard Lovell at the Jodrell Bank field station in Cheshire also joined in the hunt for radio sources. In 1950 Robert Hanbury Brown and Cyril Hazard detected radio emission from Andromeda, the nearest galaxy to our own and with a spiral structure believed to be similar to the Milky Way.[18] Detailed observations showed that the emission comes from an area larger than the visible galaxy, indicating the presence of an invisible halo of material surrounding Andromeda. Calculations also indicated that the intensity of radio emission is about the same order of magnitude as the total emission from the Milky Way, indicating that our galaxy is by no means unique as a radio emitter. The first four radio sources identified with visible objects had been unusual and rare: two supernova remnants (Taurus A and Cassiopeia A), both closeby galactic objects, and two peculiar galaxies (Virgo A and Centaurus A). The detection of Andromeda, however, a fairly ordinary galaxy, immediately raised the prospect that radio emission might be routinely expected from many of the countless number of objects already known to astronomers. Here was further evidence that the radio universe might reveal a tapestry as rich and as varied as the visible universe.

The identification of these first few radio sources aroused some interest but, as mentioned, did not set the astronomical world on fire. The event that really made astronomers sit up and notice radio astronomy was the identification of Cygnus A. Unlike the handful of radio sources which coincided with known, albeit unusual, objects in the star atlas the Cygnus source had remained an enigma. A number of objects were visible in the direction for Cygnus given by Bolton's group, but the errors in the position measurement were still too large to positively single any one out. A more accurate measurement of the celestial coordinates of Cygnus A was obviously required. In May 1949 Bernard Mills and Adin Thomas at Radiophysics took up the challenge. On this occasion they dispensed with the sea interferometer in favour of a two-aerial system similar to the instrument first used in Cambridge. Observations on Cygnus A were carried out over a six-month period with painstaking care taken to minimise all sources of error. The position measured represented a significant improvement on the earlier value—accurate to 3 minutes of arc in one celestial coordinate (declination) and to 1.5 minutes of arc in the other (right ascension)—a reduction by a factor of 4 in the area of the 'error window'.

Next Mills and Thomas examined a photographic plate of this region of the sky, sent by Rudolph Minkowski at the Mt Wilson and Palomar

Observatories in California. Minkowski had been one astronomer to develop a strong interest in the Cygnus source following correspondence from Bolton and also after hearing a first-hand account of the Australian work by Pawsey during his visit to the United States late in 1947. The examination of the photographic plate revealed a nebulous object very close to the position measured for Cygnus. Mills and Thomas believed that the object was probably closeby in our galaxy, most likely similar to the Crab Nebula. Before announcing the identification, Mills decided first to write to Minkowski and ask what the source could be. To his astonishment, Minkowski replied that on the evidence available the object appeared to be an external galaxy at an enormous distance. On Minkowski's advice Mills and Thomas decided not to risk proposing this galaxy as the Cygnus A source. It seemed highly improbable that a galaxy at such a vast distance could radiate such a huge amount of radio energy, so much energy that its signal received on earth was almost as strong as the nearby sun. Writing up their paper, Mills and Thomas played safe and stated that Cygnus was probably a faint star closeby in our galaxy, though one with exceptionally unusual properties.[19]

Radio astronomers elsewhere were now on the track of the elusive source. In August 1951 Francis Graham-Smith, carrying out research for his doctorate at Cambridge, announced a new position for Cygnus A. This position was almost identical to the Mills and Thomas value but, more important, with an error window ten times smaller in area. This was the critical breakthrough. Minkowski and fellow astronomer Walter Baade now felt that the accuracy was sufficiently good to justify a fair portion of time on the 200 inch Palomar telescope. After many hours of observation with the world's most powerful telescope locked onto the celestial coordinates sent from Cambridge, the photographic plate revealed an irregularly shaped object. Now came the big surprise. An analysis of the feeble light from the object showed sharp emission lines greatly shifted towards red wavelengths, indicating that the Cygnus source was at the astonishing distance of 1000 million light-years.[20] Mills and Thomas had been on the right track after all. The source had been their own candidate for Cygnus, but they had cautiously decided against announcing the identification.

The Palomar observations continued and by May 1952 Baade had become convinced that the strange object consisted of two spiral galaxies, each containing billions of stars, undergoing a head-on collision. A traffic accident of such violent proportions had never been detected by astronomers. Baade's evidence in support of the collision of two galaxies seemed convincing.* For over a decade astronomers believed that many

* Legend has it that Minkowski challenged Baade to prove the collision theory by showing that the spectrum of visible light from the source was consistent with such an event. Baade won the bet and with it a bottle of whisky. It is not known whether Baade returned the favour when later the theory was shown to be wrong (see note 21).

One of the interferometers constructed at Cambridge consisted of two Würzburg reflectors, former radar dishes which had been captured from German forces at the end of the war. With this interferometer, Francis Graham-Smith (pictured) succeeded in measuring the position of Cygnus A with enough precision for astronomers to identify this mysterious source. It was the beginning of an illustrious career for Graham-Smith. In 1982 he was appointed Britain's Astronomer Royal.

of the extragalactic radio sources subsequently discovered were also energy powerhouses resulting from pairs of galaxies in collision. Later, astrophysicists showed this idea to be untenable. Not even the energy generated by such a collision seemed enough to account for the intense radiation. By the early 1960s astronomers had abandoned the collision theory for Cygnus A and other distant radio sources, though for some time after the misconception was propagated in popular books on astronomy.[22] Contemporary theories are steadily making progress in understanding the physical mechanism responsible for generating the extraordinary radio emission from Cygnus and other similar sources. Its level of radio emission is approximately a million times stronger than normal galaxies such as our own or Andromeda, and up to ten billion times stronger than intense objects within our galaxy such as the Crab Nebula. Appropriately, Cygnus and similar exceptionally powerful extragalactic sources were christened radio galaxies.

The optical identification of Cygnus A in 1952 by Baade and Minkowski proved to be a major milestone in the development of radio astronomy. After the sun and the supernova remnant Cassiopeia A, by astronomical standards our near neighbours, Cygnus is the next most powerful radio source for earth-bound observers, even though its distance is estimated to be 1000 million light-years. Much weaker sources could easily be detected even with the simple aerials in use during these early days of radio astronomy. Presumably, it would be possible to detect strong sources similar to Cygnus at much farther distances. Cygnus A had not been beyond the detection limit of the powerful optical telescope at Palomar, but many hours of observation had been required to register the source on sensitive photographic plates and to distinguish it from a crowded cluster of stars and galaxies in the same field of view. The intriguing possibility arose that there might be even more distant objects radio astronomers could detect with relative ease, but which might be close to, or even beyond, the seeing power of the largest optical telescopes. Perhaps radio astronomers might be able to probe further than optical astronomers and learn more about the size, origin and nature of the universe.

In their book *Radio Astronomy*, the first scholarly exposition of the subject, Joe Pawsey and Ron Bracewell commented on the remarkable prospect raised by Cygnus:

> 'The identification of such a strong source—it approximates to the undisturbed sun—with such a distant object ... carries with it an astounding astronomical implication. A similar object, if it existed, would still be detectable by radio means at a distance which is beyond the limit of detection of the largest optical telescope .... There is little doubt that the most arresting possibility indicated by radio astronomy is that, in connection with the Cygnus 'radio star', it may be possible by radio means to extend the bounds of the observable universe beyond the optical limits.'[23]

A daring prediction, and one that essentially proved correct. The discovery during the early 1960s of remote quasi-stellar sources (quasars)—in which the Parkes telescope played a leading role—pushed back the known boundaries of the universe and led to a revolution in modern cosmology.

Budding of the New Astronomy

> No literary genius could have invented a story one-hundredth part as fantastic as the sober facts that have been unearthed by astronomical science.
>
> Fred Hoyle, 1950

The identification of radio sources with visible objects, begun by Bolton in 1947 and leading to the solution of the Cygnus mystery in 1952, forged a major link between radio and optical astronomy. The basis had been laid for an extensive astronomical program in later years, an exceptionally fruitful collaboration which by the 1960s saw the integration of the two branches of astronomy (see Chapter 9). Building on this early success, new observing programs were begun at Dover Heights by Bolton, Stanley, Slee and Kevin Westfold. A number of new aerials for sea interferometry were constructed, the most advanced consisting of an array of twelve Yagi elements. By 1953 a sky survey completed by this array had led to the detection of over one hundred radio sources, an extraordinary proliferation over the first few found at Dover Heights during the winter of 1947.[24] Despite the success of this method, sea interferometry had basically run its course by this stage, limited by a number of uncertain geophysical factors. One was the large amount of atmospheric refraction affecting sources low on the horizon, another was that choppy seas tended to smear out the reflected signals. Even uncertainties in the height of the tides altered the 'baseline' of the aerial perched on the clifftop. Another drawback was that these aerials could only operate at relatively long wavelengths (3 metres for the twelve Yagi element array), which meant that the signals were often seriously affected by irregularities in the ionosphere.

In an attempt to bypass some of these limitations a parabolic reflector 16 feet in diameter was constructed at Dover Heights in 1950. This dish could be steered to any part of the sky and was designed to operate down to the relatively short wavelength of 75 cm (frequency 400 MHz). Basically a larger version of one of the wartime radar dishes, this reflector was an early step along the road towards the Parkes telescope, constructed a decade later.

The parabolic dish produced several interesting results which encouraged Bolton to suggest building a larger version. Without official backing, Bolton and his collaborators began work in their free time on a new parabolic reflector, 72 feet in diameter and called appropriately the 'hole-in-the-ground'.[25] They did the manual labour themselves, shovelling out a shallow hole in the sandy soil. The surface of the hole was compacted with ash brought in by truck from a local power station. A mesh reflecting surface was then laid out consisting of steel strip used for binding packing crates and secured by wooden pegs every several

Two aerials at Dover Heights in 1952: *above* the first of the parabolic dishes, a 16 ft reflector built in 1950. *right* The largest and the last of the sea interferometers, the twelve-Yagi array was mounted on the turntable of a wartime radar aerial. Only the turntable still remains. Nearby, a plaque was unveiled in 1989 to commemorate the identification of the first radio galaxy.

feet. Later, with Pawsey's blessing, this makeshift telescope was upgraded by lining the hole in concrete and covering the surface with half-inch chicken wire. The size of the dish was also extended to a diameter of 80 feet by using aluminium tubes to support the overhanging chicken wire around the perimeter. With all this physical labour, the title Ph.D. among radio astronomers jokingly came to mean 'Post-hole Digger'.

The chief advantage of the hole-in-the-ground telescope was its size. With a diameter five times larger than the neighbouring 16 ft dish its resolving power—its ability to pick out fine detail—was correspondingly five times greater. Unlike the 16 ft steerable dish of course, the hole-in-the-ground was fixed in position. Its operation relied on adjusting the position of the focus on top of a movable mast, and also the natural rotation of the earth which slowly drifted radio sources through the telescope's 'beam' or field of vision. This made the telescope unsuitable for making detailed studies of particular sources, which must be tracked for long periods of time, but ideal for mapping large regions of the sky. Bolton and Dick McGee spent most of 1953 carrying out a survey of radio emission at a wavelength of 75 cm, along a wide strip of sky which included the Milky Way. The contour map of radio strength for this strip of the sky was similar in appearance to Grote Reber's earlier map, but with much better resolution. The map showed that radio emission is strongly concentrated in a very narrow band along the plane of our galaxy. In fact, never before had the galactic plane been so clearly delineated. Until then dense clouds of gas and dust throughout the galaxy had severely restricted attempts by astronomers to observe fine detail on the structure of the Milky Way. Astronomers had estimated that close to 99% of all starlight from the galaxy is obscured by these dust clouds, otherwise starlight at night would be as bright as a full moon. Although these dust clouds block out light at visible wavelengths, they have virtually no effect on the transmission of signals from distant radio sources.

McGee and Bolton found a number of radio sources along the galactic plane and the strongest they named Sagittarius A. In a bold move they proposed that this source is the core or nucleus of the galaxy—the hub from which the long spiral arms of the Milky Way flow out into space.[26] Astronomers knew that the nucleus lay in the general direction of Sagittarius, but had been unable to be more precise because of the intervening clouds of dust. After Pawsey mailed the new coordinates to the Palomar Observatory, Walter Baade replied:

> 'Frankly, I jumped out of my chair the moment I saw what it meant. I have not the slightest doubt that you finally got the nucleus of our galaxy!! … . Altogether I concluded about two years ago—after a careful examination of my 48 inch Schmidt plates of the nuclear region of our galaxy and through checking all suspicious objects at the 200 inch—that there was positively no chance whatsoever to detect the nucleus of our galaxy in the optical range and that we had to await what you radio people could do about it.' [27]

The 'hole-in-the-ground' telescope at Dover Heights, excavated and surfaced with makeshift materials by the astronomers themselves.

Radio astronomy was indeed gaining in respectability. Three years later the International Astronomical Union officially adopted Sagittarius A as the true centre of the galaxy, making its position the zero of longitude in a new system of galactic coordinates.

Sagittarius A was the last major discovery made at Dover Heights. The aerials for sea interferometry and even the hole-in-the-ground had become obsolete with the development of more sophisticated radio telescopes by the Radiophysics group. At the end of 1954 the coastal field station was closed. Until then Dover Heights had been the principal site for cosmic radio astronomy, the detection of discrete radio sources, both galactic and extragalactic, and the mapping of the general distribution of radio intensity along the Milky Way. However, cosmic radio astronomy was by no means the whole story at Radiophysics. During the period 1945–50 the radio astronomy group under Pawsey's leadership grew rapidly in size, diversifying into a number of small and relatively independent teams, operating at a variety of field stations around Sydney. At least half of the group's effort went into studies of the sun, the research program which had started the ball rolling late in 1945 and which had been continued chiefly by Pawsey himself, Paul Wild and

Chris Christiansen. Radio studies of the moon turned out to be another successful program.

Radio emission from the moon had first been detected during the war by Robert Dicke and a group at the Rad Lab in Massachusetts. The emission arises from the lunar surface being heated by the sun and then radiating some of the energy at short radio wavelengths. The radiation is extremely weak and can be detected only because of the comparatively short distance to the moon. In 1948, Jack Piddington and Harry Minnett, after having spent the immediate postwar years at Radiophysics working on radar for air navigation, began a detailed series of observations on both the bright and dark regions of the moon. An analysis of the radio 'brightness' showed that the temperature over the lunar surface varies from 210 to 290 K each month (that is from –63°C to +17°C). However, contrary to expectation, the monthly temperature cycle lagged behind the visible phases of the moon. Radio emission corresponding to the warmest temperature of 290 K was recorded not at the time of the full moon, but about four days later. Puzzling over the cause of the time lag, Piddington and Minnett eventually realised that the radio emission comes from a layer beneath the lunar surface that heats and cools more slowly than the top layer. After a study of various textbooks on heat conduction in solids they concluded that the lunar surface consists of porous rock and gravel covered by a top layer of fine dust with an average thickness of 2 cm.[28] This was a bold piece of lunar geology, carried out with a small radio dish at a distance of 400,000 km. Their discovery was eventually verified by the crew of Apollo 11 in 1969 (see Chapter 10).

'Personally I think we are pestered by enough radio stars on this earth without the CSIR finding others in the skies. But I must admit I like the discovery that the heavenly one is hissing the earthly ones. I've often felt that way myself.'
— Columnist in the *Sydney Morning Herald*, October 1948.

Radar also provided a method of studying the lunar surface, years before manned landings on the moon. In 1946 radar operators in the United States and Hungary were the first to detect radar signals bounced off the moon, but it was a Radiophysics group that achieved the first results of real scientific value. Late in 1947 Frank Kerr, Alex Shain and Charles Higgins began their own experiments using a powerful transmitter installed at the overseas broadcast station 'Radio Australia' at Shepparton in Victoria, in conjunction with a receiving station at Hornsby in New South Wales. This group succeeded in registering echoes from the moon and, by analysing the modulation of the return signal, confirmed some of the deductions made by Piddington and Minnett about the properties of the lunar surface. Another discovery was a rapid fading in the strength of the lunar echoes, caused by a swaying effect of the moon known as libration.[29] (For another unexpected outcome, see Tangent 1 below.)

TANGENT 1

Tom-toms on the Moon?

Australia's early triumphs in radio astronomy received only modest coverage by the media. The first story to make a splash was the work on radar echoes from the moon by Frank Kerr, Alex Shain and Charles Higgins. Radar and the moon, not new-fangled radio telescopes or distant radio sources, were things people could readily understand. A reporter in Shepparton picked up the story and several detailed newspaper reports appeared in November 1947.

Nowadays sizable public relations departments in large scientific establishments work hard to secure this type of media coverage. Ironically, the leaking of this story was looked on at Radiophysics as a minor disaster. Fred White wrote a consoling letter: 'I know it is terribly distressing to have this sort of thing occur, but there is very little that can be done about it'. CSIR's chief executive David Rivett also expressed his displeasure: 'Don't worry about the news hawks. After all, most people probably think it is a matter of tremendous scientific importance to get an echo from the moon. If only reporters had described it as noises from the moon, created, presumably, by the tom-toms of its inhabitants, you would have all been regarded as really great men. Has anybody, by the way, yet asked you who makes the noises in the sun?'[30]

These remarks now seem arrogant, the disdain of an ivory-tower scientist, but they need to be considered in the context of the times. For the first five of the seven years Radiophysics had been in existence, its work had been shrouded in secrecy and by late 1947 it had not shrugged off this wartime legacy.

There was another factor. Any publicity might turn out to be bad publicity in the wrong hands. Pressures at this time were beginning to mount on CSIR. A short time before the moon echo experiment Rivett had written to Bowen: 'It may be that before too long there will be a fight to ensure that CSIR is left as free as ever to work where it wishes and publish what it desires. In some directions it is apparent that 'planners' in departmental circles would rather like to be able to tell us exactly what we are to do; and that, I think, might be the beginning of the end.'[31]

A full-scale assault began less than a year later spearheaded by right-wing attacks in Federal Parliament on CSIR and on Rivett personally. The attacks were motivated partly by Rivett's resistance to a major restructuring of CSIR (the first of several since) and partly because of his reluctance to condemn the employment of several professed communists by the organisation. Australia had entered its own McCarthyist phase. These attacks on Rivett preceded by several years the eventual fall from grace of one of the leading scientists in the United States, J. Robert Oppenheimer. In 1948 David Rivett resigned defeated. The following year CSIR was reconstituted as the Commonwealth Scientific and Industrial Research Organisation (CSIRO).

David Rivett was chief executive of Australia's CSIR from 1926 to 1948. His views on science enabled new fields such as radio astronomy to flourish within CSIR.

These radar and radio studies of the moon, together with the other research in progress during the late 1940s on the sun and on galactic and extragalactic radio sources, covered the complete spectrum, from the closest to the most distant astronomical objects. A significant number of papers produced by the Radiophysics group at this time are now considered classics, the foundation stones of the new astronomy. The contribution from Radiophysics did not go unnoticed elsewhere. As one example, Edward Appleton, the noted radiophysicist and Nobel Laureate for 1947, wrote from London: 'You are turning out some magnificent work in Australia at present, and indeed have done for some years back. I am always stressing to people here how Australia, with its relatively small population, is right in the front of the advances in the field of radio astronomy.' [32]

There is little doubt that by 1950 the Radiophysics Lab had played a leading role in the postwar emergence of radio astronomy. As the 1950s progressed major research centres were formed in countries such as Holland, France and the United States, but it had been the Australian

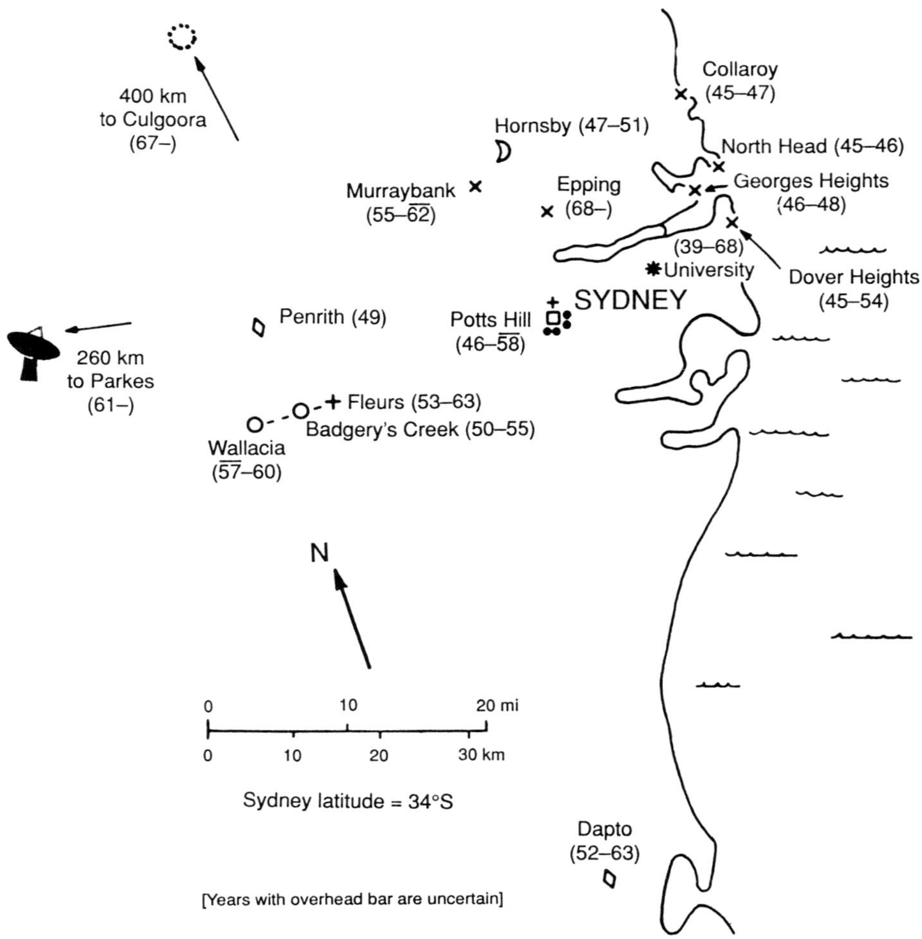

The field stations used by the Radiophysics Lab for radio astronomy from 1945 until recent times.

group that had provided a solid foundation for radio astronomy. The two principal competitors or collaborators during this early period had been the groups at the Universities of Cambridge and Manchester. However, both these English groups had consisted of small sections within their university physics departments, both had been limited to a single field station (Grange Road and Jodrell Bank), and neither had the manpower nor the financial resources to pursue all of the rapidly emerging subfields of radio astronomy to the same extent as the Radiophysics group.

We began this chapter by identifying three factors that helped launch the Radiophysics Lab into radio astronomy. One was the large and highly skilled staff, another the ample supply of modern equipment, while the third factor was morale, a potent blend of postwar optimism and enthusiasm. There is a fourth factor worth mentioning, and this involves the organisational structure of the Radiophysics group. Here the key feature was the number of field stations to which the Lab had access, some of them former wartime radar sites. These sites enabled the rapidly growing group to split up into a number of small and quite independent teams, each pursuing a particular line of investigation. On a purely practical level this had the advantage that one group carrying out observations would not be disturbed by another group needing, for example, to start up electrical machinery to carry out maintenance work. On another level, this relative isolation promoted a healthy degree of self-reliance and inventiveness. Each group had essentially a free hand in dealing with its day-to-day problems and, in the long term, in determining the direction its research would take. A large part of the time at these field stations was spent outdoors, so these new radio astronomers felt like pioneers in the physical sense as well. Chris Christiansen recalls:

> 'Each morning people set off in open trucks to the field stations where their equipment, mainly salvaged and modified from radar installations, had been installed in ex-army and navy huts. At the field stations the atmosphere was completely informal and egalitarian, with dirty jobs shared by all. Thermionic valves were in frequent need of replacement and old and well-used coaxial connectors were a constant source of trouble. All receivers suffered from drifts in gain and 'system-noise' of hundreds or thousands of degrees represented the state of the art. During this period there was no place for observers who were incapable of repairing and maintaining the equipment. One constantly expected trouble.' [33]

As another put it:

> 'People tended to explode out first thing in the morning and you didn't see them until the evening. They tended to be quite independent groups, and competitive in some ways ... and we would have this meeting once a week ... this attempt to get all the information from one group to another, just because they were so relatively isolated during most of the week ... . A person like Pawsey knew the story well, and he would move from one place to another and keep a very close contact with the various groups. But everybody ... was very involved with their own equipment and running it ... and because they were responsible individuals they were given their heads to get on with what particularly interested them.' [34]

Joe Pawsey was the one who managed to make the whole show hang together. Pawsey had effectively sacrificed his personal research career in order to guide and coordinate the activities at these various sites. He suggested research projects, helped in the design of equipment, advised on the most suitable observational methods, assisted with some observa-

Life at the Fleurs field station. A farmer comes to the aid of the mobile lab, affectionately known as 'Flo', and four of the Radiophysics staff: Bruce Slee (left), Alec Little, Les Clague and Kevin Sheridan. The poles form part of the Shain Cross, one of the major telescopes of the 1950s.

tions and, most importantly of all, became a tough and dependable critic when it came to analysing the data and writing it up for publication. Paul Wild writes:

> 'Pawsey's role at this time was quite central. Most people of his group—though perhaps there were a couple who liked to think themselves self-sufficient—looked to him for advice, encouragement and inspiration, which were freely and selflessly given. About once a fortnight the research staff would meet together with Pawsey in the chair. Each person or group would report their progress and Pawsey would ask questions and make suggestions. At other times when one ran into problems, half-an-hour's discussion with Joe tended to be both soothing and rewarding. Then on some days he would arrive unexpectedly at one's field station, usually at lunch time (accompanied by a type of sticky cake known as the lamington, which he found irresistible): or else infuriatingly near knock-off time. During all such visits one had to watch him like a hawk because he was a compulsive knob-twiddler. Some experimenters even claimed to have built into their equipment prominent functionless knobs as decoys, especially for Pawsey's benefit.' [35]

Pawsey had provided the guidance and inspiration during the formative years of the radio astronomy group. In the years ahead the group would steadily grow and consolidate its position at the cutting edge of the new astronomy.

CHAPTER FOUR

Australia Leads the Way

The number of the fixed stars which observers have been able to see without artificial powers of sight up to this day can be counted. It is therefore decidedly a great feat to add to their number, and to set distinctly before the eye other stars in myriads, which have never been seen before, and which surpass the old, previously known, stars in number more than ten times.

Galileo, 1610

Radio astronomy is the youngest child of the oldest of sciences. It is at an age at which its rate of growth is likely to be spectacular.

Joe Pawsey, 1957

The period from 1946 to 1952 marked a distinct phase in the growth of radio astronomy in Australia. A number of major discoveries had been made and, characteristic of many new branches of science in their initial period of growth, often with relative ease. The advances made during these postwar years can be likened to the rapid progress in astronomy more than three hundred years earlier. Galileo with his newly invented telescope discovered in quick succession the moons of Jupiter, the rings of Saturn, dark spots on the sun, mountainous features on the moon, and a host of new stars and nebulas previously too faint to see with the unaided eye.

The discoveries made by the Sydney radio astronomers had put the group in the forefront of this new field. It was a rare occurrence for any Australian group to have reached such international prominence. Until then, the only other occasion when Australian scientists had achieved similar recognition on the world stage was for the research in immunology carried out by Macfarlane Burnet and his group at the Walter and Eliza Hall Institute in Melbourne. Macfarlane Burnet is one of five Australian scientists to have been awarded the Nobel Prize. His award was especially notable, being the only one of the five where the prize-winning research was carried out on Australian soil.

The group at Radiophysics felt justifiably proud of its achievements and of the growing recognition this work was receiving overseas. In turn, according to Paul Wild, this created an air of enthusiasm and excitement in the Lab: 'It was a time when one could discover new things with simple, often improvised equipment; each little group had its own approach and made its own equipment. With the flow of new discoveries from a brand new science a special atmosphere developed—perhaps, though on a humbler scale, something like that at the Cavendish in Rutherford's time.' [1]

In the years between the wars the group at the Cavendish Laboratory in Cambridge led by New Zealand physicist Ernest Rutherford had laid the foundations of another new branch of science—nuclear physics. Apart from the pioneering spirit, Wild's comparison is appropriate for another reason. A brilliant experimentalist, Rutherford had become renowned for making important discoveries with simple equipment often improvised from everyday materials. This simplicity, a key feature in Rutherford's method of research, became known as the 'string and sealing wax' approach; it also essentially characterises the approach which brought such success to the Sydney group in the early postwar years. With an abundant store of wartime radar stock, almost any piece of equipment could be put together for the task at hand with a little imagination and improvisation. This meant that a range of ideas could be tried out: choose a field, rig up the aerial, get the receiving equipment operating, and the observations could begin. Because these projects were on a 'string and sealing wax' scale, each was inexpensive and so all could be accommodated within the Lab's budget. Even the hole-in-the-ground at Dover Heights—one of the most ambitious of the Radiophysics aerials during the period 1946–52—had been constructed from materials pirated from elsewhere, with the radio astronomers themselves providing the labour, all at minimal cost.

Already by 1950 there were indications that this first phase of what we might call 'little' science was coming to an end. After the initial rush of discoveries it became necessary to take stock, consolidate and attempt to understand some of these new astronomical objects and phenomena in greater depth. The early years when a radio astronomer could spend one month studying the sun and the next looking for cosmic radio sources, basically using the same instrumentation, gave way to a new period of increasing specialisation. The semi-autonomous groups operating at the various field stations around Sydney began to stake out claims on the newly emerging branches of radio astronomy, with each group defining its own set of problems and developing its own brand of instrumentation suitable for the task. As one example, for those studying the sun the discoveries of most interest were the strong radio emission from short-lived phenomena such as flares and prominences, bursting out through the solar atmosphere. The main challenge was now to study these phenomena in more detail. As a result, the solar group developed into several self-contained units with their own research programs requiring their own techniques and equipment. Once this process of specialisation had begun, it was only a matter of time before the solar group, similar to the others, would require larger and more costly radio telescopes.

Australian radio astronomy received increasing international recognition during the 1950s. A landmark was the decision by the International Scientific Radio Union (URSI) to hold its general assembly in Sydney in August 1952. This was the first occasion any meeting of an international scientific union had been held outside Europe or the United States. Fifty radio scientists made the journey to Australia, about one-third of whom were active in radio astronomy. The scientific sessions were interspersed with a busy round of concerts, dinners and tours of the Radiophysics field stations. The meeting provided a showcase to the scientific world of Australia's advanced position in radio astronomy. On a personal level, radio astronomers had the opportunity to renew friendships or meet for the first time. *above* John Bolton and Robert Hanbury Brown (Jodrell Bank). *below* At the Dapto field station (foreground): Harry Minnett (left), Francis Graham-Smith (Cambridge) and Paul Wild.

The solar radio astronomers faced a fairly well-defined task. The sun is close by and its radio emission is relatively easy to detect. Further, optical astronomers had already learnt much about its physical characteristics. The major research program was to consist of disentangling and classifying the complex and interrelated types of solar emission. By comparison, in 1950, cosmic radio astronomy remained in a more primitive state. As mentioned earlier, even though the number of known radio sources had begun to multiply, a range of fundamental questions had to be addressed: Do the number of discrete sources approach anywhere near the billions of stars and galaxies visible through the largest optical telescopes? If so, are these radio sources familiar stars and galaxies or are they some type of new astronomical object? Are most of these sources located within our own galaxy or associated with distant extragalactic objects? What is the physical mechanism responsible for the radio emission from these objects?

The answers to these questions depended largely on the ability to identify the radio source with a visible object and then to draw on the extensive body of existing astronomical information to learn more about the object. And yet, as described in some detail earlier, only four radio sources had been positively identified and the case of Cygnus A previewed the difficulties ahead. The cosmic group at Radiophysics decided to press ahead and locate as many radio sources as possible in the expectation that certain features or characteristics might be revealed by a large sample, irrespective of whether these sources could be identified with known optical objects.

The best chance of success in this search for more radio sources would be with larger and more technically refined radio telescopes, capable of improving the observations in three respects. The first was to increase the *sensitivity* of the observations so that fainter and fainter radio sources could be detected above the general background of radio noise. Basically this could be achieved by building larger telescopes with greater collecting areas and by developing more powerful receiving equipment. Second, telescopes needed to have a greater *resolving power* to distinguish between radio sources which might be clustered closely together. Third, a major improvement was required in the *pointing accuracy* of radio telescopes so that reasonably precise celestial coordinates for the sources could be measured and then passed to the optical astronomers for identification.

In Part 2 we see how the idea for the Parkes telescope arose in direct response to these demands of sensitivity, resolving power and pointing accuracy. The Parkes telescope would however take a decade to plan, design and build. Of more immediate concern to the Sydney group, circa 1950, was to settle on a new type of telescope to supersede the simple, improvised instruments characteristic of the 'string-and-sealing wax' approach and to keep the group in the forefront of radio astronomy. Most of this chapter surveys the progress of radio astronomy during the 1950s, but beforehand we look at several proposals for radio telescopes discussed by the Radiophysics group in the early 1950s. As it turned out, the principal instrument to dominate Australian radio astronomy during the decade was a new type of cross telescope devised by Bernard Mills.

☆ ☆ ☆

A search through the files at Radiophysics covering these early years reveals almost as many proposals for new radio telescopes as there were radio astronomers. One made in 1949 by Taffy Bowen grew out of the close ties that had been maintained with the RAAF. As mentioned earlier, the RAAF had placed aircraft at the disposal of the cloud and rain physics group and, in return, Radiophysics had continued to provide advice on subjects such as air navigation. Another area of collaboration was Australia's network of air-warning radar. Various schemes for improving the effectiveness of coastal radar were investigated, such as increasing the power of transmitters, the sensitivity of receivers and, what seemed a good option, increasing the size of radar aerials. Several aerial designs were roughed out, costed and presented to the RAAF for examination, the most ambitious being a parabolic cylinder approximately 24 feet high and 200 feet long which could be rotated horizontally and tilted up or down in elevation. The design called for waterpipe tubing for the structure and chicken wire for the parabolic surface. After further discussions the design was improved by adding a set of beams to strengthen the structure. Bowen had another motive in pursuing these design studies for air-warning radar. Each aerial in the network would be sited at stations such as Dover Heights and, as Bowen hoped, a time-share arrangement with the RAAF would allow Radiophysics to use these installations for sea interferometry. However, the idea was eventually abandoned. Radar technology had undergone rapid development in the United States and the RAAF decided against carrying out its own research and development in favour of a radar system which could be more readily bought 'off the shelf' overseas.[2]

In the meantime several other radio telescopes had been proposed by the Radiophysics group. Late in 1950 John Bolton suggested a new form of interferometer consisting of two large parabolic dishes, with one dish movable to vary the separation between the two. The instrument combined the good resolving power of an interferometer with the high sensitivity of a large aerial. Pawsey was one who favoured the proposal. In February 1951, in a document surveying the future radio astronomy program he pointed to the need to 'make a careful investigation of the best way of realising equipment to extend the cosmic noise radio star survey and then proceed with construction of equipment. Present ideas are in terms of two eighty-foot parabolas mounted to point in any direction, operating round 100 MHz. I regard this as the major line of development.'[3]

Even though the proposal had Pawsey's formidable support, others in the radio astronomy group argued that it would be better to build a smaller and simpler instrument. A rival proposal by Jack Piddington called for a single fully steerable dish about 70 ft in diameter. A single dish would not only avoid some of the operational difficulties associated with an interferometer, but could also be built with the existing know-how at Radiophysics and completed in a much shorter time. The proposal suddenly took on more appeal in March 1951 with the discovery by an American group at Harvard that interstellar hydrogen emits radio waves at a wavelength of 21 cm (see later in this chapter). A 70 ft dish could easily be designed to operate at this short wavelength and would allow this important discovery to be followed up in the southern hemisphere. In response to these developments Pawsey wrote to obtain technical details on a 50 ft fully steerable dish recently brought into operation by a group at the Naval Research Laboratory in Washington. A Dutch group at Leiden University, which had been involved in the discovery of the 21 cm hydrogen line, had also immediately begun planning an 80 ft dish specifically to study the new type of emission.

By the middle of 1951 Radiophysics possessed enough information and expertise to proceed with either one of Piddington's or Bolton's ideas, but for a variety of reasons neither proposal got beyond the drawing board.[4] One factor was a significant mobilisation of the Lab's resources into two major instruments for solar observations at the Dapto and Potts Hill field stations during 1952. Another factor was that a number of other proposals for even larger radio telescopes came under investigation at the time, each of which to varying degrees seemed promising (see Chapter 6). Foremost among these was Bowen's own project of building a giant parabolic dish and, though the prospects looked fairly bright in 1952, even the most optimistic estimates placed the completion date no earlier than 1956. In the fast moving field of cosmic radio astronomy, the prospect of having to wait at least four years for a new instrument raised the obvious danger that the Radiophysics group would slip from its position at the forefront. Some new type of telescope was required in the short term, until it became certain whether and when Bowen's giant dish would be built. Despite long and at times vehement discussions over the relative merits of a host of new designs, in the event an entirely new type of radio telescope emerged triumphant.

Bernard Mills had joined Radiophysics in 1942 and immediately after the war had spent two years developing new types of high voltage X-ray tubes. After transferring to the radio astronomy group in 1948 he began investigating the size and structure of several radio sources, including the enigmatic Cygnus A (see Chapter 3), trying out a variety of interferometers with simple aerials spaced at distances up to 10 km apart. This period of experimentation impressed on Mills that to achieve a high resolution the collecting areas of the aerials were relatively unimportant in comparison with the spatial extent of the array. Mills investigated a number of aerial systems before hitting on the idea of having two long arrays of small dipole aerials lying north–south and east–west along the ground in the form of a cross. The key feature was that each arm of the cross would produce a fan-shaped beam which could be fed to the central receiver through a switch which, in turn, would join the two beams alternately in and out of phase. Combining the two signals in this way would lead to a relatively narrow 'pencil' beam which would be far more effective in resolving closely spaced radio sources than other types of interferometers. The great promise of the cross was that its resolving power would be roughly equal to a parabolic dish with a diameter equal to the length of the cross arms. Because the working components housed in each arm would be inexpensive and involve no complex engineering, it seemed possible to build a cross of very large dimensions.

The theory of the cross was not exactly simple. The initial reaction to the proposal prepared by Mills in March 1953 proved far from positive:

> 'Our work was delayed somewhat by the scepticism of the powers-that-be and I remember being informed with great authority that my proposed 'cross' could not possibly work. It was first necessary to build a pilot model to demonstrate that the principle of operation was sound and that funds should be provided for an instrument of adequate size. This did have advantages, however, in enabling us to experiment with instrumental techniques and eventually produce a very reliable system.' [5]

After Mills received backing for the proposal, the pilot model with arms 120 ft in length and operating at a wavelength of 3 metres was constructed at the Potts Hill field station west of Sydney. The trials not only showed that the idea worked but also turned in enough useful astronomy to convince the radio astronomy group that, among the range of competing proposals for new telescopes, a full-scale cross was worth supporting.[6] First, a new field station had to be found because the Potts Hill site did not have an adequate area of flat land. A disused airstrip further west was leased and christened the Fleurs Radio Observatory: the new site became the main field station used by Radiophysics for the rest of the 1950s. The full-scale cross consisted of two arms 1500 ft in length, each housing two rows of 250 dipole aerials tuned to receive at a wavelength of 3.5 metres.* To scan areas of the sky the beam could be swung between north and south by making appropriate phase adjustments to the dipole aerials in the north–south arm of the cross,

* The effective 'beamwidth' of the cross was 0.8 degrees (or 48 minutes of arc). A parabolic dish operating at the same wavelength would need to be 1100 ft in diameter to achieve the same resolution—over four times the diameter of the giant dish then under construction at Jodrell Bank. However, the main drawback of the cross was that it could only operate at the wavelength 3.5 m. To change the operating wavelength would involve an immense task of replacing every dipole aerial in both arms.

Above The first full-scale cross telescope devised by Bernard Mills and built at the Fleurs field station in 1954. *below* Alec Little supervised the construction of the cross. After joining Radiophysics as a messenger boy aged 15, Little studied part-time and steadily worked his way up into the research ranks.

while rotation of the earth provided movement across the sky in the east–west direction.

Mills spent much of 1953 visiting the United States and advising the Americans on aerial design. In his absence Kevin Sheridan developed most of the complex receiving equipment, while supervision of the construction was left in the hands of Alec Little. By the middle of 1954 the full-scale cross began producing its first data. Detailed studies were made of a select number of radio sources but by far the richest scientific yield, as we see below, came from the program for which the cross had been designed—the most extensive survey yet of the southern sky.

The Mills Cross telescopes dominated research in cosmic radio astronomy by the Sydney group for the rest of the 1950s. The principle also found application in other areas such as solar studies (the Chris Cross) and in radio emissions at very long wavelengths (the Shain Cross). The construction of these crosses marked a transition from what Paul Wild has called the period of 'little' science in the postwar years, characterised by its small, inexpensive and experimental radio telescopes, to a period of 'middle' science which saw the introduction of larger and more complex equipment and more refined observing techniques.[7] The Mills Cross telescopes could, however, still be wholly designed, constructed and financed with the resources available to Radiophysics. The 1500 ft Mills Cross for example took ten men about nine months to build and the materials cost about £5000. The project remained an in-house one, though this severely restricted the possibility of backing any of the other proposals for large instruments for cosmic work put forward in the early 1950s. The decision in favour of the cross created a split within the radio astronomy group and, in fact, foreshadowed a more intense struggle for the limited resources of the Lab in the closing years of the decade.

The supporters of the 70 ft parabolic dish definitely came off second best, having to settle for a smaller 'transit' dish 36 ft in diameter erected for hydrogen-line work at Potts Hill early in 1953. However, the real losers were the supporters of the twin-dish interferometer. Disappointed at not obtaining backing for this proposal, John Bolton decided to leave radio astronomy and join the rain-making trials carried out by Radiophysics. Bolton eventually got his interferometer later in the 1950s, but that was courtesy of the California Institute of Technology at its new radio astronomy observatory at Owens Valley.

The arrival of the Mills Cross proved timely in that it served as the major type of instrument during this period of 'middle' science, before the Sydney group embarked on an era of 'big' science with the Parkes telescope. Although Bowen had already fixed his sights on building a giant radio telescope by late 1952, his plans moved much slower than hoped. The first promise of substantial funding did not come until May 1954, the same month that the Mills Cross came into operation. When the design work on the Parkes telescope did commence in 1956 it took, unexpectedly, more than three years. That story will be taken up in some detail in Part 2. For the remainder of this chapter, we survey some of the major developments in radio astronomy during the 1950s.

Sentinels of the Sun

In the early 1950s, while many radio astronomers turned their attention to the exploration of the universe at large—the Milky Way, extragalactic sources and cosmology—at least as many radio astronomers concentrated their energies on the sun, for earth-bound observers the strongest and most complex radio source in the sky. As one pioneer, Mukul Kundu, wrote:

> '... the sun still appeared an interesting and exciting object to study to many radiophysicists, because it presented a unique astrophysical laboratory, close enough to study in detail a great variety of high temperature plasma phenomena relevant to many physical processes that occur elsewhere in the universe. Consequently, there was emphasis on new instrumentation designed for the study of specific phenomena.' [8]

Solar research at Radiophysics proceeded on two broad fronts: one was to record radio bursts from short-lived phenomena such as solar flares and, the other, to map the distribution of radio brightness of the 'quiet' sun and slowly varying features such as sunspots. The first line of research, pursued by Paul Wild and his group, provided a spectacular illustration of the violent activity occurring in the solar atmosphere. The strongest outbursts are about ten thousand times as intense as emission from the quiet sun and are usually accompanied by visible solar flares. In 1947 Payne-Scott, Yabsley and Bolton had discovered that radio emission from these bursts is detected at varying times according to frequency—the shortest frequencies arrive first and the longest last. This discovery highlighted the need for a new type of radio telescope which could study these short-lived disturbances over a wide range of frequencies. The task was assigned to Wild and Lindsay McCready and, in 1949, they constructed the first radio spectrograph near Penrith at the foot of the Blue Mountains west of Sydney. The spectrograph consisted of a rhombic (diamond) shaped aerial made of wood and wire, a winch-controlled rope system for tracking the sun, a receiver which could be tuned back and forth over a range of frequencies in a fraction of a second and last, but not least, a hand-cranked movie camera. Wild recalls:

> 'We kept the aerial pointed towards the sun and photographed the cathode-ray-tube display when anything interesting happened—i.e. when the sun erupted with a burst of 'noise'. Each day we travelled to Penrith by train—neither Bill Rowe nor I had a driving licence in those days. After four

months of this we closed down the station and came back to the Lab to analyse the results, which were published as four scientific papers. It was typical of all the teams working in that group that our equipment, our results and our conclusions would start a new line of research, eventually to be taken up (usually in much more opulent surroundings and with much more substantial budgetary provisions) in laboratories and observatories around the world.' [9]

So began the study of the dynamic radio spectra of the sun, a field which has been pursued with more success in Australia than anywhere else in the world. Following the makeshift Penrith instrument another spectrograph covering a much wider span of frequencies (from 40 to 210 MHz) was brought into operation in 1952 on a dairy farm near Dapto.*

The outcome of this work at Penrith and Dapto was to unravel and classify three types of bursts, each associated with visible flare activity on the sun. Type I emission appeared as a multitude of sharp spikes on the cathode-ray tube, each lasting several seconds and superimposed on enhanced radiation from complex and stationary noise storms hovering above sunspot regions. Type II bursts were classified as intense metre-wavelength emission lasting for several minutes and often accompanying especially bright solar flares. The emission is produced by shock waves formed by the explosive force of the flare and moving out through the solar atmosphere at up to 1500 km per second. On some occasions, a day or so later, the shock wave reaches the earth producing striking auroral displays in the northern and southern skies and strong magnetic storms. Though similar in a number of respects to this second type, the type III bursts are of much shorter duration lasting only a few seconds, a radio flash caused by streams of electrons cannoning out from the sun. By measuring the rate of frequency drift Wild calculated that these electron streams travel at velocities up to one-third the speed of light and make their presence felt on earth within thirty minutes after the initial eruption. The existence of these electron streams was directly confirmed by satellite observations ten years later.[10]

Later, two more types of emission were added to the list, the next by a French group observing solar activity during the International Geophysical Year in 1957 at Nançay. This type IV burst is characterised by its long duration over a wide range of frequencies and is caused by a large chunk of corona being blown out at high velocity. Two years later Wild's group identified type V, consisting of radiation from electrons which lasts for a minute or so and which often follows type III.[11] The construction of the radioheliograph at Culgoora in 1967—after Parkes and the Molonglo SuperCross the next big science instrument for radio

* Located 100 km south of Sydney, the Dapto site at the time was the most distant and isolated of all the Radiophysics field stations, chosen to avoid sources of man-made interference which are at maximum strength during daylight hours. In contrast to the nocturnal cosmic group, however, the solar radio astronomers could at least work during normal daylight hours.

Paul Wild inspects one of the rhombic aerials forming the new solar spectrograph at Dapto in October 1951.

astronomy in Australia—consolidated the position of the Radiophysics group at the forefront of solar research (Chapter 8 has more details).

☆ ☆ ☆

In parallel with these studies of radio emission from extremely violent, transient phenomena, another group at Radiophysics led by Chris Christiansen concentrated on observations of the quiet sun and other comparatively subdued features such as sunspots. Christiansen was probably the only Radiophysics scientist who had originally considered astronomy as a profession, but then got sidetracked into radio science. In 1948 he joined the group after eleven years research in long-distance radio communication and antenna design at the AWA Research Labs in Sydney and soon became one of the most inventive designers of new radio telescopes. In 1951 he developed a new type of array with high resolving power for mapping the distribution of radio strength across the sun's disc. Known as the grating interferometer, the array consisted of

32 parabolic dishes each 6 feet in diameter and each spaced at equal intervals along one side of a reservoir at the Potts Hill field station. As Christiansen recalled, operating the instrument required a certain degree of athletic ability: 'Each of the 32 paraboloids was mounted on a very simple polar axis on a wooden pole and could be turned by hand to follow the movement of the sun. At quarter-hourly intervals and in the short time between successive scans of the sun, this involved a 200 metre run to push forward each of the 32 antennas by one 'stop'.' [12]

Christiansen and Joe Warburton carried out a series of observations on individual regions of enhanced radio brightness, measuring both the size and temperature of the emitting areas. These regions were found to correspond closely with bright patches in the chromosphere known to astronomers as plages, and so were appropriately named radio plages. Another discovery was that emission is stronger near the sun's equator than from the poles, and stronger still near the equatorial edge of the solar disc—a new effect known as limb brightening.[13]

Early in 1953 Christiansen built a second line of sixteen parabolic dishes, joining the first at right angles and running along the north–south wall of the Potts Hill reservoir. The instrument was used to confirm evidence from the French group that the sun is elliptical in shape, not circular, with a pronounced equatorial bulge. This L-shaped array in turn led Christiansen to devise an even newer type of instrument known as the crossed-grating interferometer, based on the same principle as the Mills Cross.[14] Called the 'Chris Cross', the new instrument was completed in 1957 at the Fleurs field station, each arm stretching for 1200 feet and consisting of thirty-two parabolic dishes. Operating at a wavelength of 21 cm, the Chris Cross could scan the sun's disc to produce a high-resolution two-dimensional snapshot of solar activity every 40 minutes.[15]

Variations on these solar instruments developed at Radiophysics were built around the world. In 1956 the French group at Nançay built an east–west array (used to discover the type IV solar burst) and followed it with a similar north–south line. At Stanford University near San Francisco, Ron Bracewell (a former member of the Sydney group) constructed a T-shaped interferometer operating at a 10 cm wavelength, providing maps of the sun complementary to those obtained from the Chris Cross at 21 cm. Solar radio astronomy flourished. In 1956 one writer counted over fifty separate radio telescopes engaged in solar work in countries as diverse as India, Japan, Russia and the Belgian Congo.[16]

☆ ☆ ☆

Whereas the sun had become a prime target for radio astronomers in the early 1950s, little thought had been given to studying our other near neighbours in space. Even though the weak thermal radio waves from the moon had revealed interesting properties of the lunar surface in the late 1940s, the far greater distances to the planets seemed to rule out the

Above The L-shaped grating interferometer at the Potts Hill reservoir was used to make radio maps of the sun. *right* Chris Christiansen adjusts the position of one of the 32 dishes forming the east–west arm of the interferometer.

possibility of doing any useful planetary radio astronomy. The observation of strong radio emission from Jupiter, therefore, came as a complete surprise.

Early in 1955 Bernard Burke and Kenneth Franklin at the Carnegie Institution were conducting trials of a new Mills Cross at Seneca in Maryland when they noticed unusual bursts of radio noise from a direction well away from the sun. The most likely source of the bursts seemed to be some form of terrestrial interference, possibly from something as innocuous as the ignition system on a local tractor.

Recordings of the mysterious signals continued for three months before the Carnegie group finally realised that the source was in fact Jupiter. Serendipity had struck again in radio astronomy, similar to the discovery of cosmic noise by Karl Jansky in 1932 and the discovery of solar emission by J. S. Hey in 1942.

When news of Jupiter reached Sydney, Alex Shain decided to look back through old records of a survey of cosmic radio noise at 18 MHz he had made in 1950. Sure enough, numerous series of these Jupiter bursts had been recorded at the time but disregarded as some unidentified type of local interference. Shain had enough data to make a detailed analysis of the Jupiter signals and could show, for example, that a series of bursts recorded in 1951 came from a relatively small region of the planet that coincided with a disturbed area studied in detail by optical astronomers at the time.[17]

The detection of Jupiter spurred on attempts to detect emission from other planets. For a time the hunt became the preserve of American radio astronomers. In 1956 a group at the Naval Research Laboratory in Washington detected emission at centimetre wavelengths from both Venus and Mars, while Saturn followed in 1957 through work at the National Radio Astronomy Observatory in West Virginia. The other planets followed later, with astronomers at Parkes adding Uranus to the list in 1964 (see Chapter 10). The radio emission yielded new information on each of these planets, with Venus providing the greatest surprise. The surface temperature was found to be a searing 300°C, a result of the planet's dense cloud cover. Astronomers had discovered the greenhouse effect on our near neighbour, well before it became a major concern here on earth.

In contrast to Jupiter, however, the emission from each of these planets turned out to be normal thermal radiation, and none are comparable in strength with the extraordinary non-thermal radiation from Jupiter. As we see later, a closer study of this unusual Jovian radiation became a major field of study at Parkes in the 1960s.

Away in the Milky Way

> Let us first understand what a galaxy is. The following is a recipe for making galaxies: Take about ten thousand million stars. Spread them so that on the average light takes three or four years to pass from one to the next. Add about the same amount of matter in the form of diffuse gas between the stars. Roll it all out flat. Set it spinning in its own plane. Then you will obtain an object which, viewed from a sufficient distance, will probably look more or less like a spiral nebula.
>
> Arthur Eddington, 1935

By the early 1950s radio astronomy had divided into the two quite distinct fields of solar studies and the other loosely known as cosmic studies. Sky surveys carried out at places such as Dover Heights and

Cambridge had steadily increased the number of known radio sources but, because only a handful had been identified with visible objects, no firm evidence had emerged to decide whether most sources were in our own galaxy or were more distant extragalactic objects. Further, these surveys had been carried out over a variety of observing frequencies. Radio astronomers generally believed that, because sources could be expected to emit continuum radiation over a wide frequency range, the broad features of the radio sky would not change significantly by observing at higher or lower frequencies. This picture changed dramatically in March 1951 with the discovery that hydrogen in interstellar space emits radio waves at a single sharp frequency. The early pioneers had referred to radio emission as cosmic static. In a similar way, the surprising discovery of hydrogen emission can be likened to someone turning a knob across the AM dial, through the static, then suddenly finding a radio station broadcasting a clear signal at a precise frequency.

Hydrogen is by far the most abundant element in the universe. For many years astronomers had theorised that a significant fraction of the total material in our galaxy might consist of atomic hydrogen, left in interstellar space after the earlier formation of stars. Some of this hydrogen in the vicinity of very hot stars becomes ionised and can be observed optically, but the vast bulk remains as a cold, neutral and extremely rarefied gas dispersed throughout the Milky Way, invisible to optical telescopes.

The suggestion that it might be possible to observe radio emission from interstellar hydrogen had been made as early as 1944. During the German occupation of Holland, a copy of the *Astrophysical Journal* containing Grote Reber's 1940 paper had found its way to Jan Oort, assistant director of the Leiden Observatory and a leading authority on the structure, composition and dynamics of the Milky Way. Oort had spent most of the war in hiding because of his outspoken protests against the dismissal of Jewish academics in Holland. Oort had been one optical astronomer to take an immediate interest in Reber's work, recognising the great importance it would have for his own work on the galaxy if a sharp emission line could be detected in the radio spectrum.

In the spring of 1944 Oort put the problem to one of the Leiden staff, Hendrik van de Hulst. After carrying out the calculation, van de Hulst concluded that hydrogen, the most important astronomical atom of all, just might produce the desired radio line. The electron and proton in a hydrogen atom have their own intrinsic spin: the case where the two spins are aligned, or parallel, is in a slightly higher energy state than the antiparallel case. In the parlance of atomic physics, a change from one state to the other is known as a hyperfine or spin-flip transition, which sees the emission of a photon corresponding to the radio wavelength of 21 cm. The probability of a single hydrogen atom undergoing such a transition is extremely small, on average only once every 11 million years!

And yet, despite these extremely poor odds, as well as the very low density expected for interstellar hydrogen, van de Hulst predicted that the vast distances of space would compensate for these two factors and that it might be possible to detect hydrogen's distinctive signature at 21 cm.

Immediately after the war van de Hulst travelled to Wheaton, Illinois to discuss with Reber whether equipment could be built to detect the hydrogen line. Initially, Reber was pessimistic about the possibility of building a receiver with enough stability and sensitivity to observe at, what was for the times, the high frequency of 1420 MHz (corresponding to 21 cm wavelength). Reber made a start on a 1420 MHz receiver but then abandoned the work when, in 1947, he moved to Washington to take up a new job at the National Bureau of Standards.

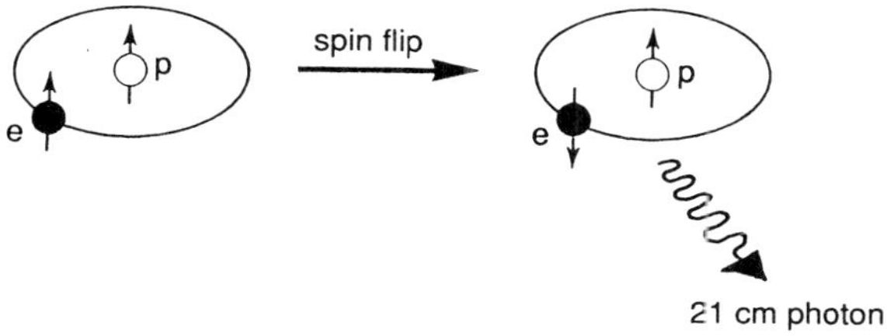

Radio emission from interstellar hydrogen occurs with a sudden flip in the spin of the electron, from parallel to the spin of the proton to the opposite direction. The reverse process can also occur with absorption of a radio photon, which 'pumps' the hydrogen atom back to the parallel spin state at higher energy. *left* Jan Oort, head of the Dutch radio astronomy group at Leiden. Oort realised the hydrogen line would provide a powerful new technique for investigating the Milky Way.

At this point the time was ripe for Radiophysics to step in and mount a search for the hydrogen line. The Sydney group had taken the lion's share of the early discoveries in postwar radio astronomy but, curiously, on this occasion no attempt was made to investigate this promising idea. Pawsey had been alerted to van de Hulst's prediction during a visit to Washington early in 1948:

> 'Mr Reber gave me some very valuable information. He tells me that there is an absorption line of hydrogen atoms on a frequency of 1420 MHz … . This is derived from theory and from laboratory work which he thinks is published, but which I have not yet seen. It may be in the *Physical Review* and is probably by the Columbia University people. If this is correct, there may be very considerable interest in searching for either cosmic or solar noise absorption or emission bands at this frequency.' [18]

Bowen too had learnt of the possibility of line radiation from the physicist Isidor Rabi, a wartime colleague at the MIT Rad Lab. Rabi's group at Columbia University in New York had extensively studied the magnetic properties of atomic nuclei and drawn up a long list of naturally occurring elements, such as hydrogen, which could be expected to absorb radiation at radio frequencies. Rabi had urged Bowen to investigate whether radio lines characterising these elements could be detected in the galaxy. Several members of the Sydney group had also become interested in the subject. One was Kevin Westfold:

> 'In some regards Pawsey was fairly conservative and he wanted at this time to concentrate the strength of our attack to certain problems. On this occasion he wasn't prepared to back the judgment of a couple of upstarts. He had respect for Bolton, but didn't like him doing things he wasn't supposed to. All we had to do was to get the receiver and the feed right. Paul Wild was also one who probably knew more about it than anyone else. He wrote up a very good report on the theory of line radiation which was later published in the *Astrophysical Journal*.' [19]

Despite the initial interest of senior staff such as Pawsey, the keenness of several younger radio astronomers, and the expertise to build the right equipment, the opportunity slipped away. After such an illustrious start, this was the first major mistake, or rather misjudgment, made by the Sydney group.

The importance of van de Hulst's prediction had not gone unheeded elsewhere. At Harvard, Edward Purcell, also a former member of the wartime Rad Lab, had extended the program begun by Rabi's group at Columbia and developed precise methods for studying atomic structure using a technique known as nuclear magnetic resonance. Rabi had already received the Nobel Prize for physics in 1944 for this work, while Purcell was to follow suit in 1952. In 1949 Purcell assigned one of his Ph.D. students, 'Doc' Ewen, to the task of building a suitable receiver and aerial to detect the hydrogen line. In the meantime Jan Oort and his group at the Leiden Observatory had decided to take up the search themselves, rehabilitating an old Würzburg radar dish abandoned by German forces in the wake of the D-Day invasion. A close race developed between the Harvard and Leiden groups, one which took on added drama when a laboratory fire destroyed most of the Dutch receiving equipment. Twelve months later, in March 1951, Ewen and Purcell announced that they had detected the hydrogen line.

Curiously, van de Hulst was visiting Harvard at this time and witnessed the discovery of the line he had predicted seven years earlier. Also present was Frank Kerr from Radiophysics, who had decided to spend a year taking the Harvard course in astronomy:

> 'Purcell called us all together on the morning after that overnight discovery. Cautious physicist that he was, he wanted to see confirmation of the

detection before publishing the result. He suggested that van de Hulst and I should cable our respective institutions in Holland and Australia to report the discovery and ask whether early confirmation would be possible.'[20]

On receipt of Kerr's cable, Chris Christiansen set aside his studies of the sun and, with Jim Hindman, embarked on a crash program at Potts Hill to assemble the necessary equipment. They were able to confirm the discovery, the product of almost two years work by the Harvard physicists, in a little over two months. In Leiden, Oort and Alexander Muller were able to announce their own detection after a similar interval of seven weeks. At Purcell's suggestion the first papers from Harvard and Leiden on the hydrogen line were published in the same issue of *Nature*, accompanied by a cable from Pawsey arriving in time to announce Sydney's confirmation.[21] In contrast to the sometimes fierce rivalry which arose between certain groups (see Tangent 3 later), the collaboration between the American, Dutch and Australian groups proved to be the first example of international cooperation in radio astronomy.

The detection of the hydrogen line proved notable in one other respect: It was the first important discovery in radio astronomy brought about by theoretical prediction, *not* serendipity. Before examining how the hydrogen line fulfilled Jan Oort's hope in 1944 that radio astronomy might provide a powerful new technique for investigating the galaxy, let us briefly summarise what astronomers had been able to learn about the Milky Way up to 1951.

Members of the twenty-one centimetre club during the 1952 URSI meeting in Sydney: Frank Kerr (left), Paul Wild, Jim Hindman, 'Doc' Ewen (Harvard), Alex Muller (Leiden) and Chris Christiansen.

The origins of our modern ideas can be traced back to 1543 when Nicholas Copernicus published the heretical thesis that the earth is not fixed in space, but revolves around the sun. The geocentric world view had been overthrown, only to be replaced by the new dogma that the sun and the solar system lie at the centre of the universe, a belief that lasted for several centuries. During the late 1700s William Herschel formed the first overall picture of the Milky Way and tried to determine its structure by making star counts in certain patches of the sky, though again in the belief that the sun was centrally placed in the star system. This egocentric belief in mankind's central position in the universe persisted until the work in the 1920s by the Harvard astronomer Harlow Shapley, using the new 100 inch telescope at Mt Wilson. By studying the distribution of star clusters Shapley showed that the stars in the Milky Way are arranged in a fairly flat disc, with the sun displaced about two-thirds of the way out from the galactic centre.

Other astronomers such as Jan Oort took these ideas further by showing that the galaxy contains about 100,000 million stars and extends about 150,000 light-years across in distance, with the disc having a maximum thickness of about 20,000 light-years at its centre. The entire galactic system was known to rotate with one complete revolution taking approximately 200 million years. And yet, though astronomers had worked out large-scale features such as these, several decades of laborious stars counts had not revealed much about the actual structure of the galaxy. Apart from the obvious reason that it is hard to decipher the structure because we are submerged within it, another major limitation are the clouds of interstellar dust that obscure the view for even the most powerful optical telescopes. In December 1951, nine months after the discovery of the hydrogen line, a group of American astronomers announced the first direct evidence for the spiral structure of the galaxy, based on a study of luminous O and B stars. However, by then, it was already recognised that radio astronomers possessed a far more powerful technique for investigating the galaxy.

Immediately after confirming the discovery of the hydrogen line, Christiansen and Hindman carried out observations of the sky along the plane of the Milky Way. Three months later they were able to announce the first radio evidence 'which suggests the possible existence of spiral arms in the galaxy'.[22] Early in 1953 a Radiophysics group led by Frank Kerr constructed a new 36 ft dish at Potts Hill and began a detailed survey of the complete southern sky. Meanwhile, the group in Leiden set about a similar survey of the northern sky. On completion of both surveys in 1957, Kerr collaborated with Oort and Gart Westerhout in Leiden on splicing the two maps together to give the first complete view of the distribution of hydrogen throughout the galaxy. At last, Jan Oort's dream had been fulfilled. Astronomers now had solid evidence for what had long been suspected—the Milky Way appears to be a galaxy with a spiral structure akin to those of other great 'island universes'.

The 36 ft dish erected at Potts Hill early in 1953 and used for the first 21 cm survey of the Milky Way.

Although the Harvard physicists had discovered the hydrogen line, the Leiden and Sydney groups were the first to use it to make a major contribution to astronomy.[23] The Leiden–Sydney collaboration provided evidence that interstellar hydrogen is not spread uniformly throughout the galaxy, but is concentrated into long spiral arms. Each arm trails slightly in its rotation about the galactic centre so that the overall spiral shape—if viewed face-on—would give the appearance of gradually being wound more tightly. The distribution of hydrogen also proved to be strongly concentrated along the galactic plane, so much so that it became possible to determine the first accurate direction of the plane. To provide a reliable grid for giving the position of stars, in 1932 astronomers had defined a system of galactic coordinates based on the optical data then available. The hydrogen-line observations showed that this grid system deviated by several degrees from the measured plane. In 1958 the International Astronomical Union adopted a new system of galactic coordinates based on these new observations, the first occasion a fundamental convention in astronomy had been decided by radio astronomy. According to Frank Kerr: 'Many conventional astronomers had been reluctant to regard radio astronomy as a respectable branch of the subject, and this was one of the key ways in which radio astronomy came of age.' [24]

Aside from these studies on the structure of our own galaxy, another set of observations by the Radiophysics group in 1953 brought a further astronomical windfall. Before starting the main Milky Way survey, Kerr, Hindman and new recruit Brian Robinson detected hydrogen-line emission from the Magellanic Clouds.[25] Visible to the unaided eye as

The spiral galaxy M81, located 10 million light-years away. The Leiden–Sydney studies of hydrogen-line emission provided convincing evidence that the Milky Way has a similar spiral shape. Astronomers still disagree, however, as to how many spiral arms the Milky Way actually has and how far they extend.

hazy patches close to the south celestial pole, these two star clouds had been observed by Ferdinand Magellan during his circumnavigation of the globe in 1519. Both clouds are essentially galactic satellites of the Milky Way, with the next nearest galaxy being M31 in Andromeda about ten times further away. They are at a sufficient distance to be viewed as isolated systems and yet close enough for their internal structure to be discerned in some detail. Studies of the clouds had played an important role in the development of optical astronomy.

The observations using the 36 ft dish at Potts Hill showed that interstellar hydrogen not only permeates both clouds but also extends well beyond their visible regions. By measuring the slight frequency (Doppler) shifts away from the precise rest frequency for hydrogen of 1420 MHz, the data also indicated that both clouds are slowly rotating. Further, the observations showed that interstellar hydrogen accounts for up to 20% of the total mass of each cloud, a much higher proportion than in the Milky Way, the first evidence that both clouds are relatively young star systems with a fair proportion of elemental hydrogen still available for star building. Compared with mature spirals such as the Milky Way and Andromeda, the Magellanic Clouds are passing through an adolescent phase in the life of a galaxy. Later, we see how the far greater sensitivity of the Parkes telescope has enabled more detailed studies to be made on hydrogen-line emission, not just from the Magellanic Clouds but from hundreds of distant galaxies as well.

Cosmology and Controversy

> Physical cosmology today is in a curious logical condition. The questions are still largely obscure, the evidence is scanty, the inferences are tenuous, and the scales are evenly weighted between the various theories. On the other hand, the intellectual stakes are so high that men naturally—and very properly—find the subject exhilarating. If it did in fact prove possible to establish, by rational enquiry, any convincing theory about the birth (or rebirth) of the material world we now know, that would be an intellectual triumph as exciting in its own way as anything which Isaac Newton achieved.
>
> Stephen Toulmin and June Goodfield, 1961

Aside from studies on the hydrogen line carried out principally at Leiden and Sydney, another major research program in cosmic radio astronomy during the 1950s concentrated on the detection and cataloguing of the discrete radio sources. As described earlier, the first sky maps drawn by Grote Reber had shown that continuum radio emission is strongly concentrated in certain regions of the sky. Then, in the late 1940s, the observations by John Bolton and the group at Dover Heights had revealed that a large number of discrete radio sources populate the sky, although at the time it was not too clear what these objects actually were. During the 1950s several new sky surveys of increasing accuracy and resolution were carried out, greatly increasing the number of known sources. These surveys were made mainly at two centres, Sydney and Cambridge. As we see in the remainder of this chapter, serious disagreements between the results produced by these two groups led to a protracted and bitter controversy. These sky surveys of radio sources added substance to the single most important issue in astronomy in the 1950s—the debate between rival supporters of the big-bang and steady-state cosmologies.

A radio astronomy group was formed in Cambridge shortly after the end of the war in circumstances similar to the origins of the groups at Jodrell Bank and Sydney. J. A. Ratcliffe had been a leading authority on the physics of the upper atmosphere before spending the war years engaged in radar research at the Telecommunications Research Establishment. On returning to Cambridge he recruited several young radar scientists to his department at the Cavendish Laboratory, among them Martin Ryle, and then later Francis Graham-Smith and Antony Hewish. Most of the first year had been spent foraging for items of surplus British and German radar equipment. Observations began with the discovery, independently of the Radiophysics group, of strong radio emission from sunspots, but from then on the interest of the Cambridge group turned from solar studies to the question of cosmic sources. Whereas Radiophysics diversified into practically every area of radio astronomy during the 1950s, the smaller Cambridge group focused its resources primarily on observations of discrete sources.

The first survey of the northern sky commenced in 1948 with an interferometer consisting of two aerials spaced a kilometre apart. Martin Ryle introduced a new technique known as phase-switching which involves combining the signals from the two aerials in and out of phase in a way that eliminates unwanted background radiation, making it easier to detect the presence of faint radio sources. Known as the first Cambridge survey, the published account in *Monthly Notices of the Royal Astronomical Society* in 1950 listed the positions and intensities of fifty radio sources. Ryle's group then embarked on a more ambitious survey and in 1952 constructed a new instrument, a double interferometer, consisting of four cylindrically-shaped aerials set out at the corners of a large field situated a few kilometres outside of Cambridge. The second Cambridge or 2C survey took three years to complete and catalogued 1936 radio sources.[26]

In the meantime the Radiophysics group had been steadily increasing the number of known sources in the southern sky. Continuing on from Bolton's initial 1948 paper which had listed six sources, Stanley and Slee published a list of 22 in 1950, Mills catalogued another 77 in 1952, and then came the Bolton, Stanley and Slee survey of 104 sources in 1954. An analysis of the Mills survey showed in particular that there are two major classes of sources, the stronger ones being confined in general to along the galactic plane (Class I), while the weaker ones appeared to be distributed randomly or isotropically over the whole sky (Class II). Evidently the Class I sources were members of our own galaxy, but for Class II there were two possibilities. Either the sources were close by in our local region of the galaxy or else they were very distant extragalactic objects. Most astronomers favoured the local interpretation, which reinforced the misleading description of these sources as radio 'stars', and only slowly did the pendulum swing in favour of an extragalactic origin for the Class II sources. In 1954 Mills began a new survey with the 1500 ft cross at Fleurs and, by the middle of 1955 and the arrival of the Cambridge 2C paper in the Radiophysics library, preliminary data had been gathered on close to 600 southern sources.

The first hint of the controversy to follow came with a letter Mills received from leading British astronomer Fred Hoyle. As the chief proponent of the steady-state theory of the universe (see Tangent 2), Hoyle was dismayed to find that the results of the 2C survey were apparently in conflict with one of the central predictions of the theory. Martin Ryle had carried out an analysis of the intensities of his 2000 sources and found a significant excess of weak, and what therefore seemed to be distant sources. According to Hoyle's theory, these radio sources ought to be uniformly distributed throughout space. In fact, a graph of the number of sources versus their measured intensities[30] ought to have produced a straight line with a slope of 1.5. However, in Ryle's

TANGENT 2

Two Cosmologies

Fred Hoyle and the Steady State. During the late 1940s in Cambridge, Fred Hoyle with Hermann Bondi and Thomas Gold developed a new cosmological theory of the universe. The theory accepted as fact the expansion of the universe—as shown by the Doppler shift in light emitted from distant galaxies studied by Edwin Hubble in the 1920s—but included a radical new hypothesis. As the galaxies rapidly recede from one another, new matter is continuously and spontaneously created in the vast distances of intergalactic space. In time, the new matter coalesces to form new galaxies, compensating exactly for the thinning of space caused by the expansion. Thus, in the future, or in the past, the universe would appear to an observer exactly as it does in the present. Consequently, according to the steady-state theory, the universe is infinite in space and time; there was no beginning, no moment of creation, and there will be no end.

Hoyle wrote in 1955: 'The steady-state theory suggests that although expansion leads to the distances between the centres of clusters of galaxies increasing, new clusters of galaxies condense at such a rate that the average number within a fairly large region of space remains effectively unaltered with time. In this way we arrive at a universe in which the individuals—the clusters of galaxies—change and evolve with time but which itself does not change. The old queries about the beginning and end of the universe are dealt with in a surprising manner—by saying that they are meaningless, for the reason that the universe did not have a beginning and it will not have an end. Every cluster of galaxies, every star, every atom had a beginning, but not the universe itself.'[27]

The hypothesis that new matter could form spontaneously from empty space became the most controversial aspect of the theory. And yet, as even the critic George Gamow was prepared to admit, the creation of new matter would be barely noticeable: 'To compensate fully for expansion no more is required than the production of one new hydrogen atom per gallon of expanding space once every 250 million years, so the creative genie would not overstrain himself doing the job.'[28]

George Gamow and the Big Bang. The Russian physicist George Gamow made the first of numerous visits to the West in 1928 and then eventually took up permanent residence in the United States, as professor of theoretical physics at George Washington University. In the late 1940s Gamow, Ralph Alpher and Robert Herman developed the evolutionary model of the universe by investigating how the basic elements in nature were manufactured during the early stage of the primeval fireball. Their theory assumed that all matter in the universe emerged at the beginning and, apart from further cooking of the heavier elements in stars since then, there has been no new creation of matter.

While 'big bang' has passed into common usage for the theory, in his early writings Gamow used the term 'big squeeze'. He believed that the present expansion might have resulted from the collapse of matter from a previous universe into one big ball causing the squeeze and subsequent violent explosion. The idea of this 'bouncing' universe is still in vogue with some cosmologists.[29]

To contrast the two cosmologies: the evolutionary model postulates that all matter instantaneously burst into existence in one big bang, whereas the steady-state theory postulates that individual atoms are slowly and continuously popping into existence over an infinite period of time.

In the 1950s Fred Hoyle in Cambridge (left) was the chief architect of the steady-state theory of the universe, the main rival to the big-bang theory developed by George Gamow in Washington. Aside from developing these fundamental physical theories, both men became renowned for their popular books on science.

analysis the graph turned out to be much steeper with a slope close to 3, a value in sharp disagreement with the steady-state prediction, but one which appeared to favour another cosmological theory of the universe known as the big bang. This rival cosmology had grown out of investigations into the consequences of Einstein's theory of general relativity, principally by Aleksandr Friedmann, Georges Lemaître and Willem de Sitter. Although a wide variety of these evolutionary models of the universe were compatible with Einstein's theory, they all possessed one feature in common: each represents the universe in a state of expansion, growing from a singular point at some time in the past when, at the moment of creation, the universe suddenly came into existence as an extremely hot and rapidly expanding fireball of elementary particles and radiation.

By the mid 1950s the question of deciding between the big-bang and steady-state theories had become of paramount importance in astronomy. The observational evidence from optical astronomers was inconclusive. Surveys using the powerful 100 inch telescope at Mt Wilson indicated that the distribution of galaxies was essentially uniform. However, no detailed, extremely time consuming 'deep field' survey had been made of very remote galaxies, regions at the edge of the universe that might decide which of the rival theories was correct. The news from Cambridge of a large number of apparently remote radio sources seemed to present astronomers with just the opportunity to explore these largely uncharted regions of space. But what then was the reason for this excess of remote sources? According to Ryle, soon after the big bang, after matter had coalesced into galaxies but before the universe had undergone significant expansion, collisions within this concentrated swarm of galaxies would have been commonplace. (Recall that the first radio source discovered, Cygnus A, was believed at this time to be just such a galactic traffic accident.) Much later, after the universe had expanded to its present size, the remnants of these galactic collisions would still be detectable as radio sources. Of crucial importance, there would be an excess of these sources in the remotest regions of the universe—exactly as the Cambridge data seemed to indicate. In Ryle's words:

> 'The apparent increase in density of the sources with distance may then be accounted for as an effect associated with the large redshifts of the distant sources. Further, it becomes possible to explain why no visible objects appear at the positions of most of the radio stars: if these stars are colliding galaxies, only a few dozens of them lie within reach of the 200 inch telescope; the rest are beyond the visible range … . Thus our present conclusions from the radio work at Cambridge support the evolutionary view.' [31]

The cosmological implications of the 2C survey were greeted with scepticism at Radiophysics. Mills examined the sources already detected in the survey underway at Fleurs and could find no compelling evidence for an excess of faint sources. His analysis of the number of sources versus their measured intensities yielded a graph with a slope of 1.8, a value in stark disagreement with 3 from Cambridge. Given the observational uncertainties this value did not rule in favour of either the steady-state or big-bang cosmologies. Furthermore, in the band of sky common to both the northern and southern surveys, Mills found very poor agreement between the celestial positions reported for sources listed in each catalogue.

In an attempt to clarify the reasons for the glaring discrepancy between the two surveys, Rudolph Minkowski from the Mt Wilson–Palomar Observatory visited Radiophysics early in 1956. After spending two months studying the Sydney data, Minkowski continued on to Cambridge where he prepared a long memorandum summarising his

Rudolph Minkowski and Bernard Mills. Minkowski visited Radiophysics early in 1956 in an attempt to clear up the controversy between the Sydney and Cambridge astronomers. At stake was the cosmological significance of distant radio sources.

analysis of both surveys. Sending a copy to Mills in June, Minkowski passed his verdict on the Cambridge observations: 'One look at the records was quite enough to show that they have tried to do the impossible by attempting to analyse completely confused patterns. The really remarkable thing was that, confronted with the picture now, they conceded voluntarily that the sources derived in this way have no validity whatsoever.' [32] Despite Minkowski's analysis and a general acceptance by the Cambridge group that their observational data had been seriously misinterpreted, Ryle pressed on with his belief in the cosmological significance of the 2C catalogue. Mills takes up the story:

> 'So far there had been a free interchange of opinions and observational results, but the battle lines were now drawn. Ryle published a popular account in *Scientific American* of the Cambridge observations and the sweeping conclusions that he drew from them. Since we now had good reason to believe that the 2C catalogue was seriously wrong and it appeared that Ryle had ignored our private criticism, it was generally felt in Sydney that the time had come to make a public statement. Accordingly I sent a letter to the editor of *Scientific American* briefly describing the discordances we found with the 2C survey and giving reasons for them; this letter was answered by Ryle. Both letters were then published in the December 1956 issue, and this was the end of any further cooperation. Our aim of publicising the uncertainties in Ryle's cosmological model had been achieved, but, in retrospect, it may have led to an easier resolution of our differences if I had ignored the article.' [33]

The rift between the Sydney and Cambridge groups referred to here had been developing for a number of years: we briefly trace its origins in Tangent 3. In response Mills and Slee prepared a paper, published in the *Australian Journal of Physics* in March 1957, which made a detailed comparison between some of their own sources and a sample area listed in the 2C catalogue. The concluding paragraph delivered the broadside:

> 'We have shown that in the sample area, which is included in the recent Cambridge catalogue of radio sources, there is a striking disagreement between the two catalogues. Reasons are advanced for supposing that the Cambridge survey is very seriously affected by instrumental effects which have a trivial influence on the Sydney results. We therefore conclude that discrepancies, in the main, reflect errors in the Cambridge catalogue, and accordingly deductions of cosmological interest derived from its analysis are without foundation.' [39]

A year later the Sydney group published the first instalment of their full sky catalogue, listing 1159 radio sources and again stressing the almost complete discordance with the 2C catalogue. According to the Sydney group, the basic reason for the disagreement was the inadequate resolution of the Cambridge interferometer. In a sky survey, the faintest and most numerous sources can blend together in the 'beam' of the telescope to produce numbers of apparently genuine sources. They can then be misinterpreted as an excess of distant radio sources. Because the beamwidth of the Cambridge interferometer was vastly greater than that of the Cross, the blending effect was much more serious, producing overwhelming numbers of fictitious sources. Technically, this effect is known as confusion. Acting as spokesman for the Sydney group, Pawsey identified the problem during the inaugural Matthew Flinders Lecture to the Australian Academy of Science in May 1957:

> 'At least one of the surveys is grossly wrong and I believe that the Cambridge one is invalid owing to serious confusion effects arising from lack of angular resolution. The Sydney statistics show no departures from the 1.5 power law greater than instrumental uncertainties. Consequently the current position is that there appears to be no reliable radio evidence showing an 'edge to the universe' effect.' [40]

Several years passed before the controversy over the two catalogues reached a settlement. In 1959 the Cambridge group published a new survey using the same interferometer, but modified to operate at a frequency twice as high (and with four times the resolution) and which catalogued a more modest total of 471 sources.[41] The 3C survey had little in common with its predecessor 2C and came much closer to the Sydney survey, though the agreement was still far from perfect. An independent third-party check came later in 1959 with data provided by a new interferometer operated by Caltech. The American results showed 3C to be just as reliable as the Sydney catalogue.

TANGENT 3

Isolation and Competition

Immediately after the war the Sydney and Cambridge groups began radio astronomy in complete ignorance of each other and both published their first results on sunspot radiation almost simultaneously. Both groups also became interested in discrete radio sources at the same time. Despite having to confront and overcome the same types of technical and instrumental problems, each group proceeded with only a vague idea of what the other was doing. Several visits were made by some of the Radiophysics group to Cambridge during these early years, yet the amount of personal contact remained low and somewhat one-sided. Not surprisingly, a degree of rivalry and competition began to arise between the two centres.

On a number of occasions Pawsey wrote suggesting closer cooperation through the exchange of information on projects planned or underway. At the very least, Pawsey hoped to avoid unnecessary duplication of increasingly expensive equipment. However, these attempts to bridge the gap met with only a half-hearted response from Cambridge. Relations then deteriorated sharply following a letter Bolton had written in which he suggested that Cambridge might attempt to verify the Radiophysics work on the nature of the fluctuations observed in the Cygnus source. The observations were duly performed in a joint experiment between Cambridge and Jodrell Bank, showing the fluctuations to be caused by ionospheric 'twinkling' and not an intrinsic property of the source. When the results were published no acknowledgment was given to Bolton. On learning the origin of the proposal, Bernard Lovell at Jodrell Bank made a public apology to Bolton. Matters only became worse when, during a tour of British radio astronomy centres in 1950, Bolton was politely informed that he was not welcome at the Cavendish Laboratory.[34]

Receiving due credit for their work became a sensitive issue for the Australians largely because, in obvious contrast to Cambridge, the Radiophysics group felt isolated from the mainstream of science in the northern hemisphere. Historian Woody Sullivan has written that things 'would have gone differently if Radiophysics had not been located 10,000 miles from its sister institutions, but instead 100 miles, or even 1000. The best airline connections to Europe required a gruelling three days (or a civilised week) and the more common passage by ship took about four weeks; moreover, the cost of a ship's berth amounted to one or two months' pay for a Radiophysics staff member. The inability to have frequent contact with colleagues from other institutions, the long interval before learning about research conducted elsewhere, the delays in publishing Australian results in the prestigious overseas journals, and the lack of foreign readership of Australian journals—these circumstances constantly bedevilled the Radiophysics staff.'[35]

In a similar vein, after his visit to England in 1950 Bolton reported: 'My trip so far has convinced me that in work done and in work being done we still have a considerable lead. But in work published and distributed to those who are most interested in it—we occupy a very bad last place.'[36]

During the late 1940s and early 1950s an increasing number of popular articles on the new science of radio astronomy appeared which, more often than not, passed over or undervalued the contributions made by the Radiophysics group. On one occasion, in response to an article in the widely read British magazine *Science Progress* which had attributed the discovery of radio sources to the Cambridge group, Bowen had felt compelled to write: 'I realise that there is nothing worse in scientific work than cries of 'we did it first', but I am sure you will understand that we are having a difficult time keeping radio work going in this rather distant outpost and it is a bit hard to find later work elsewhere getting the credit.'[37]

Though Bowen overstated the case, this isolation seemed real to the Sydney group, not just in the geographical sense. A sizable proportion of the Radiophysics group were recent arrivals from Britain. Whether it is called homesickness or psychological isolation, these feelings reinforced the belief that Sydney was a scientific outpost struggling for its fair share of recognition. Paradoxically, the reverse was equally true. In terms of the size, resources and diversity of the Radiophysics program, if anything the small European groups were the real outposts in radio astronomy.

The question of isolation, real or imagined, tells only half the story. Radiophysics established good working relationships with the groups at Jodrell Bank and Leiden during the 1950s, but not with Cambridge. Basically, it was the attitude of the Cambridge group that made the difference. Recollections by radio astronomers often use expressions such as 'an iron curtain had descended' or they 'kept their cards close to their chests' to describe the activities of the Cambridge group.[38]

Against this backdrop—the sense of isolation felt by the Sydney group, reinforced by the Cambridge group's decision to isolate itself—the disagreement between the Sydney and Cambridge catalogues of radio sources set the scene for a long and bitter controversy between the two groups.

Martin Ryle founded the radio astronomy group at the Cavendish Laboratory in Cambridge. In 1974 he and Antony Hewish received the Nobel Prize for physics, the first occasion it was awarded to radio astronomers.

The demise of the controversial 2C catalogue cut short the hope that these source surveys in the mid 1950s could provide conclusive evidence to decide between the big-bang and steady-state cosmologies. The big-bang theory later emerged victorious, but only after the discovery in 1965 of the cosmic background radiation by the Americans Arno Penzias and Robert Wilson. This background radiation could be understood in the big-bang theory, but not the steady state, and it tipped the balance decisively in favour of the big bang. Although the 2C survey proved to be a false start, radio astronomy went on to make a central contribution to cosmology in later years. By the late 1960s detailed sky surveys carried out at Parkes, Cambridge and other centres reached close agreement for regions of sky common to all observers. Graphs of the number of sources versus their intensity derived from these surveys have yielded values converging on 1.8 for the slope and provided the next piece of evidence in favour of the big bang. But here we are running ahead of our story. These source surveys at Parkes which have led to the detection of the most distant objects known in the universe—the edge of the universe sought by astronomers in the 1950s—will be taken up in Chapter 9.

To close Part 1, let us summarise this work on extragalactic radio sources. On the basis of the 2C catalogue the Cambridge radio astronomers had derived a slope of 3 for the number versus intensity graph, a value which seemed to provide decisive evidence in favour of the big-bang theory. Whereas the steady-state theory insisted on a value of 1.5, the big bang was compatible with any value greater than 1.5. Mills and the Sydney group on the other hand derived a value of 1.8, one which favoured the big bang. However, because of the usual observational uncertainties, the group decided against proclaiming this as evidence in favour of one theory or the other. Mills remarked recently: 'Something as absolutely fundamental as the origin of the universe needs a lot more than this before one can blow the trumpets!' [42]

In contrast, by favouring the big-bang theory the Cambridge group backed the right horse, but on the basis of evidence which later proved incorrect. The Sydney group produced the right evidence at its first attempt, with a slope of 1.8 which has since been confirmed in a number of later surveys. However, because of due caution over observational errors, the group declined to back either horse. Fortune favours the brave in astronomy too. In most popular accounts of astronomy, credit for providing the first observational evidence in favour of the big bang is given to the Cambridge group.[43]

Notes to Part One

CHAPTER 1

Epigraphs: Lord De L'Isle, Parkes *Champion Post*, 2/11/61; Jansky, see W. T. Sullivan in *Serendipitous Discoveries in Radio Astronomy* (Eds K. Kellermann and B. Sheets), p. 45; Reber, *J. R. Astron. Soc. Canada* **82** (1988), p. 93.

1 Parkes *Champion Post*, 2/11/61.
2 Radiophysics Archive (RPA) correspondence file A1/3/11/19, cable from B. Lovell 31/10/61.
3 An unsuccessful attempt had been made to bring Prince Philip to Australia to perform the opening. For the text of all speeches delivered see RPA file A1/3/11/19.
4 *Champion Post*, 2/11/61. This was not the only mishap. The caterer supplying afternoon tea overturned his vehicle on the way to the opening ceremony. Somehow, enough cream buns were reconstructed to serve 500 guests.
5 G. Westerhout, 'The early history of radio astronomy' *Ann. New York Acad. Sci* **198** (1972) 211–18.
6 See W. T. Sullivan, in *Serendipitous Discoveries in Radio Astronomy* (Eds K. Kellermann and B. Sheets), pp. 39–56; see also Sullivan (Ed.), *The Early Years of Radio Astronomy*, pp. 3–42. [For full details of these books and others cited elsewhere, see the Bibliography.]
7 K. G. Jansky, 'Directional studies of atmospherics at high frequencies' *Proc. Inst. Radio Eng.* **20** (1932) 1920–32. In 1973 the International Astronomical Union adopted the name 'jansky' (symbol: Jy) for the fundamental unit used to measure the strength of radio signals. A scientist can receive no higher honour.
8 K. G. Jansky, 'Radio waves from outside the solar system' *Nature* **132** (1933) 66.
9 See W. T. Sullivan, *Classics in Radio Astronomy*, p. xxi; see also P. Robertson, 'The birth of nuclear physics' *New Scient.* **93** (1982) 428–31.
10 P. Robertson, 'Grote Reber: Last of the lone mavericks' *Search* **17** (1986) 118–21. The remainder of Chapter 1 is based on an interview with Reber by the author on 6 February 1986 at Bothwell, Tasmania.
11 G. Reber, 'Cosmic static' *Proc. Inst. Radio Eng.* **28** (1940) 68–71; see also *Astrophys. J.* **91** (1940) 621–32.

CHAPTER 2

Epigraphs: Unsöld, see W. T. Sullivan, *Classics in Radio Astronomy*; Christiansen, *Daily Telegraph* (Sydney), 25/8/52.

1 F. Strebeigh, 'How England hung the 'curtain' that kept Hitler at bay' *The Smithsonian* July 1990, 120–9.
2 R. W. Clark, *Tizard*, p. 268. See also M. Barnard, *One Single Weapon*.
3 W. F. Evans, *History of the Radio Research Board 1926–45*, p. 112. See also C. S. Gillmor in *International Science and National Scientific Identity* (Eds R. W. Home and S. G. Kohlstedt), pp. 181–204 (Kluwer Academic, Dordrecht, 1991).
4 CSIRO Archives (CSA) correspondence file PH/PAW/1, Rutherford to Radio Research Board 9/3/33.
5 F. W. G. White and D. Kimpton (Eds), *Surprise and Enterprise: Fifty Years of Science for Australia*, p. 21. The Japanese eventually made over fifty air raids on Darwin, despite resistance from the hopelessly outnumbered American and Australian aircraft.

6 D. P. Mellor, *The Role of Science and Industry*, p. 441. Series 4, Vol. 5 of *Australia in the War of 1939–45*.
7 S. Davies, 'R. v. d. R. Woolley in Australia' *Hist. Rec. Aust. Sci.* **6** (1984) 59–69.
8 E. G. Bowen, *Radar Days*, p. 198.
9 The collective wisdom of the Radiophysics Lab was published in E. G. Bowen (Ed.), *A Textbook of Radar* (Angus & Robertson, Sydney, 1947).
10 R. W. Home, 'Origins of the Australian physics community' *Historical Studies* **20** (1983) 383–400.
11 CSA file KE8/1, E. G. Bowen, 'Programme for Division of Radiophysics', 2 July 1945, 26 pp.
12 Quoted in R. Rivett, *David Rivett: Fighter for Australian Science*, p. 188.
13 CSA file KE8/1, E. G. Bowen, 'A brief survey of the work of the Radiophysics Division', 23 December 1947, 5 pp.
14 RPA file F1/4/PAW/1, Pawsey to O. F. Brown 14/1/48.
15 See M. Beard and T. Pearcey, 'The genesis of an early stored-program computer: CSIRAC' *Ann. Hist. Computing* **6** (1984) 106–15.
16 J. F. Deane, 'The CSIRAC Mk I Computer 1951–64', Radiophysics Information Sheet No. 84/54.
17 E. G. Bowen, 'The influence of meteoritic dust on rainfall' *Aust. J. Phys.* **6** (1953) 490–7.
18 See F. Hoyle, *Frontiers of Astronomy*, p. 9. There are interesting parallels between Bowen's theory and Hoyle's later, and even more controversial, theory that virus or bacteria from space are responsible for some diseases that break out as worldwide epidemics: see F. Hoyle and C. Wickramasinghe, *Diseases from Space* (Dent, London, 1979). See also D. F. Martyn, *Aust. J. Phys.* **7** (1954) 358–64; W. C. Swinbank, *Aust. J. Phys.* **7** (1954) 354–8.
19 H. R. Byers, in *Climate and Weather Modification* (Ed. W. N. Hess), p. 23 (Wiley, New York, 1974).

CHAPTER 3

Epigraphs: Sullivan, *Australian Science in the Making* (Ed. R. W. Home), p. 340; Tizard, see A. L. Mackay, *Harvest of a Quiet Eye*, 148.9 (Inst. of Physics, Bristol, 1977); Hoyle, *Nature of the Universe*, p. 111.

1 E. G. Bowen, 'From wartime radar to postwar radio astronomy in Australia' *J. Elect. Electron. Eng. (Aust.)* **8** (1988) 1–11; see also Bowen in *The Early Years of Radio Astronomy* (Ed. W. T. Sullivan), p. 85.
2 G. C. Southworth, 'Microwave radiation from the sun' *J. Franklin Inst.* **239** (1945) 285; J. S. Hey, 'Solar radiations in the 4–6 metre radio wavelength band' *Nature* **157** (1946) 47.
3 RPA file A1/1/1, Part I 1945–46.
4 The Russian work carried out independently by Vitaly L. Ginzburg and Iosif S. Shklovsky was published in Soviet journals which were largely inaccessible in the West in 1946. Ginzburg's paper for example was not 'discovered' at Radiophysics until mid 1948.
5 J. L. Pawsey, R. Payne-Scott and L. L. McCready, 'Radio-frequency energy from the sun' *Nature* **157** (1946) 158.
6 J. L. Pawsey, 'Observations of million-degree thermal radiation from the sun at a wavelength of 1.5 m' *Nature* **158** (1946) 633; D. F. Martyn, 'Temperature radiation from the quiet sun in the radio spectrum' *Nature* **158** (1946) 632. The original plan had been for Pawsey and Martyn to coauthor a joint paper but, in the meantime, an argument broke out between them over who deserved priority for the discovery of the million-degree corona. Martyn's paper failed to even make reference to the

Radiophysics observations—for details see W. T. Sullivan in *Australian Science in the Making* (Ed. R. W. Home), p. 320.

7 B. Y. Mills, 'Australian contribution to the science of radio astronomy' *J. Elect. Electron. Eng. (Aust.)* **8** (1988) 12–23.

8 L. L. McCready, J. L. Pawsey and R. Payne-Scott, 'Solar radiation at radio frequencies and its relation to sunspots' *Proc. R. Soc. London* A **190** (1947) 357.

9 M. Ryle and D. D. Vonberg, 'Solar radiation on 175 Mc/s' *Nature* **158** (1946) 339–40.

10 J. S. Hey, S. J. Parsons and J. W. Phillips, 'Fluctuations in cosmic radiation at radio frequencies' *Nature* **158** (1946) 234. As it turned out, Hey and his group were right about the point-like nature of Cygnus, but for the wrong reason. Later, the Australian group showed that the fluctuations in the signal do not originate in the source itself, but are caused by the earth's ionosphere. The fluctuations are, in principle, the same as the familiar twinkling we see of stars, an effect caused by the atmosphere. The first correct evidence for the point-like nature of Cygnus came with the measurements at Dover Heights (see later).

11 RPA file A1/3/1(a), Slee to Radiophysics 4/3/46.

12 Interview with author, April 1984.

13 Pawsey decided that both sets of observations should be summarised in the one paper: R. Payne-Scott, D. E. Yabsley and J. G. Bolton, 'Relative times of arrival of bursts of solar noise on different radio frequencies' *Nature* **160** (1947) 256.

14 J. G. Bolton and G. J. Stanley, 'Variable source of radio frequency radiation in the constellation of Cygnus' *Nature* **161** (1948) 312. A more detailed report appeared in the first issue of a new science journal launched by CSIR in cooperation with the Australian National Research Council: see *Aust. J. Scient. Res.* A **1** (1948) 58. This became the established mode of publishing for the Sydney group for many years afterwards. First a brief paper announcing a discovery or important piece of work would be sent to *Nature*, while a more detailed and 'archival' report would appear in the *Australian Journal of Scientific Research*, Part A. In 1953 this journal split into two parts, chemistry and physics, with the latter (*Australian Journal of Physics*) becoming the principal outlet for Australian radio astronomy until the 1970s.

15 J. G. Bolton, G. J. Stanley and O. B. Slee, 'Positions of three discrete sources of galactic radio-frequency radiation' *Nature* **164** (1949) 101.

16 See note 12. For the paper referred to see J. G. Bolton, 'Discrete sources of galactic radio-frequency noise' *Nature* **162** (1948) 141–2.

17 M. Ryle and F. Graham-Smith, 'A new intense source of radio-frequency radiation in the constellation of Cassiopeia' *Nature* **162** (1948) 462.

18 R. Hanbury Brown and C. Hazard, 'Radio emission from the Andromeda Nebula' *Mon. Not. R. Astron. Soc.* **111** (1951) 357.

19 B. Y. Mills and A. B. Thomas, 'Observations of the source of radio-frequency radiation in the constellation of Cygnus' *Aust. J. Scient. Res.* A **4** (1951) 158–71.

20 Baade and Minkowski originally estimated the distance to Cygnus A to be 100 million light-years, based on the redshift of its emission lines and the then accepted value of the Hubble constant. Since then the Hubble constant has undergone continual revision (yes, it is still a constant!). In fact, between 1935 and 1975, the generally accepted value of the Hubble parameter has been revised downward by more than a factor of 10. In turn, this has led to a corresponding *increase* in the estimated distances to extragalactic objects. Though astronomers still do not agree on a value for the constant, we use here a contemporary value which gives the distance to Cygnus A of 1000 million light-years (see J. S. Hey, *The Radio Universe*, p. 181).

21 See R. C. Jennison, *Introduction to Radio Astronomy*, p. 77 (Newnes, London, 1966); see also F. Graham-Smith and B. Lovell, *J. Hist. Astron.* **14** (1983) 163.

22 See e.g. J. G. Taylor, *New Worlds in Physics*, p. 225 (Faber & Faber, London, 1974).

23 J. L. Pawsey and R. N. Bracewell, *Radio Astronomy*, pp. 5, 6.

24 For the eighth and final part of the sea interferometer survey see J. G. Bolton, G. J. Stanley and O. B. Slee, 'Galactic radiation at radio frequencies. VIII. Discrete sources at 100 Mc/s between declinations +50° and –50°' *Aust. J. Phys.* **7** (1954) 110–29.

25 See Chapter 4, note 4.

26 R. X. McGee and J. G. Bolton, 'Probable observation of the galactic nucleus at 400 Mc/s' *Nature* **173** (1954) 985–7. This source had in fact been discovered several years earlier by Jack Piddington and Harry Minnett [see *Aust. J. Scient. Res.* A **4** (1951) 459–75]. They noted that this peculiar source is located in a direction 'very close to the centre of the galaxy', but hesitated in claiming that the source itself is the galactic centre.

27 RPA file A1/1/1, Part IX, Baade to Pawsey 16/2/54.

28 J. H. Piddington and H. C. Minnett, 'Microwave thermal radiation from the moon' *Aust. J. Scient. Res.* A **2** (1949) 63–77.

29 F. J. Kerr, C. A. Shain and C. S. Higgins, 'Moon echoes and penetration of the ionosphere' *Nature* **163** (1949) 310–3.

30 CSA file KE5/9, Rivett to Bowen 20/11/47.

31 CSA file KE8/1, Rivett to Bowen 5/9/47.

32 CSA file KE20/1. Quoted in letter from Bowen to White 9/8/50.

33 W. N. Christiansen in *The Early Years of Radio Astronomy* (Ed. W. T. Sullivan), p. 113.

34 Quoted by D. O. Edge, see note 33, p. 352.

35 J. P. Wild, 'The beginnings of radio astronomy in Australia' *Rec. Aust. Acad. Sci.* **2** (1972) 52–61. John Bolton adds another story: 'Pawsey's failing was, in fact, known worldwide. He paid several visits to the Owens Valley Observatory in California, the last when the interferometer was in operation. A vital component in the control room that 'locked' the signals from the two dishes together was an oscillator, housed in a double cabinet for close temperature control. The outer door could be opened by turning an aircraft-type fastener a quarter of a turn with a small coin. Ahead of Pawsey's visit one of the students, Dick Read, had stencilled on the door in red 'OPEN IN CASE OF FIRE'. During the inspection of the control room it clearly was the centre of Pawsey's interest. We had previously arranged that the rest of us would be called away with a variety of excuses, telephone calls etc., leaving Pawsey alone for several minutes. When we returned he had a very sheepish grin on his face. Dick Read's message on the inner cabinet was 'NO—NOT NOW, ONLY IN CASE OF FIRE!' (Bolton to author 26/3/90).

CHAPTER 4

Epigraphs: Galileo, see S. Toulmin and J. Goodfield, *The Fabric of the Heavens*, p. 191; Pawsey, *Aust. Acad. Science Year Book*, p. 80 (Canberra, 1957); Eddington, *New Pathways in Science*, p. 206 (Cambridge Univ. Press, 1935); Toulmin & Goodfield, *The Fabric of the Heavens*, p. 264.

1 See Chapter 3, note 35.

2 In July 1952 Air Vice-Marshal Frederick Scherger visited CSIRO headquarters in Melbourne to thank Radiophysics for its design work on radar aerials. As White reported to Bowen (see RPA file Z1/7): 'While thanking us he was careful to explain that it would be impossible for the RAAF to help us financially with the work. As he put it, however, this did not detract in any way from their genuine interest in your program. We discussed the difficulty of getting money through the medium of the Services, with all its complications, and decided that there was no point in trying this, at least at present.' The visit formally ended the collaboration between Radiophysics and the RAAF begun in 1940. See Bowen, 'The prehistory of the Parkes 64 m telescope' *Proc. Astron. Soc. Aust.* **4** (1981) 267–73.

3 RPA file A1/1/1. Notes on cosmic noise programme in Radiophysics, 19/2/51.

4 Unable to get the resources to proceed with the twin-dish interferometer, Bolton and his group had gone ahead with the construction of the hole-in-the-ground at Dover Heights under their own steam—initially without Pawsey's sanction (see Chapter 3). Piddington's proposal stemmed from his observations with Minnett in 1948 on thermal radiation from the moon, and required a dish which could operate in the wavelength range 1–10 cm. However, by 1950 Piddington's own interests had shifted to longer wavelengths and so, together with the general belief that there would be very little of interest at 10 cm and below, the project was cancelled (see e.g. RPA file F1/4/BOL/1, Pawsey to Bolton 3/7/50).

5 B. Y. Mills in *The Early Years of Radio Astronomy* (Ed. W. T. Sullivan), p. 152. Bowen was one of the sceptics and Pawsey one of the supporters of the proposal.

6 B. Y. Mills and A. G. Little, 'A high-resolution aerial system of a new type' *Aust. J. Phys.* **6** (1953) 272–8. Among the results was the first detection of continuum radiation from the Large Magellanic Cloud.

7 See Chapter 3, note 35.

8 See *Serendipitous Discoveries in Radio Astronomy* (Eds K. Kellermann and B. Sheets), p. 249.

9 J. P. Wild, *Coresearch* (CSIRO staff newspaper), Aug. 1968.

10 See e.g. J. P. Wild, 'Observations of the spectrum of high-intensity solar radiation at metre wavelengths' *Aust. J. Scient. Res.* A **3** (1950) 399–408; 541–57.

11 J. P. Wild, K. V. Sheridan and A. A. Neylan, 'An investigation of the speed of the solar disturbances responsible for type III radio bursts' *Aust. J. Phys.* **12** (1959) 369–98.

12 W. N. Christiansen in *The Early Years of Radio Astronomy* (Ed. W. T. Sullivan), p. 118.

13 W. N. Christiansen and J. A. Warburton, 'The distribution of radio brightness over the solar disc at a wavelength of 21 cm' *Aust. J. Phys.* **6** (1953) 190–202; 262–71.

14 According to Christiansen, Bernard Mills conceived the idea of the cross after inspecting the L-shaped array: 'While visiting Potts Hill one morning in 1953 Mills asked me why we did not couple the two arrays to produce high resolving power in two dimensions. During the ensuing discussion it was agreed that for this to be effective the centres of the two arrays must not be separated (as they were in the Potts Hill antenna), and also that some means had to be devised to multiply the outputs of the arrays. By the next morning Mills had devised the cross antenna consisting of a pair of thin orthogonal antennas with their outputs multiplied to give a single narrow response.' See note 12, p. 122.

15 W. N. Christiansen, D. S. Mathewson and J. L. Pawsey, 'Radio pictures of the sun' *Nature* **180** (1957) 944–6.

16 J. Pfeiffer, *The Changing Universe*, p. 60.

17 See B. F. Burke and K. L. Franklin, 'Observations of a variable radio source associated with the planet Jupiter' *J. Geophys. Res.* **60** (1955) 213. C. A. Shain, 'Location on Jupiter of a source of radio noise' *Nature* **176** (1955) 836–7. Grote Reber was another to have recorded signals from Jupiter in the early 1950s, but did not recognise them for what they were. At the time Jupiter roughly coincided with the strong Taurus A source which masked the planet's emission. Later, Reber also found evidence that signals from Jupiter appeared on Jansky's original recordings from the early 1930s. Thus, it took over 20 years before the source of these strong signals was finally recognised. (Interview with Reber by the author, February 1986.)

18 RPA file A1/1/1, Pawsey to Bowen 23/1/48. Reber also provided information on another absorption line for deuterium at an expected frequency of 327 MHz (see Chapter 12, note 16). In reply, Bowen wrote to Pawsey: 'This possibility is certainly an interesting one but, in view of the present state of knowledge, I doubt very much whether we should yet devote a special effort to it. A search for the atomic hydrogen and deuterium lines could be made with the Georges Heights equipment but this

would involve a dislocation of other work which is scarcely justified at present' (RPA file F1/4/PAW/1, 18/5/48).

19 Interview with author, March 1984. In the Soviet Union the theorist I. S. Shklovsky also concluded that an observing program could be successful with equipment then available, but none of the small number of Soviet radio astronomers took up the suggestion.

20 F. J. Kerr in *The Early Years of Radio Astronomy* (Ed. W. T. Sullivan), p. 137.

21 See *Nature* **168** (1951) 356; 357; 358.

22 W. N. Christiansen and J. V. Hindman, 'A preliminary survey of 1420 Mc/s line emission from galactic hydrogen' *Aust. J. Scient. Res.* A **5** (1952) 437–65. The observations were made with a rectangular-shaped dish, 16 ft by 18 ft, hurriedly converted from the solar work at Potts Hill. For the combined Leiden–Sydney survey see J. H. Oort, F. J. Kerr and G. Westerhout, 'The galactic system as a spiral nebula' *Mon. Not. R. Astron. Soc.* **118** (1958) 379.

23 The Harvard group made no attempt to follow up the initial discovery by Ewen and Purcell, even though Harlow Shapley himself was Director of the Harvard Observatory. Purcell's primary interest was physics, and astronomy remained a sideline. Ewen became more interested in the commercial possibilities of the new technology and, after completing his Ph.D., began a successful company producing radio receivers. One was known as the travelling wave tube. These receivers were used to make a number of significant observations later, including the detection of thermal radio waves from Saturn in 1956 and confirmation of line radiation from the hydroxyl ion (OH) in 1964 (see Chapter 12).

24 F. J. Kerr in *The Early Years of Radio Astronomy* (Ed. W. T. Sullivan), p. 140.

25 F. J. Kerr, J. V. Hindman and B. J. Robinson, 'Observations of the 21 cm line from the Magellanic Clouds' *Aust. J. Phys.* **7** (1954) 297–314.

26 For the two Cambridge surveys see M. Ryle, F. Graham-Smith and B. Elsmore, 'Preliminary survey of the radio stars in the northern hemisphere' *Mon. Not. R. Astron. Soc.* **110** (1950) 508; J. R. Shakeshaft, M. Ryle, J. E. Baldwin, B. Elsmore and J. H. Thomson, 'A survey of radio sources between declinations –38° and +83° ' *Mem. R. Astron. Soc.* **67** (1955) 106.

27 F. Hoyle, *Frontiers of Astronomy*, p. 320.

28 G. Gamow, *The Creation of the Universe*, p. 32.

29 See note 28, p. 29. The term 'big bang' was first used by Hoyle as a putdown during a BBC radio broadcast. Tales about Gamow and his legendary humour are well known among physicists. As one story goes, Gamow was writing up a piece of work on the big-bang theory with Ralph Alpher and, to be fair to the Greek alphabet, he simply added the name of fellow physicist and friend Hans Bethe (pronounced beta) to the head of their research paper. The Alpher–Bethe–Gamow paper, published in 1948, became appropriately known for short as the alpha–beta–gamma theory. Bethe had no objection and, in fact, joined in some of the subsequent work. Despite Gamow's entreaties another collaborator Robert Herman refused to change his name to Delter. See R. A. Alpher and R. Herman, 'Reflections on early work on 'big bang' cosmology' *Phys. Today* **41**(8) (1988) 24–34.

30 The graph referred to is a logarithmic plot. A slope of 1.5 is expected in the steady-state theory for the following reason: The further away a radio source (at a distance R), the weaker its signal becomes, the intensity or 'radio brightness' decreasing as an inverse square law proportional to R^{-2}. However, as we look to greater and greater distances, the volume of space included in the search increases as the cube of the distance, R^3. Thus, the number of sources N expected at a given level of signal depends on the strength of the sources S to the power of –3/2. In other words, a plot of log N versus log S should yield a straight line with a slope of –3/2 or –1.5. (For simplicity the minus sign is omitted in the text.)

31 M. Ryle, 'Radio galaxies' *Scient. Am.* **195**(3) (1956) 205–14.

32 RPA file A1/1/1 Part X, Minkowski to Mills 25/6/56.

33 B. Y. Mills in *The Early Years of Radio Astronomy* (Ed. W. T. Sullivan), p. 155. For the two letters to the Editor referred to see *Scient. Am.* **195**(6) (1956) 8; 10.

34 For details see RPA files A1/1/1 and F1/4/BOL/1. Bolton has also recounted events in 'Radio astronomy at Dover Heights' *Proc. Astron. Soc. Aust.* **4** (1982) 349–58. Bolton spent three weeks in Cambridge in 1950, but as a guest of Fred Hoyle and his associates. As he reported to Pawsey: 'There is undoubtedly a visitors problem at the Cavendish but in spite of this they are both jealous and scared of Radiophysics.'

35 W. T. Sullivan in *Australian Science in the Making* (Ed. R. W. Home), p. 326.

36 RPA file F1/4/BOL/1, Bolton to Pawsey 17/7/50.

37 RPA file A1/1/1, Bowen to M. W. Ovenden (Cambridge) 13/10/49.

38 For a detailed discussion here and elsewhere in this chapter, see D. O. Edge and M. J. Mulkay, *Astronomy Transformed: The Emergence of Radio Astronomy in Britain.* Oddly, the steady-state theorists Hoyle, Bondi and Gold all held positions at Cambridge during the late 1940s and early 1950s, but none struck up a working relationship with the radio astronomy group.

39 B. Y. Mills and O. B. Slee, 'Preliminary survey of radio sources in a limited region of sky at a wavelength of 3.5 m' *Aust. J. Phys.* **10** (1957) 162–94.

40 J. L. Pawsey, 'Vistas in radio astronomy', *Aust. Acad. Sci. Year Book,* 1957, pp. 65–80.

41 D. O. Edge *et al.*, 'A survey of radio sources at a frequency of 159 Mc/s (3C survey)' *Mem. R. Astron. Soc.* **68** (1959) 37. Of the almost 2000 sources listed in 2C, all but 200 or so were later discredited: see D. O. Edge in *The Early Years of Radio Astronomy* (Ed. W. T. Sullivan), p. 358.

42 Mills to author 31/4/90.

43 See e.g. J. S. Hey, *The Radio Universe,* p. 224, or for a more recent example the bestseller by Stephen Hawking, *A Brief History of Time* (Bantam, London, 1988): 'In the late 1950s and early 1960s a survey of sources of radio waves from outer space was carried out at Cambridge by a group of astronomers led by Martin Ryle … . The Cambridge group showed that most of these radio sources must be outside our galaxy (indeed many of them could be identified with other galaxies) and also that there were many more weak sources than strong ones … . [The] explanation contradicted the predictions of the steady-state theory' (p. 47).

PART TWO

TAFFY BOWEN

THE BUILDER

CHAPTER FIVE

Finding the Funds

Socrates: Shall we set down astronomy among the subjects of study?

Glaucon: I think so, to know something about the seasons, the months and the years is of use for military purposes, as well as for agriculture and for navigation.

Socrates: It amuses me to see how afraid you are, lest the common herd of people should accuse you of recommending useless studies.

Plato's Republic

Radio astronomy flourished in Australia during the 1950s with the Radiophysics group consolidating its position at the forefront of the new science. Much of this success came about through the talent these researchers had for inventing new instruments and techniques. Five main radio telescopes laid the foundation for progress in the 1950s—the Mills Cross at Fleurs for cosmic studies, two variations on the same principle known as the Chris Cross and Shain Cross for solar and low-frequency work, the radio spectrograph at Dapto and a sizable parabolic dish for hydrogen-line studies at Potts Hill. We turn now in Part 2 to describe the main instrument for radio astronomy in the 1960s and beyond. The origin of the giant radio telescope at Parkes completed in October 1961 can be traced back more than ten years earlier, even before any of the major instruments of the 1950s had been brought into operation. The long haul leading to its completion occurred in three distinct stages: raising funds during the years 1952–55, the subject of the present chapter; the design phase 1956–58 described in Chapter 6; and the construction period 1959–61 which forms the basis of Chapter 7. In the final chapter of Part 2 we look at the major transformation in Australian astronomy at the end of the 1950s as the Radiophysics group refocused its resources in preparation for the new telescope at Parkes.

Taffy Bowen had been one of the main players in the debate early in the 1950s over the most desirable types of radio telescope to build. As chief, with responsibility for overseeing the total Radiophysics research program, including direct supervision of the cloud physics group, Bowen still found time to keep a close watch on the radio astronomy group. In 1949 he had been the architect of an attempt to interest the RAAF in a large radar aerial and, although the joint project came to nothing, the idea of building a telescope with a very large collecting area began to take shape in Bowen's mind. Largely as a result of his own efforts, Bowen eventually marshalled official backing and sufficient funding to proceed with the Parkes telescope. No scientific project of this type had ever been undertaken in Australia, so in the present chapter we also look at some of the implications this had for Australian science at large.

Before turning to the question of funding we begin by describing three other developments in the 1950s, each of which strengthened Bowen's resolve to build a giant radio telescope. The first came with the announcement by Australia's leading astronomer Richard Woolley in February 1951 that the Commonwealth Observatory planned to build a large radio telescope at Mt Stromlo near Canberra. The announcement not only threw out a direct challenge to Radiophysics, but also called into question whether radio astronomy should in fact continue at Radiophysics. The second development was the plan by Bernard Lovell at the University of Manchester to build a giant radio telescope at the Jodrell Bank field station. In Bowen's view, if the Jodrell Bank group got in first it would be virtually assured of picking up many of the important discoveries that would follow with such a large instrument.

The third development came with the surge of interest in radio astronomy by the United States. With the Americans planning a massive investment in radio astronomy, at one stage Bowen seemed set to get his giant radio telescope, not in New South Wales, but in California. Before taking up that story we begin with the challenge to Radiophysics from the Commonwealth Observatory.

The View from Mount Stromlo

After the war the Radiophysics Lab did not have radio astronomy completely to itself in Australia. In 1946 a small group of physicists in Perth led by Syd Williams began observations of solar radiation using a Yagi aerial set up on the campus of the University of Western Australia. The observations led to four research papers (three published in *Nature*) and a modest amount of international recognition, before the work was discontinued two years later.[1] A group at the Commonwealth Observatory in Canberra entered the scene even earlier. In November 1945, only weeks after the first Radiophysics observations of solar radio emission at Collaroy, Richard Woolley wrote to Joe Pawsey stating that the Commonwealth Observatory planned to carry out radio observations and provide data complementary to the work of the Sydney group. Pawsey replied indicating the willingness of Radiophysics to collaborate and immediately made arrangements to lend the Observatory suitable equipment.

Since its establishment in 1924 by the physicist Walter Duffield, the Commonwealth Observatory had become the principal centre for astronomy in Australia, overtaking the pioneering role played by observatories set up in each of the state capitals during the 1800s. In their day these observatories had been prominent in the scientific and cultural life of the colonies. Their directors had been acknowledged as the chief scientific advisers to their governments, while the world's second largest telescope, a 48 inch reflector, was part of the Melbourne Observatory. By the turn of the century a combination of factors, economic, political and scientific, saw the observatories go into a long decline. At the same time, Mt Stromlo's star began to rise. Initially Mt Stromlo concentrated on the investigation of solar and related geophysical phenomena. Upon Duffield's death and the eventual appointment in 1939 of Woolley as director of the Observatory and Commonwealth Astronomer, the research program turned abruptly to more challenging areas of astronomy. Similar to Radiophysics, the Observatory had become engaged in work for the armed services, playing a major role in the design and production of optical devices such as anti-aircraft gun sights. At the peak of its activities, over sixty people were employed in war-related work in the Mt Stromlo workshops. A recent history noted:

> 'World War II marked a great watershed in the development of the Observatory at Mt Stromlo. It occurred right at the start of Woolley's directorship, and, during the war, the size of the staff and the quality of the facilities both increased markedly to support the optical munitions work that was undertaken. Up until that time the main emphases of the research work had been on solar and geophysical phenomena; but following the war, Woolley steered the Observatory in quite a different direction—that of stellar and galactic astrophysics. This has remained the broad research area of the Observatory ever since, and the area in which its world-class reputation has been established.' [2]

Early in 1946 a steerable array of four Yagi aerials with a 200 MHz receiver was installed at Mt Stromlo, courtesy of Radiophysics, and used each day to monitor the sun's emission. Clay Allen and David Martyn had immediate success with the discovery of a high degree of circular polarisation in the emission from a group of sunspots (see Chapter 3). However, by the middle of the following year the collaboration between Mt Stromlo and Radiophysics had begun to sour. The Mt Stromlo astronomers had no expertise in radio engineering, relying totally on Radiophysics to patch up the equipment when anything went wrong, not an infrequent occurrence with the often temperamental equipment during these early days of radio astronomy. In response, Woolley contemplated establishing a radio engineering workshop at Mt Stromlo to design and build its own equipment, but then decided against a straight-out duplication of the work already being carried out in Sydney. Besides, at this time Woolley was far too preoccupied with negotiations for the acquisition of a new 74 inch optical telescope, to be equal largest in the southern hemisphere. It seemed that this project would occupy the resources at Mt Stromlo for the foreseeable future.

Despite some tension this early collaboration with the Commonwealth Observatory undoubtedly benefited the Radiophysics group. The radio scientists, turned radio astronomers, came into contact with Australia's leading astronomers at a time when the Sydney group was only learning the basics of the science. Clay Allen in particular provided a steady flow of information to Radiophysics on solar phenomena and astronomical objects. The association between Mt Stromlo and Radiophysics was the first major collaboration between optical and radio astronomers anywhere in the world.

In 1951 Allen was appointed to the chair of astronomy at University College in London. Combined with Woolley's earlier decision not to establish a radio laboratory on Mt Stromlo, Allen's departure seemed to spell the end of the Commonwealth Observatory's involvement in radio astronomy. It therefore came as a surprise when Woolley wrote to Fred White at CSIRO's head office in Melbourne to announce that plans had been made to establish a major radio astronomy group in the nation's capital and to erect 'as soon as possible a radio telescope of large resolving power in the paddocks at the foot of Mt Stromlo.' [3] The instrument would be used to carry out studies of the galaxy in conjunction with the main optical research program of the Observatory. White immediately summoned Bowen to Melbourne to discuss the Mt Stromlo proposal at a meeting of the CSIRO Executive in February 1951. The next day White drafted a letter to Woolley cautiously welcoming the news but adding firmly: 'I think Radiophysics has established, as far as Australia is concerned at least, a degree of priority for this particular project, and you will agree, I hope, that it would not be in the interests of collaboration for the Commonwealth Observatory to enter precisely the same field.' [4]

Woolley's proposal to establish a rival radio astronomy group puzzled White and Bowen. On the one hand, they were aware of Woolley's commendable ambition to promote the Commonwealth Observatory and also his belief that the properly constituted centre for astronomical research in Australia, whether optical or radio, was at Mt Stromlo and not in one of CSIRO's divisions. On the other hand, Woolley had personally shown a strange ambivalence towards radio astronomy. He became interested in radio emission from the sun and the galaxy, publishing several papers on the subject. Yet, when it came to the mysterious cosmic sources, Woolley like many other optical astronomers found them incomprehensible. To make matters worse, they had been discovered by people with no knowledge of astronomy. On one occasion Woolley had antagonised several radio astronomers at a scientific meeting in Perth. In an hour-long address on 'Opportunities for astrophysical work in Australia', he failed to make a single mention of radio

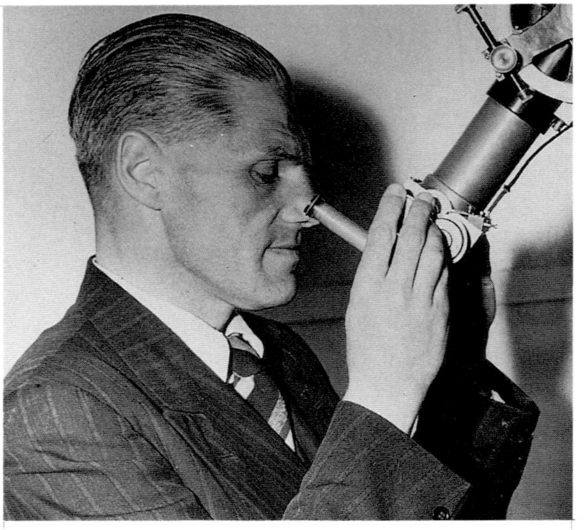

Right Richard Woolley was director of the Commonwealth Observatory at Mt Stromlo from 1939 until his appointment in 1955 as Astronomer Royal in England. *below* Clay Allen with a camera for observing solar eclipses at Mt Stromlo. Early in 1946 Allen began radio observations of the sun, gaining the distinction of being the first professional astronomer to personally turn his hand to radio astronomy.

astronomy. At the end of the lecture, in response to a question on where radio astronomy was likely to be in ten years' time, Woolley had flippantly stated that it would be forgotten.[5]

In view of Woolley's apparent shortsightedness, White and Bowen at first believed that David Martyn had been the instigator of the radio telescope proposal. Since his chequered wartime career at Radiophysics Martyn had continued research at Mt Stromlo, developing a strong interest in solar radio astronomy. In 1950 Martyn had been elected a Fellow of the Royal Society of London, the first Australian-based physicist to gain this high distinction in twenty years, and he had become a powerful and respected figure in the scientific circles of Canberra. However, Martyn was quick to disclaim any responsibility for Woolley's proposal. As he explained in a letter to White, the governing body of the Commonwealth Observatory—known as the Board of Visitors and consisting of several of the most influential figures in Australian science—had 'pressed rather strongly' for the idea. In Martyn's opinion, Woolley had been coerced into supporting the proposal.[6]

As one member of the Board of Visitors, Mark Oliphant made no secret of his desire to see the Commonwealth Observatory develop as the main centre for astronomical research in Australia. Born and raised in Adelaide, Oliphant had pursued a brilliant career in nuclear physics, first at the Cavendish Laboratory in Cambridge where he had risen to be Rutherford's second-in-command, then at the University of Birmingham where among his achievements his laboratory had invented the magnetron. One of the most outstanding physicists of his generation, Oliphant returned to Australia in 1950 and became a driving force in the planning of the Australian National University in Canberra. Appointed foundation director of the Research School of Physical Sciences, Oliphant then played a part in transforming the Commonwealth Observatory into the principal arm of the ANU Department of Astronomy. Postgraduate students at the ANU would have, if Oliphant's plan came to fruition, the opportunity to enter either optical or radio astronomy at Mt Stromlo. In the long term, this steady stream of new talent would ensure the healthy development of astronomy in Australia. Gifted, at times outspoken, Oliphant possessed enough political sense to know that CSIRO would not willingly relinquish the pre-eminent position it had established in radio astronomy. As he explained to a senior government official in March 1951: 'I hope that these doubts will be dispelled and that the work in radio astronomy will be established on Mt Stromlo.' While acknowledging the pioneering start at Radiophysics, Oliphant maintained nevertheless that 'it would be a grave mistake to allow this temporary state of affairs to prevent the natural growth of radio astronomy in the Commonwealth Observatory.' [7]

Fred White responded with a forceful letter, pointing out that exploratory studies on a large radio telescope had already been undertaken at Radiophysics (referring to the abortive RAAF radar aerial) and that the idea had already been supported in principle by CSIRO. Ending on a conciliatory note, White offered a closer degree of collaboration between Radiophysics and Mt Stromlo provided a clear division remained between the radio and optical studies. At best Radiophysics would consider establishing a new field station at Mt Stromlo to supersede the 'lend and mend' equipment installed in 1946, one which would specialise in observations on the galaxy to complement the optical research program.[8] Bowen proved far less willing to compromise. Informed of these developments during a trip to London, he staunchly opposed setting up any radio equipment at Mt Stromlo. With typical

The 74 inch telescope at Mt Stromlo near Canberra, commissioned in 1955. Together with a sister instrument in Pretoria, South Africa this telescope was equal largest in the southern hemisphere until the construction of a new generation of 150 inch telescopes in the late 1960s.

candour, Bowen replied to White that the proposal by the Canberra astronomers was a blatant attempt to jump onto the Radiophysics bandwagon.[9]

The issue dragged on for over a year until a meeting between the Mt Stromlo Board of Visitors and the CSIRO Executive in May 1952 decided against enforcing a partnership between the Commonwealth Observatory and Radiophysics. Even with this truce, Mark Oliphant continued to press for the idea of a national radio astronomy facility. Following a visit to Jodrell Bank he informed Bowen of his personal support for a giant radio telescope in Australia but:

> 'An undertaking of this kind is too revolutionary and too expensive to receive proper consideration as an item in the budget of a division of CSIRO however well conceived it may be. Under prevailing conditions it can succeed only as a national undertaking and as a matter of national prestige .... It is necessary to think in terms of a continued program which might far outlast the interest of those of your team who are at present skimming the cream in this field.' [10]

Oliphant's lobbying for a national facility was an idea ahead of its time. Indeed, it was not until the inauguration of the Australia Telescope in 1988 that a national radio astronomy centre came into existence.

The possibility that Mt Stromlo might establish its own radio astronomy group or, going further, that radio astronomy might be wrenched from CSIRO and relocated in a national facility controlled directly from Canberra set the alarm bells ringing at Radiophysics. The challenge emphasised the need for Radiophysics to break with the free-wheeling approach of the 1940s and to formulate long-term objectives, in particular to decide the types of radio telescopes required for future work. A second and no less important spur to this need to define future directions came with the announcement early in 1951 by the Manchester group of a plan to build a giant parabolic dish. The proposed diameter of 250 feet dwarfed any similar telescope contemplated elsewhere and went far beyond existing engineering knowledge. With such a powerful instrument it seemed only a matter of time before the Manchester group, and not Radiophysics, would begin 'skimming the cream' in radio astronomy. In retrospect, there are several interesting parallels and contrasts in the planning and construction of the two large telescopes at Jodrell Bank and Parkes, so it is well worth looking at the Jodrell Bank story in some detail.

Foggy Field in Cheshire

> Astronomy has marched forward with the growth in size of its telescopes. The need is always for more light-gathering power and more resolving power. In radio astronomy history will doubtless repeat itself; the building of radio telescopes with more sensitivity and more resolving power should yield striking advances in our knowledge of the universe.
>
> Bernard Lovell, 1953

Lovell dates the start of his career in radio astronomy to a foggy day in December 1945 when a trailer loaded with ex-Army radar gear was towed onto a horticultural field station in Cheshire, known as Jodrell Bank.[11] Most of the assorted equipment had been in operation only a few months earlier to direct anti-aircraft guns to enemy planes. After years of wartime radar work Lovell had returned to academic life at the University of Manchester eager to investigate a number of scientific applications of radar. Lovell's first experiment was an attempt to detect radio waves reflected from trails of cosmic rays—a highly penetrating radiation consisting of very energetic gamma rays or subatomic particles passing through the earth's atmosphere and producing long shower columns of ionised particles. In the attempt, Lovell succeeded instead in detecting meteors, which produce a similar but much larger trail of ionised particles in the atmosphere. In October 1946 during the annual Giacobinid meteor shower he confirmed the discovery made by J. S. Hey's group a year earlier of the radar echoes produced by meteor trails.

Lovell had already begun dreaming of large radio telescopes. In 1947 he and John Clegg constructed a fixed parabolic aerial 218 feet in diameter consisting of a spider's web of wire suspended on upright tubular supports. Although Lovell initially had radar studies of cosmic rays and meteors in mind, the 218 ft aerial was used increasingly as a radio telescope to study cosmic emissions. This led to two important discoveries, both due to Robert Hanbury Brown and Cyril Hazard. One

A sketch prepared early in 1952 of the 250 ft dish planned by Bernard Lovell at Jodrell Bank.

was the radio detection of the Andromeda galaxy in 1950, the other the remnant of the supernova first observed visually by the Danish astronomer Tycho Brahe in 1572. After the Crab Nebula and Cassiopeia A, this was the third supernova remnant to be found by radio observations.

Soon after the completion of the 218 ft telescope Lovell set about planning a fully steerable instrument which would be at least the same size. Early in 1948 he discussed the idea with Patrick Blackett, head of physics at the University of Manchester. Lovell could not have had a better department head than Blackett, a scientist with an outstanding record and who, later in the same year, received the Nobel Prize in physics for his research on cosmic rays and nuclear physics. Blackett believed that research groups with access to the largest and most advanced instruments would lead the way in physics, a point of view he argued vigorously in the scientific circles of postwar Britain. With Blackett's encouragement, Lovell sketched out a proposal for a large telescope and in September 1949 approached H. C. Husband, a consulting engineer who prepared a preliminary design for the telescope. Lovell then lodged an application with the Department of Scientific and Industrial Research (DSIR) in London which, in June 1950, granted a sum for Husband to proceed with a detailed design and cost estimate of the project.

Lovell and Husband knew that the maximum size, and thus the final cost of the telescope, would depend to a large extent on the surface accuracy specified for the dish. Almost all radio astronomy in the late

1940s had been carried out at wavelengths of several metres or longer, so Lovell specified one metre as the shortest operating wavelength with approximately ten metres as the maximum. For the telescope to operate with sufficient accuracy, any part of the dish would have to remain within 10 centimetres of a true parabolic shape, whatever the orientation of the dish. To avoid leakage of the signal, the wire mesh forming the reflecting surface could be no coarser than 5 cm square. Given these constraints, Lovell and Husband chose a working diameter of 250 feet (76 m). The dish itself would be supported on two huge towers, each fitted with powerful motors to tip the dish in elevation and, if need be, turn it completely upside down. The dish and towers were in turn to be supported by a large circular steel frame, fitted with several undercarriages which would rest on a circular railway track. Additional motors would then drive the whole telescope structure in azimuth (rotate it about the vertical axis) to within an accuracy of 0.2 of a degree. To help shield the receiving equipment from man-made interference, a decision was also made to have the focal point of the dish lying in the plane of the aperture. This required a sixty-foot tower rising from the centre of the dish upon which the receiving equipment would be mounted.

Detailed blueprints incorporating these ideas were completed in February 1951 and, the following month, Lovell applied to the DSIR for a grant of £259,000 to fund the project—£200,000 for the telescope itself with the remainder for a separate control room, power generators and other sundries. Husband estimated that the project could be completed within three years.

The Americans Enter the Scene

The progress of Lovell's plan to build the world's largest radio telescope was followed with great interest at Radiophysics. Several of the Sydney group visited Jodrell Bank and returned home with glowing appraisals. If a small university department could embark on such an ambitious project, then why not Radiophysics with its much larger staff and technical resources? Bowen first had to confront the obvious problem. A project on the same scale as the Jodrell Bank telescope could not be funded from CSIRO's normal annual budget, as Oliphant had been quick to point out. As Bowen knew from informal discussions with White, the possibility seemed remote of the Australian government providing special funds for a project on this scale, especially one which would not bring any clear economic benefit to the country. Although Bowen continued to think more about the idea, the possibility of Radiophysics building a giant radio telescope looked for most of 1951 like it would remain in the realm of wishful thinking.

Events took a new twist later in the year during a visit by Bowen to the United States, one of several made since the war largely in connection with his own research in cloud physics. Bowen always made a point of looking up old colleagues, many of whom had held wartime posts at the MIT Rad Lab or the Office of Scientific Research and Development—among them Vannevar Bush, a former president of MIT and in 1951 president of the Carnegie Institution in Washington; Alfred Loomis, a trustee of both the Carnegie Corporation and the Rockefeller Foundation, a multi-millionaire and a physicist in his own right; Lee DuBridge, president of Caltech; and Robert Bacher, a leading figure in the Los Alamos atomic bomb project and then head of the physics department at Caltech. These were men of formidable stature and power in the American scientific establishment, impressed by Bowen's descriptions of the flourishing radio astronomy program in Australia, and concerned at how the United States had steadily fallen behind in this rapidly growing area of science.

Even though Jansky, Reber and Southworth had pioneered the field, the Americans had failed to capitalise on their early lead. It was a curious situation in view of the massive government funding of American science after the war—a level of support which saw international leadership in most fields of science shift from Europe to the United States. The potential for the Americans to add the radio astronomy string to their scientific bow had certainly been there. After the war many of the thousand or more scientists and engineers at the MIT Rad Lab returned to academic posts but, oddly, no major radio astronomy group emerged in any of the universities. A number of important, yet isolated discoveries in radio astronomy were made by physicists such as Robert Dicke at Princeton and Ed Purcell and Doc Ewen at Harvard, but these never developed beyond interesting sidelines.[12] A promising start was also made at Cornell by Bill Gordon and Marshall Cohen, but again this group did not have a major impact until the construction of the giant Arecibo dish in 1963. The only significant radio astronomy program carried out in the immediate postwar years was by the group at the Naval Research Laboratory in Washington led by John Hagen and Fred Haddock. This group developed a strong interest in radio astronomy at short wavelengths (in parallel with its principal area of research in radar), which helped set the pattern of subsequent radio astronomy research in the United States. The Washington group constructed a 50 ft parabolic dish, the first sizable telescope built specifically to operate at wavelengths as short as 1 cm.

Lastly, mention must be made of Grote Reber. After joining the National Bureau of Standards in 1947, he began designing a new telescope with a structure similar to the successful Wheaton dish and with a projected diameter of 220 feet. Reber lobbied long and hard in Washington circles to get support for the project, but to no avail. In 1951 he resigned his position and returned to radio astronomy the way he had started, by himself, first in Hawaii and then in Tasmania. Paradoxically, Reber believes the problem was not a shortage of funds:

> 'The situation was different in the United States. There was tremendous wealth and we could embark on huge projects such as nuclear energy. After the war that's where the real empires were made in science. Instead of tens of thousands of dollars, there were tens of millions to spend. The scientific community was totally mesmerised by these sums of money available to build big machines. Unlike Australia, England and Holland, the American scientists simply failed to see the potential in radio astronomy.' [13]

Bowen's recipe for his American friends was straightforward: set up a research establishment with radio astronomy as its prime objective and a large steerable dish as its principal instrument. His advice fell on receptive ears. In December 1951 Robert Bacher invited him to work out a broad specification for the type of telescope likely to pay dividends in radio astronomy, an invitation repeated shortly afterwards by Lee DuBridge in Pasadena: 'The idea of creating in conjunction with the Mt Wilson and Palomar astronomical observatories and in connection with Caltech, a new laboratory or observatory of radio astronomy, is attracting the enthusiastic interest of many people.' DuBridge asked for a detailed proposal specifying the types of problems to be tackled and the instrumentation required. Symptomatic of the times, he encouraged Bowen 'to let your imagination run wild as possibly several million dollars' could be provided.[14] In short, DuBridge proposed building the radio equivalent of the Mt Wilson–Palomar Observatories which possessed the finest collection of optical telescopes in the world. The Mt Wilson Observatory had been founded in 1904 by the dean of American astronomers, George Ellery Hale, who had recognised the viewing advantages of this site in the San Gabriel mountains. A 60 inch reflector had been completed in 1908, followed by the 100 inch Hooker instrument in 1917, the world's largest telescope for the next thirty years. In the late 1920s Hale began planning an even larger instrument, leading eventually to the completion in 1949 of the famous 200 inch telescope at Palomar, near San Diego. For DuBridge, the rightful complement to this outstanding collection of telescopes would be the establishment of the world's finest centre for radio astronomy.

Bowen set to work preparing a proposal for the new observatory. Initially he suggested building two telescopes, one at least 200 feet in diameter for work in the wavelength range 1–10 m (the same as the Jodrell dish), the other 100 ft in size to operate at wavelengths down to 21 cm or lower which could exploit the recent discovery of radio emission from interstellar hydrogen. It soon occurred to Bowen how the two types of telescope could be combined into one. With a single large dish in the range 200–250 ft, the parabolic surface of the central portion out to a diameter of 100 ft could be made accurate enough to allow observations of the hydrogen-line emission, with the engineering tolerances relaxed out to the edge of the dish but still providing sufficient accuracy for observations at the longer wavelengths. The telescope would be driven on an altitude–azimuth form of mounting and, similar to the Jodrell Bank dish, cost in the region of £A250,000. For the observatory's personnel, Bowen suggested a staff about the same size as the radio astronomy group at Radiophysics, which consisted at this time of ten scientists plus a further thirty technical and engineering support staff. The annual salary and operating budget could be expected to be in the vicinity of £A50,000. These proposals were summarised in a report 'A Large Radio Telescope for Radio Astronomy' dispatched to DuBridge in May 1952.[15]

Bowen felt reasonably confident that the project would go ahead. Two of DuBridge's most powerful backers, Vannevar Bush and Alfred Loomis, had pledged to promote the idea and seek sources of funding. Both had enthusiastically supported the appointment of Bowen as director of the new observatory; indeed, John Bolton had already agreed to accompany Bowen as his second-in-command.[16] And yet, despite such a rosy future, Bowen had reservations about the idea. Although he would fulfil a growing ambition to build and take charge of possibly the world's finest radio telescope, it would of course be on American soil. It would also mean starting a new research centre from scratch, not to mention leaving the group he had played no small part in guiding to a position of international leadership.

Fred White kept an eye on these developments from Melbourne, careful on the one hand not to stand in the way of Bowen pursuing his career overseas. On the other hand, after having already fended off the challenge from Mt Stromlo, White did not welcome this American raid on the most successful of Australia's postwar research programs. The 'brain-drain' of Australian scientific talent was developing into an issue of national concern. White counselled Bowen to be patient and to explore other avenues before abandoning hope that a giant radio telescope could be built in Australia. One interesting possibility had arisen earlier in 1952 when the Nuffield Foundation, one of Britain's main philanthropic organisations, had agreed to fund half the cost of the Jodrell Bank telescope. Nuffield might be persuaded to fund a similar telescope in the southern hemisphere, especially if this was presented in the context of strengthening scientific ties within the British Commonwealth. In July 1952 Bowen wrote to sound out his former mentor Henry Tizard, who had just retired as scientific adviser to the British Ministry of Defence. Tizard felt pessimistic about the likelihood of the Nuffield Foundation agreeing to fund a second large telescope, especially one outside Britain. Instead he suggested that Bowen try the British Dominion and Colonies Fund, part of the giant Carnegie Corporation in New York. Buoyed by these developments Bowen informed DuBridge that he still wished to take up the post at Caltech but, nevertheless, he felt duty bound to investigate the possibility of having a large dish built in Australia, even if it meant having 'a foot in both camps'. At last, Bowen's transatlantic network of wartime colleagues had begun to yield results.[17]

As the name implies, the British Dominion and Colonies Fund provided financial support for countries in the British Commonwealth, other than Britain. The Fund formed only a small part of the philanthropic empire established by Andrew Carnegie in his attempt to systematically give away one of the world's great personal fortunes. Among the legacies of Carnegie's philanthropy were over 2500 public libraries across the United States, the Carnegie Institution in Washington established to conduct basic scientific research, and the Carnegie Corporation in New York founded in 1911 with a charter to promote the advancement and diffusion of knowledge.

Taking Tizard's advice Bowen wrote in August 1952 to Vannevar Bush, president of the Carnegie Institution in Washington, to ask whether in addition to the Caltech instrument Carnegie would consider granting further funds for a similar dish in the southern hemisphere.[18] Both radio astronomy centres would benefit as the design of the telescope could be a collaborative effort and the costs shared. Bush responded that he was very much in favour of the idea. Although the Carnegie Corporation had for several years been redirecting its support to the social sciences, mainly in reaction to the massive support given to the physical sciences by postwar governments, for the British Dominion and Colonies Fund it was a different story. Grants had been suspended during the war years and with interest steadily accruing on the original $US12 million endowment from the parent Carnegie Corporation, the

Vannevar Bush (left), president of the Carnegie Institution in Washington, and Lee DuBridge, president of Caltech. Both were strong supporters of the proposal to build a giant radio telescope in Australia.

Fund was flush with money. In 1951 nearly $1 million had been released to support individual study awards and to fund research projects, a substantial proportion going to Australian cultural and educational institutions. In Bush's view, the timing of an approach to the Fund for a substantial grant in the physical sciences could not have been better.

For most of 1953 the funding issue was held in abeyance while several firms in the United States and Australia carried out studies on the feasibility and approximate cost of building a large dish (see Chapter 6). A final opportunity to lobby for the proposal came in January 1954 when Bowen and several members of the Radiophysics group attended a symposium on radio astronomy in Washington, sponsored by the National Science Foundation, the Carnegie Institution and Caltech. The symposium had been organised in direct response to the growing recognition by American astronomers and physicists that the United States had been left far behind in this important new field. The main purpose of the meeting was to examine ways to sponsor the establishment of major new facilities and instruments; in turn, this led directly to the proposal to form the National Radio Astronomy Observatory, founded at Green Bank in West Virginia in 1957. The Australian contingent came away from Washington with renewed assurances from Bush and other influential figures at the Carnegie Institution that its funding application would receive a favourable hearing at the parent body in New York. Confirmation came in April. From his Fifth Avenue office, the president of the Carnegie Corporation, Charles Dollard, informed Bowen:

> '… our program is concerned chiefly with the social sciences and since the war we have made only a few scattered grants in the natural sciences. The chief question we had to debate, therefore, was whether the character and quality of your program would justify an exception to our general policy. Our final conclusion was that it would.' [19]

On 21 May 1954 the trustees of the Carnegie Corporation formally approved a sum of $US250,000 towards a giant radio telescope in Australia.

The announcement of the Carnegie grant concluded a curious sequence of events. Late in 1951 Caltech had invited Bowen to propose a new astronomy centre consisting initially of a single large radio telescope. Bowen was to be director of the facility, Bolton his deputy and no doubt others in the Radiophysics group would have joined them along the way. Later, the proposal was broadened to involve two large telescopes, one at Caltech in the northern hemisphere, the other in Australia. As it turned out, the proposal to build a large American telescope staffed partly by Australians had changed, two years later, to an Australian telescope funded partly by American money!

Caltech eventually devised a new plan and in 1955 began construction of an interferometer consisting of two 90 ft dishes at Owens

How Sydney's *Daily Telegraph* viewed plans for a giant Australian telescope, shortly after the announcement of the Carnegie grant in July 1954.

Valley in California, funded by the US Office of Naval Research. John Bolton became the first director of the new Caltech observatory, assisted by his colleague Gordon Stanley. We can only speculate on the course Australian radio astronomy might have taken if the original Caltech plan for a giant telescope had gone ahead.

Search for Further Funds

The Carnegie grant meant that the first important funding hurdle had been cleared. Initial estimates at Radiophysics put the cost of a telescope, at least as large as the one under construction at Jodrell Bank, to be in the region of $US1 million (£A500,000) or about four times the amount promised by Carnegie. During the latter half of 1954 Bowen continued to marshal support for the project. Part of the strategy consisted of enlisting the backing of two of the most prestigious organisations in astronomy. The International Astronomical Union sent a statement of support in July signed by its president Otto Struve and its general secretary Pieter Oosterhoff. A second statement came from the Royal Astronomical Society in London, drafted by Bernard Lovell and signed by twenty leading British astronomers:

> 'The following members of the Council of the Royal Astronomical Society strongly support the movement for the building of a giant radio telescope in Australia, which can do for the southern hemisphere what the 250 ft radio telescope, now under construction at Manchester, will do for the northern hemisphere. A large instrument is needed to complete the survey of the southern hemisphere especially of the Milky Way, our local galaxy. The centre of the galaxy is in the southern hemisphere. It is obscured by interstellar dust and is not accessible to observation by photography. Radio observations do not suffer from this limitation, and the large telescope proposed would give us information about the central regions of the Milky Way of the greatest importance in astronomy. ... It is most desirable that Australia, where so much pioneer work has been done in radio astronomy, should keep in the van in the advancement of this young and rapidly growing field of knowledge.' [20]

In April 1955 a formal appeal for government funding went through the official channels in Canberra. First, Ian Clunies Ross (CSIRO chairman) wrote to Richard Casey (CSIRO minister) and stressed the proven record of the Radiophysics group:

> 'A major part in the development of radio astronomy has taken place in Australia, and the country is in the very forefront of this field of science. It has a group of highly qualified and experienced workers skilled in the difficult techniques involved. The work deserves the utmost support so that the country's position can be maintained.' [21]

The funding application had Casey's wholehearted support and he took the matter to the prime minister, Robert Menzies. Although Menzies endorsed the project in principle, he remained guarded on the question of government funding:

> 'I agree with you as to the value of the project in the national interest, and I am prepared to agree to subsidise private contributions to the cost of this instrument, on a pound-for-pound basis, on the understanding that one-half of the cost of the project must be in sight from private sources before any Commonwealth grant becomes payable.' [22]

Menzies' stipulation that one-half the cost be in hand from private sources before the government would come to the party meant that a sum as large as the Carnegie grant still had to be found. The two options open were either to raise the money within Australia or to again approach philanthropic foundations overseas. Bowen and White had already begun exploring the first option in the middle of 1954, anticipating that this would strengthen their hand when approaching the government for funds. A number of large companies were contacted in the hope that one, or possibly several, would agree to make a substantial donation to the project. The idea proved futile. As one example, Essington Lewis, deputy-chairman of Broken Hill Proprietary Ltd, replied that he would be prepared to make a private contribution, but that Australia's largest company was not in a position to act as a local version of Carnegie.[23]

With the unlikely prospect of attracting a single large benefactor, a more promising option seemed to be to cast the net wider and attempt to catch a large number of small donations. Early in 1955 White and Casey formed the Radio Astronomy Trust to raise funds. Four prominent figures were invited to act as trustees.[24] The purpose of the trust was to solicit donations and also to manage an investment folio for these funds, including the Carnegie grant, until the design and construction of the telescope got underway. Despite numerous appeals to companies and wealthy individuals (admittedly very few of those approached had any real appreciation of radio astronomy), the response was disappointing. After the first year of fund raising, donations pledged to the Trust amounted to slightly more than £20,000, far short of the £125,000 required which combined with the Carnegie grant would make up half the projected cost of the giant radio telescope.

With only one option remaining Bowen flew to the United States in June 1955 to approach cap-in-hand a number of other large philanthropic organisations. He drew a blank with the Sloan Foundation, basically because of a clause in its charter which severely restricted the award of funds outside the United States. Initially the Ford Foundation appeared promising, but then turned down Bowen's application on the grounds that it would be inconsistent with its policy of concentrating on issues of human welfare, leaving the support of scientific research to governments and other organisations. Ford funded five main areas—peace, democracy, education, human behaviour and human relations—five ingredients aimed at strengthening the American way of life. The support of a somewhat obscure scientific program halfway round the globe did not quite fit into the picture.[25]

Bowen's approach to the Rockefeller Foundation proved far more profitable. With a history similar to its Carnegie counterpart, the Foundation had been established in 1913 by John D. Rockefeller with the lofty aim of promoting the well-being of mankind throughout the world. By the early 1920s the Foundation had grown to be the largest philanthropic organisation of all with an annual expenditure of $8 million, most of this being injected into medical research in the United States. One of the Foundation's offshoots, the International Education Board, was established in 1923 largely to provide assistance to scientists outside the United States, particularly those in the economically depressed countries of Europe. Almost all of the major figures involved in the revolutionary new quantum theory in the 1920s, physicists such as Niels Bohr in Copenhagen and Max Born in Göttingen, received substantial Rockefeller support either to build up their own institutions or to establish a network of travel scholarships.[26] Astronomy also benefited with the International Education Board in 1928 making the first of a series of grants, eventually totalling $6 million, to build the 200 inch Palomar telescope.

In addition to this tradition of support for the physical sciences, another factor in Bowen's favour was his personal acquaintance with Warren Weaver, director of the Natural Science Division and a key figure in the Rockefeller hierarchy. Originally a mathematician, Weaver had followed the rise of postwar radio astronomy with keen interest. One further factor was the personal intervention of Richard Casey, on this occasion in his capacity as Australian minister for foreign affairs. In September 1955 during a visit to New York, Casey called on Dean Rusk, president of the Rockefeller Foundation and later to become US secretary of state during the 1960s. Casey spoke persuasively in support of the telescope project and at the end of their meeting Rusk himself suggested the sum of $250,000. The official announcement of the grant

Marshalling funds for the telescope: Fred White (left), Richard Casey and Taffy Bowen in January 1958.

was made on 17 December 1955.[27] Although later in the project there were unforeseen escalations in the cost, the Rockefeller and Carnegie grants together with Menzies' pledge meant that, by the end of 1955, enough funds were in hand to proceed with the Parkes telescope.

Big Science Comes to Australia

The promise of funds gave the green light to the telescope project, one of several large ventures in Australia in the 1950s qualifying under the title of big science. The first had been the construction of the 74 inch telescope completed at Mt Stromlo in November 1955 at a cost to the government of £100,000 and which, for the next two decades, was the equal largest optical telescope in the southern hemisphere. Several factors, however, put the radio telescope project into a whole new league. Whereas the Mt Stromlo instrument was based on tried and true design principles, the radio telescope involved a journey into unknown areas of engineering; this story forms the subject of our next chapter. Other aspects, such as a projected cost five times greater than the optical telescope and the fact that half the funds came from overseas, raised some interesting issues. For the remainder of this chapter, we look briefly at the telescope project in the wider context of CSIRO and Australian science.

One point worth making is the way the telescope project highlighted the different attitudes to science in Australia and the United States in the early 1950s. Although the government in Canberra supported the project, it did so with considerable caution, only promising to put up half the money on condition that the other half was in hand from private sources. It was as though the government first needed reassurance from the Americans that the idea was worth supporting. The Carnegie and Rockefeller foundations on the other hand showed no such hesitation, clearly recognising the Radiophysics group as world leaders. Both were intent on supporting the project even though it would bring no direct benefit to the United States. As in most facets of American life in the 1950s, science had an abundance of funds, so much so that there was enough to support major centres in other countries as well. The large philanthropies recognised that science funding could play a key role in helping to broaden and strengthen American cultural and political influence in the postwar world, particularly in countries such as Australia where a power vacuum had developed as a result of Britain's declining influence. The president of the Carnegie Institution, Vannevar Bush, proclaimed the desirability of strengthening these ties shortly after the announcement of the Carnegie award in 1954:

> 'Nothing, I am sure, would bring our two countries closer together more effectively than for Australia to lead the way in an important area of fundamental research. I am sure that this would have a salutary effect in Australia and throughout the world. We look to Australia for many things, and admire a great deal we see from this distance. But if we could also see the Australians as the recognised leaders in an intellectual effort of a high order in the field of science, I am sure that the respect for Australia would grow in this country and that this would be helpful indeed. Also I feel sure that this accomplishment on your part would lead the Australians to view this country increasingly as a place where good fellowship is found in the discussion of mutual interests.' [28]

In contrast to the large sums dispensed by the American foundations, Australia had no comparable tradition of philanthropic support for science, particularly in the physical sciences. It came as no real surprise when the appeal to several large Australian companies in the hope of finding a major benefactor produced no response. The Radio Astronomy Trust set up to canvass funds on a wider level hardly fared much better. After a considerable effort, particularly by Casey, only 37 companies and individuals contributed, with one of the trustees Andrew Reid himself donating £10,000 towards the final total of £27,000. This sum represented only 3% of the eventual cost of the Parkes telescope: in fact, the interest earned by investing the Carnegie and Rockefeller grants until construction began amounted to slightly more.[29]

The poor response in raising funds could not be blamed solely on the lack of philanthropy in Australian science. Though nowhere near the same level as the United States, a modest amount of private support did exist, but in most cases it was directed towards more utilitarian research in fields such as medicine, biology and agriculture. To illustrate the point, there is one example which has several interesting parallels with the telescope project. In the late 1950s the CSIRO Division of Plant Industry led by Otto Frankel began planning a new facility in Canberra known as a phytotron, a special laboratory with rooms and cabinets where environmental factors such as temperature, humidity, light intensity and daylight-length can be controlled to study plant germination and growth under a wide variety of conditions. With an estimated cost of £500,000 for the facility, a Phytotron Trust was proposed in April 1958 and within a month donations totalling £30,000 had been pledged. This was more than the sum raised by the Radio Astronomy Trust over a period of several years. In the meantime Casey lobbied in Canberra for the project, arguing that the phytotron would yield substantial economic benefits to Australia. In contrast to the telescope project, Casey's political colleagues could easily appreciate this line of argument and the government had little hesitation in agreeing to meet the full amount of £500,000. Those who had pledged funds to the Phytotron Trust were politely informed that their donations were no longer required! [30]

Although it proved easier to tap into sources of both government and private funding in fields such as medicine, biology and agriculture, examples of large-scale support in the physical sciences in the 1950s were not unknown. One which benefited from substantial government funding, the other from even greater amounts from the private sector, are highlighted in Tangent 4. Whereas nuclear physics provided Oliphant in Canberra and Messel in Sydney with a powerful lever in their funding campaigns, the relevance of a giant telescope to the economic or social well-being of the nation proved a difficult case to argue. Most scientists at the time believed, and many still do, that attempts to establish such a connection are pointless because benefits eventually flow, unpredictably but inevitably, from almost any type of basic research. Attempts to identify or predict the immediate gain from basic research, so the argument goes, prove shortsighted and possibly even counterproductive. In response to a reporter's question on the value of a giant telescope, Bowen summed up this philosophy:

> 'No instant value if that's what you're looking for. This is pure research. It doesn't give immediate results. But long term—ah—that's different! This is exploration, this is Columbus voyaging over the western seas. I suppose when Columbus set out, a Scottish shepherd, if he'd heard about it, might have said, 'What's all this to me? How many more sheep will this bring me?' The discoveries of Columbus didn't affect him directly perhaps. But they certainly did have a tremendous effect on his descendants.' [33]

Though most would agree or at least sympathise with this justification for curiosity-driven basic research, when it came to raising funds for a radio telescope the argument did not carry much weight. Other more pragmatic figures such as White and Casey often went to considerable lengths to stress the practical benefits of radio astronomy in their appeal for the Radio Astronomy Trust, for example an increase in our understanding of radio communications or of plasma physics. But in the event they failed to convince the Australian private sector to dig deeply into its pockets. Beneath these attempts to justify the telescope in terms of its practical benefits lay a more important concern for White and

TANGENT 4

Oliphant and Messel

Mark Oliphant accepted the new post of director of the Research School of Physical Sciences in 1950 only after a firm assurance by the government that it would provide large sums of money to build a powerful new particle accelerator at the Australian National University. Oliphant called his dream machine a 'cyclosynchrotron', combining features of the cyclotron and synchrotron accelerators in use overseas and a power source known as the homopolar generator.

By late 1953 it had become clear that the accelerator would be extraordinarily complex and costly to build, and probably be no great step forward on the Birmingham synchrotron which Oliphant had constructed before his return to Australia. Oliphant's group then formulated a far more ambitious plan which incorporated the half-built cyclosynchrotron into a new and larger proton synchrotron. This accelerator would be ten times more powerful, generating proton pulses at energies up to ten thousand million volts (10 GeV), and would supersede the generation of accelerators under construction in Europe and the United States. Again, design problems and escalating costs bedevilled the project.

A biography of Oliphant noted: 'The constant arguments over money were a trial for Oliphant, who continued to feel that his project was a test case for government support for science. The sums involved were likely to total about £1 million. While these were large by Australian standards, they were not when seen in an international context. If the government failed to honour its commitment to see his project through there would be little hope of support for other significant efforts such as the large radio telescope then being talked about. Other observers saw the matter differently. They feared that if Oliphant failed to deliver the goods within a reasonable time governments might be scared off and money harder to obtain. In retrospect it seems fair to say that Oliphant's enterprise neither helped nor hindered other large projects in their quest for funds.' [31]

In 1961 Oliphant finally abandoned his dream after £500,000 had been spent on the accelerator project (£300,000 on materials and £200,000 on salaries). In the meantime the Berkeley Bevatron had come into operation, after seven years of effort by the Americans and at a much higher cost of £7 million. The Bevatron began harvesting the discoveries which might have otherwise gone to the proton synchrotron. Throughout the long episode Oliphant was subjected to the peculiar Australian trait known as the tall-poppy syndrome—criticised for planning a project large and ambitious by Australian standards and, what was worse, failing in the attempt. Inevitably his strongest critics branded the accelerator the 'White Oliphant'.

If nothing else Oliphant succeeded in making the government realise that if Australia was to play in the big league of contemporary science, large sums of money were required. Another physicist, Harry Messel, played a similar consciousness-raising role among some of Australia's industrial leaders. Shortly after his appointment as professor of physics at the University of Sydney in 1952, Messel established the Nuclear Research Foundation, a fund-raising organisation consisting of a number of leading business and political figures. In its first eight years the foundation raised a phenomenal £3.2 million to support the research program in Messel's department, over £2 million of which came from overseas. A number of the foundation's benefactors were rewarded by receiving honorary degrees from the University of Sydney. The theoretical physics department was later named after the media magnate Frank Packer. Another benefactor, wealthy Sydney jeweller Adolph Basser, donated most of the £50,000 prize money after his horse Delta had won the Melbourne Cup! [32]

The physicists Mark Oliphant (left) and Harry Messel established vigorous research centres at the Australian National University and the University of Sydney in the 1950s. Both centres challenged CSIRO's traditional position as the leading research organisation in Australia.

Much of the credit for this extraordinary success in raising funds goes to Messel himself, a flamboyant Canadian with formidable powers of persuasion and with the knack of getting good media coverage of his work, talents also possessed in no small degree by Oliphant. A more important reason was the purpose for which these funds were to be used. Rather than build large and expensive particle accelerators, Messel planned to use cosmic rays, nature's own particle beams, to investigate the fundamental structure of matter.

The debate over accelerators versus cosmic rays often saw the Canberra and Sydney groups at loggerheads, but when it came to attracting generous benefactors both groups were on the right track. Nuclear physics and the search for the fundamental laws of nature—the holy grail of physics—were seen as vital to Australia in two respects. First, with the arms race in the 1950s nuclear physics and its applications became central to long-term strategic planning of the nation's defence. Second, atomic energy (or more correctly nuclear energy) promised to bring major economic benefits to the country. Like most of the world in the 1950s, Australia was in the honeymoon stage with atomic energy, believing it would one day provide an unending supply of cheap, safe power.

Later, Messel changed the name of his fund-raising organisation to the Science Foundation for Physics and broadened its scope to support other areas of research, including a new department of astronomy in 1960. In the early 1950s Oliphant had challenged Radiophysics' dominance in radio astronomy by proposing a new government-funded national centre at Mt Stromlo. Oddly, the first real competition came instead from Messel at the end of the decade, not with government but with private funds (see Chapter 8).

Casey—the effect the telescope would have on the existing balance between pure research and applied research in CSIRO. Since the reconstitution of the organisation in 1949, increasing emphasis had been placed on research to support secondary industry reflecting (somewhat belatedly) the continuing diversification of the Australian economy away from its traditional 'ride on the sheep's back'. By 1955 CSIRO had grown to an organisation of about 1400 staff and 17 separate divisions, concentrated largely in Sydney and Melbourne but with isolated sections scattered around the country. These divisions separated into two main groupings, one consisting of eight divisions operating mainly in the field of primary industry, with the other nine divisions basically serving the requirements of Australian secondary industry. As part of this second group, Radiophysics accounted for 7% of the CSIRO budget of £5 million in 1955, ranking third in size behind the Division of Industrial Chemistry (11%) and the National Standards Laboratory (10%).[34]

In size and in its style of operation, Radiophysics had much in common with these other two divisions. Located at Fishermen's Bend in Melbourne, Industrial Chemistry carried out a wide range of applied chemical research under the leadership of Ian Wark. It also contained a large section headed by Lloyd Rees devoted almost entirely to pure research in atomic and molecular physics. Similarly, the National Standards Laboratory (NSL) carried out its own blend of pure and applied research. Established in 1939 and sharing with Radiophysics the same rambling complex of stately buildings on the University of Sydney campus, the NSL had begun basically in the role of a technical consultant, providing a testing and calibration service to Australian industry gearing up for wartime production. Later, the NSL expanded its operations to include a broad range of precise standards of physical measurement, an essential technical foundation for Australia's postwar industrial reconstruction. Though much of its work was fairly routine (some would call it pedestrian science), a substantial part of the NSL research effort was devoted to fundamental problems in physics. As one example, a small group led by Ron Giovanelli began research in 1946 on the physics of the outer layers of the sun and, by the late 1950s, had become a world authority on the subject.[35]

Even though divisions such as Industrial Chemistry and the NSL channelled a substantial part of their resources into fundamental problems in chemistry and physics, this investment could be readily defended because of the close links this work had for the applied, practically oriented research program. To this extent a certain amount of pure and uncommitted research could be seen as justifiable and desirable. But in this one important respect, radio astronomy was different. Admittedly the techniques of radio astronomy found application in related areas such as radio engineering and physics, yet in hard economic terms the connection seemed tenuous and shaky. Until plans for the Parkes telescope had been announced, it had been possible to conveniently hide the problem. Radio astronomy had been able to flourish because of the patronage of CSIRO's top management: a strong and enduring interest by White in the activities of his former division, the role of CSIRO chairman Clunies Ross as a warm but somewhat detached admirer of these achievements, and then Casey the minister who had become fascinated by the work.

Until the mid 1950s radio astronomy had been funded entirely within the Radiophysics budget and in several respects tucked away from outside scrutiny. Various instruments such as the Mills Cross and the radio spectrograph were situated at isolated field stations, and each had been designed and constructed with relatively inexpensive materials by the Radiophysics staff itself. Further, with Radiophysics sharing the same site as the NSL, any casual visitor would walk away with the impression of a laboratory bustling with applied research work. Even so, what made the CSIRO management increasingly nervous was that by appealing for funds for a giant radio telescope, either publicly through the Radio Astronomy Trust or direct to the government treasury, CSIRO had exposed itself to the charge that radio astronomy was an activity outside its charter.

Compounding the problem was the extent to which radio astronomy had come to dominate the research program at Radiophysics. A report prepared at Fred White's request in August 1956 revealed that radio astronomy had grown from an unlisted item in 1948, to 38% in 1952 and then to 51% of the Radiophysics budget of £300,000 in 1956. The cloud and rain physics group had decreased slightly to 33%, with the air navigation and computing groups being the two areas to suffer at the expense of radio astronomy, both dropping to 8% of the total budget.[36] Even with his close interest in the activities at Radiophysics, White admitted to being startled by these figures. In response Bowen argued that if CSIRO wished its most successful program to continue, it could not be done on the cheap:

> 'The fact that radio astronomy absorbs just over one-half our budget is no accident. It reflects the very deliberate policy we have followed of putting the best research and technical officers on to this work .... The annual expenditure of £154,000 on radio astronomy is about ten times that at Cambridge or Manchester (excluding the big telescope) and far greater than that of any individual group in the USA. I would regard this kind of expenditure as thoroughly justified if—as was certainly the case in 1951—we had a clear lead over the rest of the world. However, in 1956 we no longer have this lead and, in my view, are trailing the field badly in many respects.
>
> Our radio astronomy program is canvassed around as one which shows what can be achieved by relatively simple and inexpensive devices. String a few wires up in a field and, lo and behold, a miraculous result! Nothing could be further from the truth. Taking the Chris Cross as an example, this represents about one-fifth of the activity of the group, that is about £30,000 a year. When finished it will have taken three years, that is an expenditure of £90,000 before any research results have been achieved.' [37]

The radio astronomy program was beginning to make Radiophysics burst at the seams. With the construction of the giant radio telescope in the not too distant future, the rest of the astronomy program could not continue to grow unchecked. A meeting to discuss the issue was arranged for senior Radiophysics staff and the CSIRO Executive in October 1956. The meeting set the scene for the future.[38] Support would go to the telescope, not by further cuts to the other Radiophysics programs, but by winding back parts of the radio astronomy program. Later, in Chapter 8, we see that by the end of the decade the pressures mounting from this decision eventually tore the radio astronomy group apart.

CHAPTER SIX

Turning Concepts into Blueprints

In my experience most major programs of scientific research would never have got started if the people who proposed them had not greatly underestimated the cost, time and amount of work involved.

Robert Hanbury Brown, 1987

As time would tell, raising funds to build the Parkes telescope turned out to be the easy part. The decision by the US foundations and the Australian government to back the project had largely been an act of faith, a pledge made before any detailed design work or realistic cost estimate had been carried out. Unlike the large optical telescope at Mt Stromlo constructed in the early 1950s, based as it was on decades of experience by telescope builders around the world, Parkes was a venture into an unexplored area of engineering. The project had more in common with the unsuccessful attempt by Mark Oliphant to build the world's largest particle accelerator in the 1950s, a bold gamble to transform untried physical ideas into an engineering reality. In this chapter we trace the evolution of the design concepts for the Parkes telescope, beginning in 1951 with the first proposals at Radiophysics for a telescope with a very large collecting area, through to the completion of the blueprints for a giant parabolic dish in 1959.

Earlier in Chapter 4 we saw that among a variety of competing proposals for new telescopes at Radiophysics, the new cross devised by Bernard Mills emerged victorious and became the principal type of instrument in cosmic radio astronomy during the 1950s. Two proposals passed over at this time were plans to build a single 70 ft dish primarily for hydrogen-line studies, the other to build an interferometer consisting of two 80 ft dishes. To some extent these proposals were the forerunners of the Parkes telescope, basically a scaled-up version of these smaller parabolic dishes. However, there were two other proposals also taken quite seriously at this time which illustrate how the Radiophysics group was already contemplating instruments with very large collecting areas. One proposal for a new sea interferometer put forward in 1951 by John Bolton called for a 'standing paraboloid', consisting of a series of upright posts about 20 ft high and set into the ground at intervals to form a parabolic arc 200 ft long. Resembling a long and curved tennis court fence, the telescope would focus radio waves to a receiver set 150 ft in front of the structure, allowing the position of radio sources to be measured to within a minute of arc, an accuracy not then achieved by any other instrument.[1]

Bowen went one step further in planning a much larger version of the parabolic cylinder he had earlier tried to persuade the RAAF to build as a radar aerial. The cylindrical structure would be 200 ft wide with a total length of 1000 ft, consisting of five sections each 200 ft long and each weighing about 20 tons. This massive cylinder would lie on its back along an east–west direction and be steered by an arrangement of cables and winches tilting the structure in either the north or south direction. One appealing feature of the proposal was that the telescope could be built in stages, each section being added one at a time so that over a period of five years the estimated total cost of £125,000 could have been met within the normal Radiophysics budget.[2]

Though the proposals for these two large aerials remained on the drawing board, they provided additional stimulus to the early plans for a giant parabolic dish at Radiophysics, formulated in response to the external developments at Mt Stromlo, Jodrell Bank and Caltech discussed in Chapter 5. Bowen first sounded out engineering firms for their advice early in 1953, asking for design concepts for a telescope as large as possible with particular attention paid to the way the dish would be mounted and steered across the sky. One idea investigated by the Chicago Bridge & Iron Company was to rest the parabolic dish on a section of a large sphere, which in turn would float in a pool of water or oil. This 'eyeball' design had the advantage that stresses would be more evenly distributed around the dish structure, rather than localised as in the Jodrell Bank telescope where the dish was supported at just two points. The Hughes Aircraft Company also formulated a proposal which included the idea of enclosing the telescope in a huge protective bubble known as a radome. Similar to the dome enclosing a large optical telescope, the radome would guard against deformations in the dish surface caused by strong winds or large temperature variations. The Fuller Research Foundation in Wichita, Kansas, established by the engineer Buckminster Fuller, put forward another design idea based on Fuller's famous idea of the geodesic dome. Architects and engineers at this time had started to discover the potential of the geodesic dome, in particular its exceptionally high strength to weight ratio. However, as one astronomer noted, not all were impressed with the telescope proposed by Fuller: 'Most people think that in a decent wind the thing will blow away down the valley.'[3]

Closer to home a group of engineers under the direction of Arthur Wills at the Aeronautical Research Laboratories (ARL) in Melbourne also agreed to carry out a design study on the structural features of a large dish. Radiophysics had forged close links with the ARL during the war when the ARL itself had been part of CSIR. In 1949 the ARL had been transferred to the Department of Supply following the decision that the type of classified military research partly conducted at ARL would no longer continue in the new CSIRO. Nevertheless, it remained the best qualified government laboratory to carry out an engineering study of a giant radio telescope. Late in 1953 the ARL issued a report which concluded that it should be feasible to build a dish 250 ft in diameter, weighing approximately 100 tonnes and supported on an altitude–azimuth type of mount, with a surface accuracy which would allow observations at wavelengths as short as 21 cm.[4]

Although the ARL and the American engineering firms concluded that a telescope as good as the one under construction at Jodrell Bank seemed possible, these preliminary investigations were still a long way from a full design study. Immediately after the announcement of the Carnegie grant in May 1954, Bowen and Pawsey set up a planning committee to evaluate the various design ideas and to draw up a set of general technical specifications the telescope would be expected to meet. The committee's deliberations were summarised in a publicity booklet issued in May 1955 and distributed to several dozen astronomers, as well as engineering firms which had expressed interest in the project. Seven possible designs were featured in the booklet, each having in common a fully steerable parabolic dish 250 ft in diameter and differing only in the type of mount used to support and steer the dish.[5]

The minutes of the planning committee reveal that support for the type of large parabolic dish featured in the publicity booklet was by no means unanimous. Some committee members argued that other types of large reflectors should be investigated which might avoid the formidable design problems involved with a movable dish. One particularly attractive alternative was a proposal made by ARL scientist Alan Head for a new and more refined version of the hole-in-the-ground telescope used with such success at Dover Heights. Head's idea involved two reflectors, one a very large fixed reflector set in the ground, hemispherical and not parabolic in shape, the other a small movable barrel-shaped reflector suspended above ground. Incoming radio signals would be reflected from the large hemisphere into the barrel-shaped reflector, which would then focus the signals onto the receiver or feed. According to Head's calculations, the performance characteristics were as good as a movable parabolic dish of comparable size. Importantly, the large fixed bowl was a straightforward and inexpensive exercise in civil engineering, especially if a ready-made geological formation of roughly the right shape could be found.[6]

Several members of the radio astronomy group including Pawsey developed a strong interest in Head's idea, realising it might provide a

What might have been—four structures investigated at Radiophysics during 1954: *top left* The dish is mounted on two movable towers, a design proposed by Hughes Aircraft in California and similar in concept to the Jodrell Bank telescope. *top right* Jacks on a turntable, where the dish is tilted by two jacks mounted on a rotating platform. *below left* The eyeball design devised by the Chicago Bridge & Iron Company, where the dish is fixed to a section of a sphere which in turn floats in a pool of water. *below right* The dish moves on tracks inside a hemispherical hole in the ground, a design that avoided a number of structural problems.

relatively simple and inexpensive alternative to a giant steerable dish. Even if the Head telescope did not fit the bill, Pawsey had fundamental reservations about building a dish on the scale envisaged by Bowen. A smaller telescope would allow much higher surface and pointing accuracies, and could also be brought into operation much sooner, no small advantage in a field progressing as rapidly as radio astronomy. In Pawsey's view, a smaller and cheaper dish would release funds for other aerials which could be connected to the main dish and used for interferometry. At a meeting of the planning committee in March 1955, with Bowen absent and Pawsey in the chair, the issue was discussed at length and led to the resolution: 'A 150 ft aerial would be quite good enough for many types of investigation, a bit doubtful for some, but in these cases a 250 ft dish might not be much better. In any case, a 150 ft in existence could do more than a 250 ft on paper. For these reasons, there was a general feeling in favour of reorienting the program towards a small dish.' [7]

As head of the radio astronomy group, Pawsey's views carried considerable weight and so it is worth looking at the reasons for his opposition to a very large dish in some detail. Since pioneering the technique of sea interferometry in 1945, Pawsey had remained first and foremost an interferometer man, attracted by the theory of the technique and the challenge of developing new combinations of aerials, such as the ingenious arrays devised by Chris Christiansen for solar work at Potts Hill. Pawsey certainly did not share Bowen's view that interferometry in radio astronomy would soon be outmoded. On one occasion, Bowen's foresight deserted him when he stated that 'the interferometer has now been exploited to the utmost. It will not be of much use for the discovery of new stars or extension of our knowledge beyond the present frontiers, although it will certainly be of use for a long time for further studies of the sun and some of the stronger radio sources.' [8]

Aside from the obvious danger that a single large dish would reduce the resources available to develop interferometers, Pawsey harboured a deeper fear about the project. Because of the competition by astronomers for observing time on a large and expensive instrument, subtle pressures come into play where basically 'safe' research projects are pursued, not the 'wildcat' ideas that might lead to fundamental and unexpected discoveries. For this reason the research programs at well-established optical observatories often tended to be staid and predictable. During the heady times since 1945 Pawsey knew that most of the important discoveries in radio astronomy had been made by those who had been flexible in their approach and constantly prepared to experiment with new types of instrumentation. A total commitment to a single large dish hardly seemed the most convincing way to maintain this momentum. On a more personal level, the sheer size of the project went against the grain for Pawsey. More than any other, Pawsey epitomised the 'string and sealing wax' approach to research that he had learnt so well early in his career at the Cavendish Laboratory. In sharp contrast to Bowen, he did not have the instinct for projects under the name of big science, with their large-scale engineering works and million-pound budgets. Pawsey had been in his element in the days when radio astronomers had worked in the open fields stringing up simple aerials with their own hands. After a visit to Jodrell Bank, Pawsey summed up his position to Lovell in a nutshell: 'I am still rather appalled at the magnitude of your project, but I am very interested indeed at the possibility of being able to construct an aerial of about half the size.' [9]

Despite these reservations, Pawsey gave the project his unqualified support once the design study for the giant telescope was underway in 1956. Nevertheless, for Bowen, with his principal deputy half-hearted about the project and with several others quite vocal in their opposition (see Chapter 8), there were occasions when he had serious doubts himself about pressing ahead. Combined with this dissent from within Radiophysics, there were also times when Bowen received discouraging advice from influential figures elsewhere. By the middle of 1954, when Lovell had run up against a number of formidable engineering problems, he urged Bowen to learn from his own experience: 'there is a very strong case for building a somewhat smaller telescope—say 150 ft or 200 ft with very much tighter tolerances.' [10]

Similar advice also came from Merle Tuve, director of the Department of Terrestrial Magnetism in Washington (the Carnegie Institution's largest laboratory), and one who had played a key role in securing the Carnegie grant for the telescope. In 1954 Tuve had been appointed chairman of a National Science Foundation committee responsible for the distribution of government funds for radio astronomy throughout the United States. Together with the noted radiophysicist Lloyd Berkner, Tuve had also been a prime mover in establishing the National Radio Astronomy Observatory (NRAO), the first genuinely national facility of its kind. The NRAO was funded by Tuve's committee and operated by an umbrella group known as Associated Universities Inc. (AUI), a non-profit consortium of nine powerful universities on the eastern seaboard.[11] The AUI had already established the Brookhaven National Laboratory on Long Island near New York, which had developed into one of the world's leading centres for nuclear science with accelerators and reactors devoted to a multipurpose program in biology, chemistry, engineering, medicine and physics. Presumably, with this same high-powered backing and high-level funding, it would only be a matter of time before the NRAO became a leading force in radio astronomy.

Immediately after its formation, the new national organisation announced its intention to begin long-range plans for two telescopes 300 and 600 ft in diameter to be built at Green Bank, West Virginia. For the short term, NRAO stated it would begin with a dish of 150 feet and then use the design experience to scale up to the larger instruments. As he informed Bowen in March 1955, Tuve believed it would be jumping the gun to start with a dish as large as 250 feet because of the unknown engineering problems involved: 'I would rather have a 130 ft dish on a polar axis [than] a 200 ft dish on an alt–az mount. It seems to me that an irrational ambition for the 'biggest' is behind all this discussion of 250 ft dish designs, and very little of the foundation rests on anything related to astronomy. It's a kind of engineer's paradise.' [12]

For Bowen this was the straw that almost broke the camel's back. A few days later in a letter to CSIRO chairman, Ian Clunies Ross, Bowen gloomily forecast that it would only be a matter of time before the Americans built the biggest radio telescope in the world and that as far as Radiophysics goes 'we just can't compete'. Bowen proposed that plans for a large telescope should be dropped and instead Radiophysics should collaborate with AUI on the design of a smaller American dish, which could then be replicated at some future date in Australia.[13] On the same day he confided to Tuve:

> 'I have become very despondent in the last few months about the possibility of making progress here along the lines we originally considered. There are a number of reasons for this, all of which make sad reading. In the first place, the response outside this Laboratory to the possibility of constructing a large radio telescope has not been good. The attitude of people in industry and in financial circles is that this is the responsibility of the government: meanwhile the government is quite apathetic. Even in research circles there has been a disappointing tendency to say that sheep are more important and that radio astronomy is all right for other countries. Finally, purely from the point of view of constructing a large device I have been surprised to find how scared the rugged Australians are of going one better than anyone else in the world.' [14]

Bowen's despondency did not last for long. The following month the British engineer Barnes Wallis wrote to express his strong interest in the challenge of designing a giant dish. The possibility in April 1955 of recruiting an engineer of Wallis' stature proved a crucial turning point. Several days later the Australian government agreed to match all funds raised privately and from there the telescope project steadily gained momentum. Before continuing that story, we turn now to catch up on progress at Manchester.

Back to Jodrell Bank

In the previous chapter we left Jodrell Bank soon after the engineer Charles Husband had completed a detailed design of the 250 ft telescope in February 1951. Two months later news broke of how Ewen and Purcell at Harvard had succeeded in detecting line emission from interstellar hydrogen at 21 cm, a discovery quickly confirmed at both Leiden and Sydney. Until then, there had been what seemed solid reasons for believing there would be very little of interest in the radio sky at short wavelengths. As a result the Jodrell Bank telescope had been designed for use at wavelengths of 1 metre or longer. The surface tolerances of the dish were far too rough for observations at 21 cm—radio waves reflected from the surface would arrive at the focus in a scrambled, incoherent fashion. Similarly, with the coarse wire mesh planned for the surface, most of the 21 cm radiation would leak through the dish like a sieve. To enable observations at these shorter wavelengths Husband redesigned the dish with a higher surface accuracy, replacing the 100 ft central section with a mesh of finer weave. This change required strengthening the back girder and the two supporting towers to take the extra weight, and also the extra wind loading on the structure.

Excavations began on the foundations in September 1952. Over 150 piles were sunk to reach the underlying bedrock and 4000 tonnes of concrete poured to ensure the telescope would have stable footings. Next a circular railway track was bolted to the concrete foundations. Special rails strong enough to take the weight of the telescope had to be manufactured and laid so that the track would be level over the entire circumference to within 3 millimetres. This stage of the project took almost two years to complete, four times longer than the original projection. As Bernard Lovell recalled, this was 'a ratio of promise to achievement with which we were to become wholly familiar.' [15]

In August 1954 the erection of the steelwork began but, in the meantime, Husband continued to have doubts about the design of the dish. The decision in 1951 to use a finer mesh over the central 100 ft section was overturned early in 1954 by a new plan to have an even finer mesh over the whole surface. A problem then arose in finding an economical way of joining the panels together which would also provide sufficient rigidity to the dish. Eventually this led to a decision to do away with mesh altogether in favour of solid steel sheets which would give the necessary rigidity without any additional drawbacks—for example, the wind loading on the structure would remain the same as the calculations had already assumed that occasionally during the northern winter the mesh would completely ice over. Although this was the fourth type of surface proposed, a further unforeseen problem emerged. Wind-tunnel tests on a scale model of the telescope revealed that the solid sheet dish underwent an effect known as structural oscillation—a phenomenon where a structure can be destroyed by swaying to and fro in winds of relatively low speed, even when the structure has been designed to withstand the static force of strong winds. In 1940, structural oscillation had caused the spectacular destruction of the Tacoma Narrows Bridge in the United States. To guard against the same phenomenon, Husband added a large girder to the back of the dish, fitted with hydraulic brakes to damp the effects of these oscillations.

Some of these numerous design changes were made in response to increasing military interest in the instrument, others came about because of the very nature of the project. No structure of this type had ever been built and experience had to be gained as the project progressed and new problems arose. Calculations underpinning the design had to be ground out the hard way by slide rule and elementary calculating machines; often a change in one feature chain-reacted and meant a redesign of other parts of the telescope structure. The cumulative effect of all these changes was a serious delay in the construction timetable and, of increasing concern to Lovell, a major blowout in the cost of the project. The original estimate in 1951 had been £UK259,000 and only one year later this had risen to £373,000, a sum the Department of Scientific and Industrial Research (DSIR) and the Nuffield Foundation had agreed to share equally. As time went by the costs steadily escalated and by late

1955 had reached a projected £630,000, leaving a deficit of £257,000 with no guarantee that additional funding would be provided. Bowen summarised the situation after visiting Jodrell Bank in June 1955:

> 'The most intriguing news is that they have scrapped the original dish and have redesigned it completely. This accounts for the delays, the increased cost and the covering up which has been going on. As far as I can gather Lovell has not told his sponsors about this, which I should imagine is a serious tactical mistake on his part.' [16]

Early in 1956 a special DSIR committee formed to inquire into the project criticised the number of design changes and the way they had been implemented. The committee also identified defects in the management of the project where the lines of responsibility had often become confused. Again, Bowen:

> 'An extraordinary contractual situation exists with the Manchester project which we should be careful to avoid. The three parties principally concerned are Lovell, the [University] Bursar and Husband, all very excellent people. Their roles in simple terms are that Lovell states the requirements, Husband gets the design work done either by his own staff or by outside contract and the Bursar then calls for tenders for each part of the structure as the design is completed. There are a large number of contractors and no one of them has overall responsibility. In effect the Bursar not only holds the money but is also acting as the main contractor. I do not think this is what they intended at the start, but this is the way it has worked out.' [17]

While it is debatable whether the construction would have moved more smoothly had it been managed differently, the absence of a main contractor with overall responsibility for the project added up to a daily workload for Lovell which at times became a crushing burden. As he confided to Bowen:

> 'If I could give one piece of general advice may I strongly urge you to find somebody who can handle all the details of this telescope with no other distractions and preoccupations. I have found the task of carrying on with the normal life of Jodrell Bank and the University together with the worries of the radio telescope quite back-breaking.' [18]

In response to a key recommendation in the DSIR report, a site committee was formed in April 1956 to provide overall coordination of the project. By September most of the towering steel framework had been completed and before the end of the year work began on fitting and welding over 7000 steel sheets to form the surface of the dish. In June 1957 the telescope moved under its own power for the first time. Shortly afterwards, Lovell could joyfully record in his diary: 'we shan't forget August Bank Holiday 1957 ... we got our first recording. 160 MHz with the bowl in the zenith. A remarkable run completely free of any sign of interference.' [19]

The 250 ft Jodrell Bank telescope in mid 1957, near to completion of the surface panelling.

As we see later in Chapter 10, Jodrell Bank soon shot into the international limelight, not for its radio astronomy but for its role in tracking the Russian rockets used to launch the Sputnik satellites. Lovell's elation at the successful completion of the construction was, however, tempered by financial and legal problems which continued to besiege the project. In April 1956 the DSIR had agreed to fund half the deficit but several attempts to persuade the government to provide the remaining £130,000 proved unsuccessful. To make matters worse, early in 1958 the telescope's designer Charles Husband announced that his reputation as a consulting engineer had been unfairly damaged by certain allegations and that he planned to serve a writ on Lovell for a quarter of a million pounds. Lovell's biographer wrote:

> 'Lovell hit the depths of despondency. The radio telescope was an instrument that had become the symbol of Britain's continuing greatness in scientific and technological achievement, and her contribution to space research. It was the only equipment in the Western world capable of detecting and tracking ICBMs, and of providing the basic know-how for the speedy development of military equipment for such national security purposes. It was the sole system available and capable of providing the USA the cooperation essential to its newly developing space program. And for enabling the British nation to possess this superb and scientifically important instrument, the only major scientific research instrument produced in Britain since the war, the British government was ready to leave Lovell to face a court case, brought by Husband, which might well result in imprisonment!' [20]

Fortunately the dispute with Husband was settled privately and Lovell managed to stay out of prison. The question of who would pay for the telescope took longer to settle. Late in 1958 the University of Manchester launched a Radio Telescope Fund and began wooing large companies and wealthy individuals with letters of appeal and a series of lavish charity functions. By the end of 1959 the fund had raised £65,000, over three times the amount raised by the Radio Astronomy Trust for the Australian telescope, but still only half the sum required to cover the deficit. It was not until well into 1960, three years after the telescope had begun operations, that the Nuffield Foundation bailed out the project by agreeing to pay all outstanding debts.

The events at Jodrell Bank were followed with considerable interest at Radiophysics. Early in the project it had seemed that the best option for building a telescope in Australia would be to take the Jodrell Bank design, perhaps not lock, stock and barrel, but at least as the basis for a new and more refined version. Lovell himself had been anxious to see this happen and had generously offered to provide the blueprints: 'It would be a great pity to duplicate the years of design work and the investigation of alternative types which have already gone into our

Bernard Lovell shortly after the completion of the Jodrell giant. Later he wrote that probably no other scientific project ever survived so many crises to reach ultimate success.

program.'[21] On more than one occasion Husband had also offered his services as a design consultant but, as the Manchester project slid further into trouble, the option of copying the Jodrell Bank telescope became less and less attractive. By 1955 the Australians had decided to go their own way.

Venturing into the Unknown

Although plans to build a large dish in Australia had firmly taken shape in 1952, by early 1955 there was little to show for the effort. For one, the funding issue had still to be resolved. Further, the various feasibility studies on large dishes carried out by engineers at home and abroad, favourable though they seemed, were still only ideas on paper. The turning point in the project came with the recruitment of Barnes Wallis, widely regarded as the foremost British inventor–engineer of his generation (a brief biography is given in Tangent 5).

TANGENT 5

From Dirigibles to Dishes

The illustrious career of Barnes Wallis began in 1913 in the airship division of the British firm of Vickers. In 1917 he completed the R80 for the Admiralty and during the 1920s designed the most successful dirigible ever to fly, the R100. By the 1930s airships had fallen from favour with both military and civil authorities. Wallis turned to aircraft design and applied some of the same principles that he had developed for airships—geodetics and new alloys which allow the construction of light structures of exceptional strength. The results were the Wellesley bomber which broke several long-distance records during its test flights (including the non-stop England to Australia record in 1938), followed by the Wellington bomber which went on to become the main workhorse for the RAF during the war. The Wellington endeared itself to many airmen for its ability to limp home after parts of its structure had been shot away and fabric burnt from large areas of its frame.

Early in the war, Wallis devised a way for carrying out air attacks on heavily fortified dams, an idea for which he is most widely remembered. His proposal was to drop a spherical bomb from a low-flying aircraft a fair distance upstream from the dam. The bomb would avoid defences such as steel torpedo nets in the water by ricocheting across the surface, impact on the dam wall, and then sink in close contact with the wall, detonating by hydrostatic pressure. Wallis first demonstrated the feasibility of the idea, assisted by his children, using marbles sent skipping across a tank of water in the garden of his home. The culmination came in May 1943 when several large dams important to the German war effort were damaged or destroyed in the series of famous Dambuster raids. Tizard who had championed this somewhat eccentric idea against considerable opposition called the bouncing bomb 'the finest individual technical achievement of the war.'[22]

Peacetime. After 1945 Wallis became involved in a variety of new projects such as submarines, bridges and high-rise buildings, mixed with his abiding interest in aircraft design (variable geometry or swing-wing aircraft was another of his ideas). Even telescopes began to interest him. In 1952 the Astronomer Royal, Harold Spencer Jones, invited him to join the design team for the Isaac Newton Telescope, Britain's bid to challenge American dominance in the field of large optical telescopes. Given responsibility for designing the drive mechanism, Wallis branched out and proposed several new features to make the telescope structure absolutely rigid. However, while the design work was still in progress Spencer Jones retired and, with the elevation of Richard Woolley from Mt Stromlo to the position of Astronomer Royal in 1955, a new design committee was appointed to start the project afresh. Wallis was excluded from the new committee and his ideas ignored. The 2.5 m (98 inch) Isaac Newton Telescope was eventually completed in 1967 at the Royal Greenwich Observatory at Herstmonceaux. Later, in 1981, it was shifted to a far better observing site on La Palma in the Canary Islands.

The approach to Wallis late in 1954 on the subject of radio telescopes could not have been better timed. Wallis was annoyed at the way he had been treated by the telescope committee, another occasion he believed where his farsighted ideas had gone unappreciated. If he was to be shunned by Britain's optical astronomers, then he would respond by demonstrating his talents to the Australian radio astronomers. Apart from the obvious differences between optical and radio telescopes Wallis knew that many of the design problems are common to both. The questions of how to build rigidity into a large structure subject to varying gravitational distortions and of how to track astronomical objects with pinpoint accuracy were problems to which Wallis had already given considerable thought.

The Monument Remains. Although some of the basic design features of the Parkes telescope stemmed from his fertile mind, oddly, Wallis received no financial reward for the work. At the commencement of the design study Wallis had arranged for his consultant's fee of £1000 to be donated on his behalf to an organisation known as the RAF Foundation at Christ's Hospital School in London, the school where Wallis himself had been educated. Wallis had established the foundation immediately after the war using a government inventor's award of £10,000 received in recognition of his war work, especially the idea of the bouncing bomb. Wallis had been deeply affected by the loss of RAF personnel in the Dambuster raids, many of whom he had known personally. The foundation was set up to help educate the children of RAF crew killed in action.

Halfway into the telescope design study Wallis left, disgruntled by the course the project had taken. Despite several attempts to settle the issue, made as late as 1962, Wallis stubbornly refused to accept his £1000 consultant's fee. Although his involvement in the project ended on a sour note, the Parkes telescope remains as an enduring monument to Britain's outstanding inventor of the twentieth century.

British inventor–engineer Barnes Wallis. One biographer wrote of Wallis that he served his country better than it served him.

The first contact with Wallis had been made during a visit by Joe Pawsey to London in September 1954. Pawsey met with Henry Tizard and described the uncertain future surrounding the telescope project. Two years earlier, Tizard had helped to start the project by suggesting the Carnegie Corporation as a possible source of funds. Once again he provided sound advice by encouraging Pawsey to meet with Wallis, head of the research department in the aircraft division of Vickers Armstrong.

Wallis immediately warmed to the idea. In April 1955, after receiving a copy of the Radiophysics publicity booklet, he confidently informed Bowen:

> 'I think that I shall find it possible to meet, at least, the 'compromise' figures of the specification given ... and possibly to better them slightly, with a dish of larger diameter than the 250 ft which you have in mind. Before spending too much time on this, it would help me if you could give me some indication of how *much* larger you would like the dish to be *if you could have it?* I expect your first reaction will be that the cost is greatly influenced by the structural design and that, for a given cost, the size will depend on the efficiency of the design. I am not proposing to alter the form of your dish which is of course essential for your purpose, but I believe from my experience in building large rigid airships that we can adopt methods of construction which will be cheaper than any of those illustrated in your booklet.' [23]

After agreeing to act as a consultant, Wallis began work on a conceptual design of the telescope, concentrating initially on the structure of the dish. Years earlier Wallis had patented a device known as an incompressible column, a rod of steel or aluminium incorporating a servo-controlled screw jack which automatically resists any change in length when the rod is stretched or compressed under stress. Wallis proposed using these incompressible columns in the structure of the dish so that the reflecting surface would retain a true parabolic shape, whatever the orientation of the dish or the operating conditions. An earlier proposal by Wallis to use them in the structure of the Isaac Newton Telescope had been flatly rejected by the design committee. Another radical idea investigated by Wallis was to set the dish spinning like a top around its major axis. The spinning motion would stabilise the structure against wind gusts and also offset the effects of solar heating which can lead to differential rates of expansion across the dish's surface. In effect, the spinning dish would act as a huge gyroscope and, as required, remain pointing in a fixed direction in space. In September 1955 Wallis summarised his ideas in a report entitled 'Giant Radio Telescopes'. The foreword proclaimed: 'The design of a large radio telescope involves many of the same problems as did in the past the design of large rigid airships, for the construction of a giant radio telescope is a venture into the unknown which certainly involves the creation of a large and costly structure.' [24]

Not all of Wallis' ideas stood up to closer examination but a number of basic features from his preliminary design are the same or closely related to the final design worked out for the Parkes telescope. Several features are worth commenting on. First, the dish is supported at its centre, which immediately avoided a major problem at Jodrell Bank where the two towers supporting the bowl at its perimeter gave an uneven distribution of stress. Further, the dish incorporates the same type of geodetic structure which Wallis had developed for his famous airship and then applied with brilliant success in the Wellington bomber. Interestingly, in his report Wallis deliberately avoided specifying a diameter for the dish. He could see no intrinsic structural limitations in building an exceptionally large dish and on more than one occasion he urged Bowen to shoot for a diameter of 1000 feet, regardless of the cost. By far the most successful feature of Wallis' design was his solution of how to drive and point the dish with great precision. According to Bowen, Wallis conceived the idea of the master equatorial control system one day at the exclusive Athenaeum Club in London:

> 'About half way through lunch, Barnes Wallis asked 'How do you propose to drive this machine?' I gave him the conventional answer, which was that we would do it, analogue fashion, from a small equatorial mount built alongside the main telescope. It took him about thirty seconds—certainly not more than a minute. He pointed a finger at me and said 'You're wrong. The place to put the equatorial unit is at the intersection of the two axes of rotation. You derive an error signal and servo this back to the main telescope. It's perfectly obvious that's the way to do it'.' [25]

Wallis later patented the idea and the master equatorial unit became a feature not only of the Parkes telescope, but several other large dishes built during the 1960s (see Chapter 11).

Wallis' unbridled optimism breathed new life into the telescope project. In June 1955 an advisory committee was formed at Radiophysics which included several of Australia's most distinguished engineers and which met to decide how and, of equal importance, where to carry out the design study.[26] At this stage Wallis had formulated several ingenious concepts but this was only one step on the road towards turning these ideas into a thorough design study—the detailed analysis required to demonstrate the technical and economic feasibility of a proposed design. The lesson of Jodrell Bank had already underlined the dangers of embarking on the construction prematurely.

Up to this point Wallis had acted in a private capacity but there was a possibility that the design study might be done through his employer Vickers Armstrong at Weybridge in Surrey. The most appealing option was to assemble a design team in Australia. Radiophysics staff could have been assigned to the project (reducing costs) and the Aeronautical Research Labs in Melbourne had also expressed its willingness to continue with its earlier studies on large telescopes. After considerable

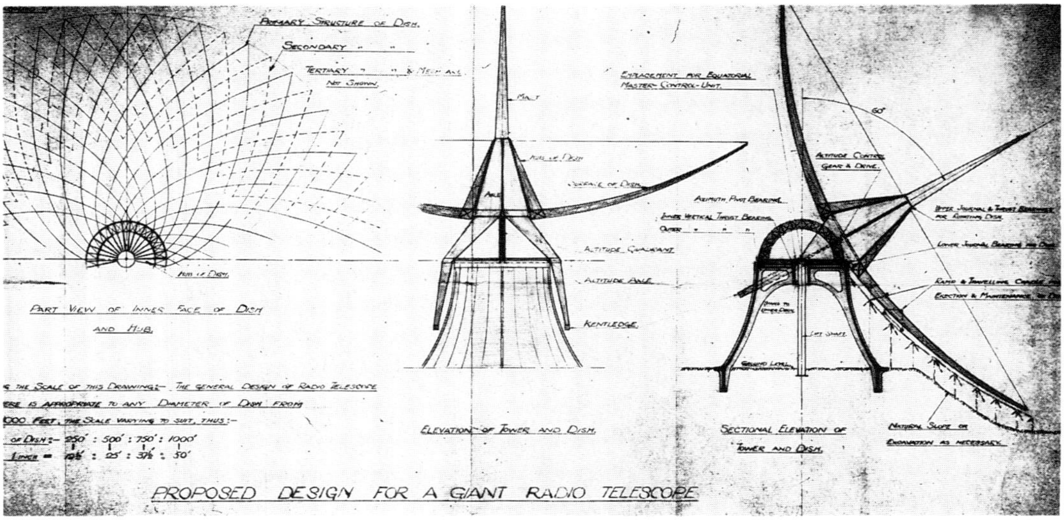

In his first drawing from October 1955, Barnes Wallis deliberately omitted a scale for the telescope. Wallis believed the design would be suitable for any dish up to a diameter of 1000 feet.

discussion the idea was shelved, partly because of doubts on whether a local group with sufficient expertise could be assembled and also uncertainty as to whether Wallis could be lured to Sydney for an extended period. Another possibility was the United States where there seemed to be no shortage of suitable engineering firms. However, with the American economy booming in the mid 1950s the main drawback seemed to be the likely cost, not just for the design study but later in the project when firms would be approached to tender for the construction. For these reasons the decision was made to limit the search to Britain. On the advice of the advisory committee at Radiophysics, three engineering firms were approached and each offered to undertake a design study for a suitable fee. Easily the best offer made by the three candidates came from the London firm of Freeman Fox and Partners.[27]

London Bound

Freeman Fox had already made a name for itself in Australia. The firm's founder, Sir Ralph Freeman, had designed the best known landmark in the country, the Sydney Harbour Bridge. Since the 1930s the firm had expanded and ran three separate offices in London dealing with bridges, motorways, power stations and a range of other projects. In November 1955 Freeman Fox agreed to undertake the design study which, if acceptable to Radiophysics, would then provide the basis for the full-scale engineering design. Freeman Fox divided the task into five main areas, which it stated as follows:

(1) to examine the structural forms of the dish, supporting structure and foundations, including both rigid and self-compensating frameworks, and the use of steel and/or light alloys;
(2) to examine the altitude–azimuth and/or equatorial systems for steering the telescope, together with appropriate drive and control mechanisms;
(3) to give approximate estimates of the overall cost of the project according to the diameter of the dish;
(4) to carry out wind tunnel tests and other experimental studies as required; and
(5) to study the stress, deflections and temperature changes on the telescope structure.

Freeman Fox estimated that with four or five engineers full-time on the project the design study would be completed in six months; failing that, the firm guaranteed at a minimum to supply a detailed interim report by June 1956. The cost of the study was reckoned at £10,000 (or £A14,000) with £5000 as Freeman Fox's fee, £1000 for Barnes Wallis who would act as a design consultant, and the remaining £4000 for outside firms engaged to conduct the experimental studies.

In the meantime a document had been drafted at Radiophysics setting out the main telescope specifications and the role Freeman Fox was expected to play.[28] As one example, Radiophysics would take responsibility for the development and construction of the receiving equipment, but Freeman Fox was required to provide a cabin at the telescope's focus which could house a certain minimum weight of equipment and be readily accessible to telescope staff. Aside from spelling out the division of labour between Sydney and London, the bulk of the report listed the main technical specifications of the telescope. The most important, as expected, was for a dish as large as possible and limited only by the funds available. Next, it was considered essential that the telescope could operate down to a wavelength of 21 cm, preferably even lower. In turn this meant no point on the dish surface could deviate, on average, from a true parabolic shape by more than half an inch (or 1.3 cm). Further, when tracking radio sources across the sky the pointing error of the telescope could be no larger than one minute of arc (i.e. one-sixtieth of a degree). Given that the pointing error would be the cumulative effect of a host of factors such as structural deflections, errors in the drive and control mechanisms, and buffeting winds, this specification alone presented a formidable engineering challenge.

The design specification also proposed a number of compromises which took into account the effect meteorological factors would have on the telescope's operation. The telescope would not be operated in winds exceeding 50 km per hour and Radiophysics would take on the task of finding a site with a low average windspeed. As well, Radiophysics guaranteed that the site would be free of snow all-year round which meant that the wire mesh used for the reflecting surface would be in no danger of icing over. The porous wire mesh would in turn reduce the maximum projected windload on the dish (recall that less favourable climatic conditions at Jodrell Bank eventually led to a solid steel surface). As one further limitation, the dish would only be required to tilt to a maximum angle of 60° from the vertical or zenith, a decision which was made for two reasons. First, observations low on the horizon (90° tilt) are of poor value because of the uncertain effects of atmospheric refraction. Second, even with a 60° limitation the telescope would still be able (at Sydney's latitude) to point to the south celestial pole and as far north as +27° declination. Telescopes in the northern hemisphere were better placed to cover the remainder of the sky north of +27°. This limitation greatly eased the design problem by reducing the maximum projected gravitational and wind stresses, and also gave greater flexibility in designing the layout of the turret structure supporting the dish.

The Radiophysics design specification was completed in November 1955 and then taken to London by Jack Roderick, professor of civil engineering at Sydney University. Roderick spent two weeks at Freeman Fox finalising arrangements for the design study and then toured several centres in Britain and the United States where work was under way on other large radio telescopes. To offset Freeman Fox's lack of experience in radio engineering a decision was also made to post Harry Minnett in London for the duration of the design study, not only to act as the resident radio consultant but also to assist in the flow of information. One of the most experienced aerial designers at Radiophysics, Minnett had made his reputation as a radio astronomer through his pioneering studies of the moon in the late 1940s (see Chapter 3).

To develop a design to meet the set of specifications laid down in Sydney, Freeman Fox first had to tackle two important questions: What should be the structural form of the dish? What type of mount should be used to support and steer the dish? Both questions required an answer before decisions could be made on a myriad of other design features—not least the size of the dish itself.

Structure of the dish. Basically there are two ways to counteract the variety of effects which lead to a telescope dish distorting from an ideal parabolic shape. One is to incorporate a mechanism where the dish automatically adjusts its shape to compensate for the distortion. This is the approach Barnes Wallis strongly advocated with his incompressible columns. The other approach is to design a dish so rigid that any distortion occurring during normal operation remains small and within the specified tolerance. Here the main drawback is that the larger the dish, the amount of structural material required increases out of proportion, the dish becomes heavier and heavier, and finally the approach becomes self-defeating.[29]

Freeman Fox opened its account by investigating the structural characteristics of a 150 ft rigid dish and a 300 ft compensated dish. The calculations indicated that a compensated dish would indeed be feasible but it required the invention of some new device to automatically monitor distortions across the entire surface of the dish and then feed the information back to allow adjustment of the incompressible columns. In all likelihood the technique would be expensive to develop and comprise a substantial proportion of the construction costs. As a result, Freeman Fox carried out an investigation of a 150 ft rigid dish with Wallis' geodetic structure and then extended the calculations to determine the maximum size possible without using any form of compensation. Surprisingly, the diameter turned out to be in excess of 300 ft, and from there it was a straightforward decision to choose a rigid, uncompensated structure. Steel was selected for the skeleton of the dish because of its superior rigidity compared with aluminium or other light alloys.

Driving the dish. The question of how to mount and drive the dish proved the most difficult of all. Again there were two possibilities, but with neither mount appearing to have a clearcut advantage over the other. One type known as the equatorial or polar-axis mount has the dish supported along an axis chosen to be exactly parallel to the earth's axis of rotation. Such an axis is usually quite tilted; in fact, over 50° from the vertical at Sydney's latitude. The great appeal of this form of mount is that tracking a celestial object is simply achieved by making the dish rotate about this telescope axis at a rate exactly equal but opposite in direction to the earth's own rotation, that is once every sidereal day (23 hours 56 minutes). Polar mounts were standard in all optical telescopes and, if that was not recommendation enough, in most of the smaller radio dishes built at this time. The other form of mount investigated was the alt–az already chosen for the Jodrell Bank dish. With this mount a radio source is tracked by simultaneously driving the dish about two axes—the horizontal axis to bring the telescope to the correct angle or *altitude* above the horizon, and the vertical axis for the correct point on the compass or *azimuth*.

Although structurally much simpler, the main disadvantage of the alt–az mount is that the dish has to be driven at continuously variable rates of rotation about both axes. Consequently, the drive and control system required is much more complex than for a polar mount. Even if this problem could be overcome, it seemed that the alt–az mount would take funds away from the design feature that mattered most, a dish of maximum size. On the other hand, studies of the awkward polar mount revealed its own set of perplexing problems. The larger the dish the more structural and mechanical difficulties come into play and the harder it becomes to maintain a good pointing accuracy over most of the sky. Whether or not these difficulties with the polar mount were more serious than the problems inherent in the alt–az mount with its complex drive features developed into an issue of some controversy. Supporters of the alt–az mount included Barnes Wallis and the engineer in charge of the design study at Freeman Fox, Gilbert Roberts, while aligned on the other side were several consultants Bowen had approached personally. One was Bruce Rule who had been responsible for much of the design work on the 200 inch Palomar telescope and then the Owens Valley interferometer. Merle Tuve and the designers of the 140 ft dish at Green Bank also argued forcibly for the polar mount.

Only unusually complex calculations could decide the issue. Eventually, the conclusion was reached that for dishes exceeding approximately 150 feet, the alt–az mount would meet the stringent specifications for the pointing accuracy better and more cheaply than the polar. Two factors proved crucial. One was the advantage offered by Wallis' master equatorial unit, the other was a novel solution worked out by Harry Minnett for the servo-drive problem. Minnett's work had shown how a large dish could be controlled with an accuracy ten times greater than the Jodrell Bank telescope. The decision in favour of the alt–az was a landmark in the development of large telescopes. No longer was the polar mount considered a serious option. Even the designers of large optical telescopes eventually adopted the alt–az mount.

☆ ☆ ☆

Freeman Fox's original projection that the design study would take about six months proved far too optimistic. As promised, an interim report was delivered by September 1956 recommending a rigid, uncompensated structure for the dish but the central issue of the mount, not to mention a host of other design features, took much longer to resolve. It was not until November 1957 that copies of the long awaited 'Design Study on Proposed Radio Telescope' arrived at Radiophysics, delivered in person by Harry Minnett. As expected the report discussed at length the relative merits of both types of mount before coming down solidly in favour of the alt–az type, utilising the master equatorial unit to control the pointing of the dish. Freeman Fox had been able to meet and in some cases better the design specifications set out in the Radiophysics document of November 1955. As one example, with the dish tilted at 20° from the zenith and operating in windspeeds of 15 km per hour, the surface deviations would be no more than 3 millimetres near the centre of the dish, increasing to 9 mm at the perimeter—well within the original specification of 13 mm. The pointing accuracy would also be comfortably within the required one minute of arc.[30]

The only major issue left unresolved by the report was the size of the dish. Freeman Fox felt confident that a dish meeting all the design specifications could be built as large as 250 ft, but on this score the matter became a financial, not an engineering issue. Using a formula which estimated that the cost would rise as the 5/2 power of the dish diameter, Freeman Fox calculated that the price for diameters of 230 and 210 feet would be £A644,000 and £500,000 respectively. Gilbert Roberts visited Sydney in January 1958 and decided the issue at a series of meetings at Radiophysics. At this stage, with funds in hand slightly in excess of £500,000, common sense demanded that the size of the dish be limited to 210 feet.

The engineer Gilbert Roberts was known as Britain's master bridge builder. While Barnes Wallis contributed several brilliant ideas, Gilbert Roberts was the one responsible for developing the overall conceptual design of the Parkes telescope.

The response to the Freeman Fox report at Radiophysics and by several local engineering consultants was exceptionally favourable. Numerous minor features in the design had yet to be worked out but Freeman Fox had clearly come up with a proposal which, in terms of both design and cost, looked superior to either the Jodrell Bank dish or the 140 ft dish proposed by the Americans at Green Bank. In February 1958 Freeman Fox was given the green light to begin the full-scale design of the telescope and to engage contractors for its construction. To assist in the next stage of the project, a group of six senior radio astronomers at Radiophysics met to discuss matters such as the most appropriate layout of the telescope's tower, the instrumentation and control panels, the best way to gain access to the receiver cabin—all issues aimed at ensuring that the telescope would be as 'user friendly' as possible. The group's chairman Frank Kerr relayed this information to Harry Minnett in London who, in turn, continued to send a steady flow of reports on progress at Freeman Fox.[31]

Maintaining the Momentum

At the beginning of 1958 it seemed all would be smooth sailing. With the most difficult problems overcome the timetable ought to have been busy but brisk: flesh out the design study by drawing up a set of blueprints and technical specifications and then place contracts for the construction in the latter half of the year. As it turned out, the project began to lose momentum. One reason was that unforeseen contractual problems arose in negotiations with the firms chosen to construct the telescope. Other problems resulted from design changes. In February 1958 a decision had been made to reduce the minimum operating wavelength from 21 cm to 10 cm, but this meant extra time would be needed for another round of wind tunnel tests and the recalculation of a range of telescope parameters. Quite unexpectedly, Gilbert Roberts was also forced by rising costs to make major simplifications to the mechanical drive system. These problems were compounded when Freeman Fox was unable to strengthen the design team with enough engineers to cope with the extra work. As mentioned in Tangent 5, Wallis had already abandoned the project, peeved that some of his ideas had not been implemented exactly as he wanted. Since then he had become immersed in a new defence project at Vickers Armstrong on the design of supersonic bombers. Harry Minnett provided most of the drive and control expertise, but he was in the difficult position of having a foot in each camp—responsible both to Radiophysics and to Freeman Fox. His heavy workload was at odds with his other role of pressuring Freeman Fox to see the project through to completion. Primary responsibility for progress rested with Gilbert Roberts, but the telescope was only one of several major projects under his direction, one that could not receive his undivided attention. His reluctance to delegate responsibility or to communicate effectively with the design team did little to maintain the momentum. Twelve months into the design study Minnett had reported:

> ' ... how difficult it is to get a statement of intentions from Gilbert Roberts or to alter his course of action. He gets quite upset at any suggestion that Freeman Fox are not doing the right thing, and some sharp words have been exchanged in the course of our talks. Ralph Freeman [son of the founder] is much more informative and amenable to argument, but unfortunately he takes only a very general interest in the work.' [32]

These management problems were accentuated by difficulties in finding firms genuinely interested in taking on the construction of the telescope. Earlier, in 1956, Freeman Fox had approached a number of British firms but most had shied away from the prospect of journeying into unfamiliar territory with a project involving so many untried engineering features and, not least, one which promised little financial reward. As Gilbert Roberts explained, those firms with the resources and know-how required to tackle the project were far more interested in chasing lucrative defence contracts: 'Clearly the development of a special piece of apparatus, of which there are not likely to be any repeats, and, in fact, not even the certainty of a single order, is a most unattractive proposition for any firm here with adequate staff. They are practically all fully engaged on work associated with our defence requirements.'[33]

In view of this disappointing response in 1956, Freeman Fox felt there would be little to be gained by throwing the project open to a competitive tender and, instead, decided to divide the project into three contracts and make its own selection of the contractors it believed best suited to the task. Early in 1958 three firms were chosen: MetroVick (Manchester) to construct the mechanical and electrical drive systems, as well as to act as the main contractor; Grubb Parsons (Newcastle) to develop the master equatorial system; and Sir William Arrols and Partners (Glasgow) to fabricate the heavy structural components and to construct the telescope at a site halfway around the globe. MetroVick and Grubb Parsons had been closely involved in the design study during 1957, so both firms seemed a logical choice.

As Britain's largest engineering firm and one of its major defence contractors, MetroVick unquestionably had the expertise to take on the project. Its collaboration with Freeman Fox however took a turn for the worse when both sides reached a stalemate over the terms of the contract, and then deteriorated as MetroVick stalled in providing an estimate of the cost and likely duration of the construction. Not even the personal intervention of CSIRO minister Richard Casey had much effect. During a visit to London in September 1958, Casey received an assurance from its chairman that MetroVick was anxious to participate in the project, but then a further six months went by before the firm was prepared to commit itself to a quotation.

Other problems arose with Grubb Parsons, a firm renowned for its construction of optical telescopes. The original firm of Grubb had built the Great Melbourne Telescope at its Dublin works in 1868 and, in more recent times, the 74 inch telescope brought into operation at Mt Stromlo in 1955. The firm's task was to translate the master equatorial concept by Barnes Wallis into a fully engineered system. However, even with assurances that the project would receive top priority, Grubb Parsons maintained that the control system would take at least two years to construct.

Months of slow progress and discouraging news from London added up to mounting frustration in Sydney. Writing to Fred White in November 1958, Bowen expressed the blackening mood at Radiophysics:

> 'It is now 2 years and 11 months since Freeman Fox first accepted responsibility for this project and as of today not a single piece of hardware exists, technical specifications on which a manufacturer may quote have not been written, no firm quotations of costs have been obtained, and there is no firm quotation of delivery dates. We regard the situation as quite intolerable ...'.[34]

The United States planned two major instruments in the mid 1950s in an attempt to narrow the lead held by Australian and British radio astronomers: *above* The Caltech interferometer built at Owens Valley and completed in 1958. *below* The polar-mounted 140 ft dish built by the NRAO in West Virginia and eventually brought into operation in 1965.

Encouraged by Casey and White, Bowen flew immediately to London. After assessing the situation he hammered out an agreement that MetroVick would submit its quotation by February 1959. There were however ominous signs on the likely cost and completion date which strengthened Bowen's decision to investigate alternative arrangements. Early in 1959 he and Harry Minnett toured the United States and between them held discussions with a number of firms, among them some of the giants of the American corporate world such as Bethlehem Steel, General Electric and Westinghouse. The interest and enthusiasm shown by most of these firms, together with their rough estimates of the cost and completion date, all pointed in the same direction. The earlier recommendation by Freeman Fox to rely on handpicked British firms, with their questionable commitment to the project, was no longer the only option—the telescope instead could be put out to international tender.

Back in London by March, Bowen's worst fears had been confirmed. MetroVick estimated that the telescope would cost £A715,000 (£200,000 more than the funds available) and that the construction time would be almost four years. Both estimates were plainly unacceptable. The time had arrived to move to international tendering. When Freeman Fox seemed reluctant to support the idea, Bowen could not contain his frustration:

> 'I can quite understand why this project has taken such an appalling time. The procedure is never under any circumstances to cross a bridge when you come to it, but to camp on the near side, indulge in a round of lunches, telephone calls and trips to the country, by which time the bridge has probably washed away anyhow.' [35]

In the end Bowen prevailed. After a period of intense activity at Freeman Fox, the documentation required for calling international tenders was completed in April 1959.

The problems encountered with two of the prospective British contractors should not have been completely unexpected. Freeman Fox had approached both firms, not vice versa, and so neither had much incentive to quote rock-bottom prices or to promise rapid delivery dates. Freeman Fox had been over the proverbial barrel. Admittedly, the various British firms approached originally in 1956 had shown little or no interest in the project. However, the situation had been dramatically transformed when, in October 1957, the Soviet Union stunned the world with the launch of Sputnik. The only instrument in the Western world capable of tracking the Russian launch rocket was the newly completed telescope at Jodrell Bank. From that time on large steerable dishes—until then expensive hardware used only by a handful of astronomers—were to become an integral part of military and civilian communications technology around the world. As we see in the next chapter, when the telescope project went to international tender in April 1959 a large number of American and European firms expressed interest in this new space-age technology and there was no shortage of competitive bids. If advantage had been taken earlier of this dramatic change, the construction of the Parkes telescope might have been completed well before October 1961.

CHAPTER SEVEN

Erecting the Acre of Ear

> As far as I am concerned the construction of radio telescopes seems to be like a repetitive series of gramophone records: delays and financial problems calculated to drive the astronomer to despair and then, in the end, success beyond expectation.
>
> Bernard Lovell, 1973

Putting aside the frustration and anger which crept in during the final stages of the design work, the three years at Freeman Fox had undoubtedly been time well spent. The final design had been thought through in such careful detail that once construction got underway in late 1959 only two years were required to erect the telescope. As well as a rapid and virtually trouble-free construction stage, another bonus of the Freeman Fox design was that the project kept to a realistic budget. Although the eventual cost rose to 50% more than the 1957 predictions (as we see later), the increases were certainly not unreasonable given the unusual nature of the project and given the difficulties encountered with other large telescopes. By way of comparison the Jodrell Bank telescope, which underwent major design changes during construction, eventually cost over twice its original estimate. Similarly, the 140 ft dish at Green Bank faced long delays and a rapid escalation of costs, caused largely by severe design problems associated with its polar mount. The episode vindicated the hard-fought decision at Freeman Fox to choose the alt–az mount for Parkes. The final cost of the Green Bank telescope rose to six times the original estimate or, put in another light, over six times the cost of the Parkes dish.

In this chapter we look at the construction of the Parkes telescope, not just the on-site erection of the structure, but the fabrication of many of the components beforehand in European factories, the development of new types of receivers for the telescope and the host of tasks involved in commissioning the instrument in readiness for its first astronomical observations. To begin we take up another aspect of the story which occurred in parallel with the design study in London—the search for a site.

Cliffvale, Canberra or Over the Mountains?

The selection of the area near Parkes in central New South Wales for the new telescope took the best part of four years to settle. Preliminary discussions on a suitable site began in mid 1954, prompted by press reports on the award of the Carnegie grant which saw a large bundle of mail arrive at Radiophysics from real estate agents, shire councils, chambers of commerce, even a politician lobbying to have the telescope built in his electorate. By early 1956 a list of over thirty possible locations had been compiled, with an area close to the main Radiophysics field station at Fleurs heading a shortlist of the ten most promising sites. Several technical requirements were taken into consideration in short-listing a site. The ideal location would need to be geologically stable to provide a solid foundation capable of supporting a structure weighing close to 2000 tonnes. The site would also need to have a mild climate free from ice and snow, with a low average windspeed all year round. Above all, the site had to offer a very low level of radio interference. This meant being well away or at least well shielded from industrial sources of noise such as factories and power plants, a requirement crucial in the same way that an optical telescope must be sited well away from the night-time light 'pollution' of towns and cities. To add to his troubles at Jodrell Bank, Lovell had been drawn into a protracted battle trying to persuade a local electricity authority to reroute a new high-voltage line away from the new telescope. Though the route originally planned came no closer than 6 km, the radio noise generated from these lines would have made Britain's premier scientific instrument virtually useless.[1]

While even this rural pocket of Cheshire proved an uncomfortably noisy part of England, from the start it was clear that a country as large and as sparsely populated as Australia would have no shortage of sites free of radio noise, and which would also fulfil the other geological and meteorological requirements. Consequently, other logistical and political factors came into play. The issue settled down into a three-way contest—a site close to Sydney, one near Canberra, or a third 'over the mountains' well to the west of Sydney.

On grounds of sheer convenience the site ideally would be not too distant from the centre of Sydney. From the beginning the Radiophysics field stations had been no more than a couple of hours' drive away and the radio astronomers had grown accustomed to a pattern of frequent, often daily visits to carry out their observing programs. The longer periods of data analysis and writing up research papers, or the development and construction of equipment, could then be carried out in comfort in the well-appointed labs back at Radiophysics. A site too distant would disrupt this routine and create problems of travel, accommodation and technical support. For these reasons most of the sites examined initially fell within a comfortable radius of Sydney, the best candidate to emerge being an area known as Cliffvale, on the Nepean River at the foot of the Blue Mountains, 80 km south–west of Sydney. Three sites in this area were selected and tests begun to gauge the levels of radio interference. In February 1957 Bowen wrote to Fred White recommending that the telescope be built at one of these sites: 'Their low height above sea level, together with the existence of higher ground to the east and north–east, will almost certainly combine to provide an adequately low noise level at the site, which is, of course, the vital and overriding requirement.' [2]

Oddly enough, the problem with the Cliffvale site turned out not to be radio noise, but political noise. About 20 km south of Cliffvale, CSIRO had recently established an Upper Atmosphere Section at Camden under the direction of David Martyn. A major part of the research program planned for Camden was a study of the properties of the ionosphere using powerful radar equipment. Upon learning of the plan to site the telescope so close to Camden, Martyn let it be known that 'nearly all our work demands and uses powerful pulse transmissions on many frequencies, which would be almost certain to interfere with the operation of a giant radio telescope placed so near to us as is apparently contemplated.' Further 'it would be the height of folly to erect a radio telescope within a few miles of a centre concerned with pulse sounding of the ionosphere.' [3]

In view of his topsy-turvy relationship with Radiophysics in the past, Martyn was understandably wary of being asked to share his patch of turf with the Sydney radio astronomers. If the future work at Camden did lead to unacceptable levels of interference at Cliffvale, Martyn felt sure that because of the prestige and the massive investment in the new telescope, pressure would inevitably be brought to bear on him to alter his own research program. In short, in any conflict of interest it was a fair bet that the Upper Atmosphere Section would come off second best. Martyn voiced the underlying reasons behind his opposition to the Cliffvale site to CSIRO headquarters in August 1957:

> 'There is no need to remind you of the circumstances which forced us to transfer our work from Radiophysics to Stromlo; of the decision of the Executive to oppose the development of radio astronomy at Stromlo, which naturally killed interest in radio work there and again sent us 'into the wilderness'; of the decision of the Executive that I should not come back to live and work in Sydney for fear of causing complications with Radiophysics; and of the hard but finally successful struggle to convince the Executive that the site and existing buildings at [Camden] provided an economic and suitable site for our work.' [4]

In view of its recent decision to establish the new section for Martyn, the CSIRO Executive naturally wished to steer clear of a situation which seemed certain to inflame the long-standing feud between Martyn and

Radiophysics. In turn this precarious situation cast doubt over the wisdom of choosing any site relatively close to Sydney.

The row developing over the Cliffvale site in fact played into the hands of Fred White, who personally favoured locating the telescope close to Canberra. White's reasons were largely political and stemmed from the events of the early 1950s when plans had been proposed to establish a national radio astronomy facility in Canberra, centred around a large dish at Mt Stromlo (see Chapter 5). Although this attempt to board the Radiophysics bandwagon had been repelled, this was not the end of the mistrust which had arisen between CSIRO and the Canberra group. Shortly after the announcement of the Carnegie grant in May 1954 Mark Oliphant—in his capacity as foundation president of the new Australian Academy of Science—decided to establish a committee to assist in promotion and fund raising for the new telescope. Among those on the committee Oliphant nominated himself, Martyn, Woolley and from CSIRO White and Bowen. When White learnt of the new committee and suggested that the courteous thing might have been to have discussed the plan beforehand, Oliphant abruptly withdrew the Academy's offer of assistance. In an attempt to defuse the issue, White explained to Oliphant that he wished 'to avoid getting this project mixed up with the several personal incompatibilities which seem to exist in Australian physics today. This is a project which Bowen and Pawsey have regarded for some time now as their own.' [5]

In White's view the best way to mend these fences would be to locate the telescope near Canberra. Such a move would be seen not only as a gesture of goodwill towards the circle of influential scientists in the national capital, but also one that would lead to a greater degree of cooperation between Radiophysics and the astronomers at Mt Stromlo. White's argument had been strengthened with Woolley's departure for England in December 1955 to become the eleventh Astronomer Royal, together with a plan to reconstitute the Mt Stromlo Observatory as a new department within Oliphant's Research School of Physical Sciences at the Australian National University. As part of this reorganisation, the new director of the Mt Stromlo Observatory would also wear the hat of Professor of Astronomy at the ANU. The prospect of having the new telescope, as well as the optical instruments at Mt Stromlo, readily accessible to a new generation of astronomers training at the ANU clearly had some appeal. A further indication that the time was ripe to heal the rift between the radio and optical astronomers came with the appointment of Bart Bok as Woolley's replacement. Bok had studied at Leiden and Groningen under leading astronomers such as Jan Oort, moved to the Harvard Observatory in 1933 and, during the early 1950s, started up a successful program in radio astronomy at Harvard. In contrast to Woolley's indifference to radio astronomy, Bok stood out as one of the first classically trained astronomers to appreciate the potential of the new science, particularly for his own field of interest, the Milky Way. The appointment of Bok had in fact been a deliberate attempt by Oliphant to develop this new integrated brand of astronomy in Canberra.

In response to these encouraging developments the region was surveyed late in 1957 and a candidate site chosen near Hoskinstown, about 35 km south–east of Canberra. However, for all the apparent advantages of siting the telescope near Canberra, two factors worked against the idea. One of Bok's first tasks after taking up the new post was to organise the search for a new site for optical astronomy. Increasing light pollution at Mt Stromlo from Canberra's sprawling suburbs meant a far more remote site would be required for any future optical instruments. Bok privately favoured bringing the new radio telescope to Canberra but was unwilling to be too outspoken in his support, fearing it might only confuse the more important issue of convincing government officials that optical astronomy had to get out of Canberra. (The new site was eventually chosen at Siding Spring near Coonabarabran in northern NSW.) A second and more telling argument against Canberra proved to be its distance from Sydney. If the telescope was to be operated in the same way as planned for the Cliffvale site, then the radio astronomy group would have to be relocated in Canberra, a prospect the group opposed almost unanimously. Alternatively, the Canberra site could be operated as a remote field station—fly staff to and from the telescope and provide on-site accommodation. Yet, in this case the potential benefits of Canberra hardly seemed to outweigh the various logistical problems of having most of the telescope's users in Sydney. Why in this case choose a Canberra site ahead of numerous other possibilities at a similar distance from Sydney?

Reservations about both the Canberra and Cliffvale sites led to an extensive search during 1957 for a suitable location further west of Sydney, west of the Great Dividing Range. Large areas in the sparsely populated region bounded by the towns of Bathurst, Cowra, Forbes and Parkes were inspected, with a site near Parkes chosen as the best of these 'over the mountains' candidates. Situated in the Goobang Valley 20 km to the north of the Parkes township, this site was particularly well shielded from radio interference by a surrounding wall of hills. Though not as accessible as Cliffvale, the site about 400 km west of Sydney could be reached in five to six hours driving time or by a twice-daily air service to Parkes. (For a snapshot history of the area see Tangent 6.)

Early in March members of the radio astronomy group met to discuss the relative merits of the Cliffvale, Canberra and Parkes sites. Tests at each site showed that Parkes had by far the lowest noise levels and, in view of the industrial and residential expansion forecast near Cliffvale and Canberra, on the technical grounds Parkes had a clear advantage. Paul Wild recalls the meeting:

TANGENT 6

The Bushmans Radio Telescope?

The *Kildare* property and the district surrounding the new telescope site had been settled by Europeans about 120 years earlier. Major Thomas Mitchell was the first European to explore the area in 1835 and also the first to sight the Goobang Valley from a hill north–east of the site. Mitchell encountered some of the Wiradjuri people and recorded with sentiments typical of the times: 'The natives whom we met here were fine looking and enjoying contentment and happiness within the precincts of their native woods. Their enjoyment derived so directly from nature that it almost excited a feeling of regret that civilised men enervated by luxury and all its concomitant diseases should ever disturb the haunts of these rude happy beings.'[6]

Only months later an enterprising squatter following in Mitchell's tracks established the first cattle station on a creek on the southern end of the Goobang Valley, attracted by the 'country with beautiful open grassy flats'. By the early 1860s settlers from the crowded coastal plains around Sydney had begun to move west and claim large tracts of land, transforming the open grasslands and rolling woodland into fertile farming country. In 1871 the population of the area accelerated sharply with the discovery of gold, which saw thousands rush to pan the alluvial deposits in the creek beds of the Goobang Valley. Over the next two years an estimated 1500 kilograms of gold was extracted from the area, including a 4 kilogram nugget from the Welcome field.

In August 1873 one of the colony's most prominent figures, Henry Parkes, toured the area ending the visit with an address to a large crowd gathered in the new goldfield town known as Bushmans. Parkes had arrived from England as a young man and drifted through a variety of jobs before establishing in 1850 the *Empire*, a newspaper which became renowned for its radical views, until its financial collapse eight years later. By then Parkes had already entered politics and rose to be prime minister of the Legislative Assembly in 1872, the first of five occasions he held the highest political office in New South Wales. Referred to by the London *Times* as 'the most commanding figure in Australian politics', Parkes later went on to champion the idea of uniting the colonies under a single national government. To honour his visit the people of Bushmans decided in December 1873 to change the name of their town to Parkes.[7]

The population of Parkes varied wildly at first, falling to just 500 early in 1874 as prospectors and their families flocked north to new strikes in Queensland, and then jumping to 10,000 only months later as more gold was discovered south of the town. With much of the new found wealth invested in property the 1880s saw a land boom and a steady expansion in farming. The coming of the railroad in 1893 consolidated the Parkes area as one of the major wheat and wool producing regions of Australia.

News of the giant radio telescope with the obvious prestige it would bring to the town was greeted enthusiastically in Parkes. With the announcement of the Goobang Valley site in August 1958 the town's *Champion Post* ran an editorial welcoming the new 'acre of ear': 'The selection of a site near Parkes for the largest telescope in the southern hemisphere is the biggest event in the history of Parkes since the discovery of gold brought the town into being. The radio telescope will make Parkes one of the major centres of the world in the study of radio astronomy. Parkes can be proud that, in a day when science has produced monsters capable of annihilating whole cities in one blast, it will be a centre where pure science—man's probing to discover more about his surroundings in God's universe—will be practised with the aim of benefiting all mankind.' [8]

Sir Henry Parkes (1815–96), known as the father of Australian federation. In 1873, a gold mining town was named after him, in 1961, a radio telescope.

'To his credit, Taffy Bowen did not attend the meeting. At the beginning, the general mood of the meeting was that Cliffvale would inevitably be the site, partly because Bowen favoured it and partly because people did not like the idea of having a residential field station where they would have to spend days away from home comforts. I personally spoke very strongly in favour of choosing the best possible site from an astronomical point of view, which would mean 'over the mountains'. I also assured the meeting that my solar colleagues and I were very used to working at a residential field station (Dapto) and that such a mode of work was entirely satisfactory.' [9]

Wild's intervention helped turn the tide. By the end of the meeting, the decision in favour of Parkes was unanimous.

☆ ☆ ☆

Several days later negotiations began with Austie Helm, the owner of a 360 hectare (900 acre) property known as *Kildare* in the Goobang Valley. The area chosen initially consisted of a 50 hectare parcel but, to allow for the possible addition of other instruments in the future, two adjacent paddocks were included in the transaction to bring the total area of the site to almost 170 hectares. At £100 per hectare the negotiated price of £16,400 amounted to less than 2% of the projected cost of the project. A small team under the supervision of Lindsay McCready immediately began work at the site. First the area was surveyed and pegged out. Test bores sunk through the loam top soil indicated a hard clay layer at a depth of about 12 metres. A study of local meteorological records also confirmed the area's favourable climate. Although snowfalls, as well as wind gusts reaching 150 km per hour, had been recorded in the district, the occurrence of such freakish weather was extremely uncommon. Anemometers installed on the site showed that windspeeds were generally low, exceeding 15 km per hour 10% of the time, and 30 km per hour only 1% of the time. Only on rare occasions would unfavourable weather conditions shut down operations and force the telescope to be stowed in its safety position.

In the meantime tests of radio noise levels had continued. With a radio transmitter specially set up in the Parkes township, a DC3 aircraft with a sensitive receiver on board made several low-flying sweeps through the Goobang Valley. These tests showed that the site was exceptionally well shielded from surrounding sources of interference, in fact as good as any radio observatory in the world. To protect the area, negotiations began with several organisations. The department of civil aviation agreed to prohibit low-flying aircraft, not only because of the danger of joy-flights crashing into the telescope. Besides their own direct radio transmissions, even the metallic body of an aircraft can reflect or 'backscatter' radio noise into the dish. Similarly, the local shire and Parkes town council agreed to establish a quiet zone around the site. Industrial development within 10 km of the site would be restricted, and to within 30 km for any factory using radio transmitters or radio-frequency processes. Plans were also made to issue local farmers with ignition suppressors for their cars and farm machinery, but this turned out to be an unnecessary precaution.

Out to Tender

We mentioned earlier how the launch of Sputnik in October 1957 and the birth of the space age created a sudden interest in large steerable dishes. Many companies particularly in the United States were eager to get in on the ground floor of this rapidly growing technology. The interest shown by these companies in the proposed Australian dish had then led to the decision to open the project to international competition, rather than rely solely on British firms. In April 1959 Freeman Fox completed the tender documents and bids were invited from a shortlist of companies that had expressed strong interest in the project. In view of the complex interplay between the structural, mechanical and control features of the telescope, bids would only be accepted from companies prepared to take on the whole job as prime contractors. Freeman Fox would retain responsibility for overall supervision of the construction, including inspection and testing of the multitude of components during manufacture. By the close of the tender period early in June bids had been received from seven companies, with the lowest being £A611,000 from the West German firm of Maschinenfabrik Augsburg Nurnberg (or MAN for short). The second lowest came from the John Brown Construction Group in England, the third from Loewy Hydropress in New York, while the remaining four came from American firms which priced themselves out of the project with bids exceeding £900,000. Each of the seven companies quoted a completion time of under 27 months, with the MAN firm again leading the field with a promise of completion by April 1961, an extraordinarily short period of 21 months.[10]

MAN with three major works in West Germany had established itself as one of the world's leading producers of a wide range of steel products. About half the 3000 bridges constructed in Europe since the war had come from its workshops. Following a visit in April 1959, first to the Krupps factory and then to the MAN Gustavsburg plant near Frankfurt, Bowen came away convinced of MAN's ability to take on the project, and also impressed by West Germany's spectacular postwar reconstruction:

> 'I have not been on the Continent since 1951, when most places were still feeling the effects of the war. Now, in 1959, the contrast between Germany and Great Britain is startling. Comparing Dusseldorf with a provincial city like Manchester is like comparing Fifth Avenue with Footscray. The country is bursting with activity and productivity, and one does not have to look at a table of statistics to see why Germany is the foremost shipbuilding country in the world. I have no great love for the Germans, but there is no doubting their vigour and efficiency.
>
> The MAN story is much the same as Krupps. Their Gustavsburg plant (one of three) has 5000 employees and turns out about 2000 tons of fabricated steel a month. Their products range from light steel pressings for cars and buses, heavy equipment for the chemical industry, penstocks, hydraulic equipment, cranes, fixed and moving bridges, etc. (Their Augsburg works manufactures one third of the world's output of diesel engines for ships.) I had not learned the obvious lesson from my talk with Krupps, and repeated our requirement for getting the steel work out of their shops in less than twelve months. They treated this as a rich joke.' [11]

In July MAN was awarded the job as prime contractor, with three other firms as main subcontractors. One was Associated Electrical Industries (AEI) of Manchester, a conglomerate which had recently absorbed MetroVick. AEI was given responsibility for the servo-control system as well as all cabling, lighting and communication systems for the telescope. The other two subcontractors were Askania-Werke of West

Berlin which would take over development of the master equatorial system, and Concrete Constructions of Sydney which would lay the foundations and erect the reinforced concrete tower.

Despite strong interest in the project by American firms, it is not too surprising that MAN walked away with the contract. With postwar recovery well advanced and West Germany's re-emergence as a major economic power, labour and production costs had remained low compared with most other industrialised countries. In contrast, the boom economy in the United States saw living standards bound well ahead of any other country. Its high production costs made it difficult for American firms to compete openly on the international scene. German or American firms aside, the fact that the project had not remained in Britain raised a few eyebrows. The magazine *New Scientist* in an article 'German Radio Telescope for Australia' mourned how no British firm had been able to compete effectively either in terms of price or in terms of the way the productivity of British workers measured up against their German counterparts: 'When MAN was asked whether it could have the work done within two years, it replied that it would be uneconomic to have the job hanging about in the shops as long as that.' [12] A project begun four years earlier as a fine example of Anglo-Australian collaboration had now largely passed into the hands of a firm belonging to Britain's arch rival.

Even though MAN had bid 20% lower than any other firm, it still created financial problems for the project. The £611,000 price tag applied only to the telescope itself, not costs associated with on-site developments such as laboratory and residential buildings, roads, landscaping and the provision of essential services. Once these additional outlays were included the total cost of the project climbed to almost £800,000, well above a number of estimates made during 1958. More importantly, the actual funds in hand by mid 1959 totalled only £560,000, consisting of the Carnegie and Rockefeller grants made in 1954 and 1955, the modest sum collected by the Radio Astronomy Trust, plus the government pledge to match these funds pound for pound.

With the project facing a substantial shortfall, Bowen began another round of visits to the American foundations. The Carnegie Corporation refused point-blank to supplement its original grant, but the Rockefeller Foundation again provided a sympathetic ear.[13] In June 1959 its vice-president Warren Weaver gave Bowen an assurance that a further $US107,000 (£A50,000) would be forthcoming. Equally encouraging was his suggestion to apply for a grant from the Sloan Foundation. Weaver was in the process of transferring to a new post as Sloan vice-president and he felt confident that this foundation could match the amount promised by Rockefeller. And so, with a combined £100,000 from these two foundations, matched in turn by Treasury, it seemed certain that the projected shortfall of £240,000 would almost be covered. However, for reasons which are now unclear, Weaver had to go back on his word.

Shortly after taking up office at the Sloan Foundation, he personally rejected the grant application. Following lengthy negotiations in Canberra, the treasurer Harold Holt finally agreed in January 1960 to cover the shortfall; by and large the last piece of the financial jig-saw had fallen into place. Although the shortfall anticipated in mid 1959 eventually blew out by a further £100,000 (giving a final total of £900,000), the excess was covered by CSIRO's capital works budget, without the need to again seek out sources of external funding.

Whispers from the Deep

So far we have discussed only the structural and control aspects of the Parkes dish and have neglected one feature of central importance in any radio telescope. Before looking at the construction and commissioning of the telescope in the remainder of this chapter, we turn now to the receiver systems developed for Parkes.

In bald terms, a radio telescope is a device for collecting, focusing, amplifying and recording a steady hiss of radio noise. In a radio telescope the part played by the receiver is similar to that of the photographic plate at the focal point of an optical telescope. The energy of the radio signal is extremely feeble. In fact, the total energy collected by the Parkes telescope during its first thirty years of operation would light a torch bulb for only a tiny fraction of a second! To produce a signal that is strong enough to be recorded, the receiver must amplify the incoming radio waves over ten billion times.

Aside from the astronomical objects of interest, radio noise comes from a wide variety of sources. First, regardless of how well a telescope is sited, noise infiltrates from a range of modern technological devices. Then, there are various natural sources such as the background sky radiation, the earth itself, even the most unlikely objects. Radio telescopes are so sensitive that even natural radio signals from something as innocuous as a galah can interfere with observations. Astronomers often refer to a strong source of interference as a 'birdie'.

The most pervasive source of unwanted noise originates in the telescope's own receiver, a kind of electronic fuzz intrinsic to the device itself. Combined with the other types of extraneous noise, this sets a limit on the strength of an astronomical source which can be detected. Sources too faint simply drown in the background sea of noise. One way round this limitation is to increase the sensitivity of the telescope by making its collecting area as large as possible. (Economics had of course limited the size of the Parkes dish to a diameter of 210 feet.) Another way to minimise the problem is to use receivers which have very low intrinsic noise levels. The quality of receivers is therefore just as crucial in determining the performance of a telescope as the physical dimensions of the structure.

During the 1950s several major breakthroughs occurred in the development of receivers designed to operate in the microwave region of the electromagnetic spectrum (that is, over the wavelength range from one millimetre to one metre, the principal region of interest to radio astronomers). Two important devices to emerge were the solid-state maser and the parametric amplifier, both of which promised substantial reductions in receiver noise levels. In addition to their applications in radio astronomy these new low-noise receivers created a revolution in communications technology, comparable in importance with the great leap forward in radar brought about by the wartime invention of the magnetron.

Since the start of radio astronomy at Radiophysics, receivers had usually been improvised as the need arose, often assembled by a group with a particular aerial and a particular observing program in mind. In preparation for Parkes, Joe Pawsey decided in August 1959 to establish a specialist group to concentrate solely on new receiver technology. The group would have the dual role of keeping a close watch on technical developments elsewhere and of taking responsibility for the construction of a range of advanced receivers for the new telescope. By then the development of greatest interest was a revolutionary new device invented in 1954 by Charles Townes and his group at Columbia University in New York and christened the maser, an acronym for microwave amplification by stimulated emission of radiation.[14] The maser was the forerunner of the first laser (developed at Bell Labs in 1960) which, as the name suggests, amplifies light rather than microwaves. The first maser had used ammonia gas but, by 1956, the Columbia group had developed a solid-state device suitable for application in radio receivers. The physicist Gib Bogle was the first to hear of the breakthrough in Australia:

> 'It seemed almost certain that the solid-state maser would be a thousand times more sensitive than previously existing microwave receivers, and this represented a tremendous jump forward in the process of refinement which had been going on slowly over the previous decade. It is true that a type of gas maser had already been demonstrated a few years beforehand; but it was a feeble device tied to one unalterable frequency and was as much use on its own to communications technology as a tuning fork to the conductor of an orchestra. In contrast, the new solid-state maser packed a solid punch and could be tuned to any frequency. Its importance was so obvious that few Australian physicists doubted the rumour that the inventor of the new principle had been paid $100,000 by a big corporation not to patent his invention.' [15]

Close on the heels of the solid-state maser came another device known as the parametric amplifier, based on new types of nonlinear circuit elements and which promised under certain operating conditions to lead to noise levels as low as the maser. The maser was however the first to be actually used in radio astronomy. Successful trials conducted by Townes' group with a maser receiver fitted to the 50 ft NRL telescope

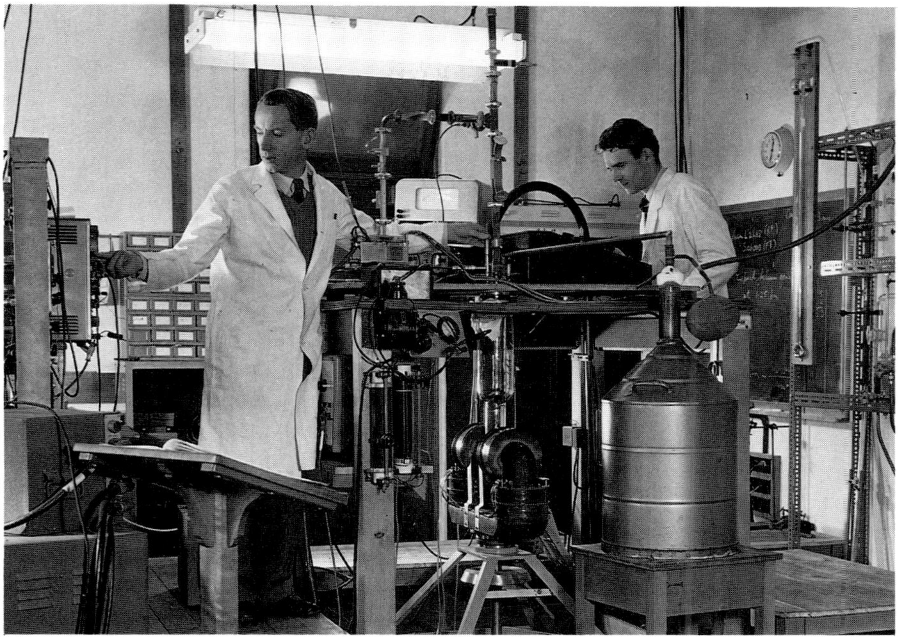

Gib Bogle (left) and Doug Milne with the maser built by Bogle in 1958. Milne's task was to develop a maser package small enough for use as a receiver on a radio telescope. The main drawback of the maser was that it required cooling with liquid helium to temperatures close to absolute zero.

in Washington and operating at a wavelength of 3 cm soon led to plans to install masers on a number of other American telescopes. Before his appointment as head of the new receiver group at Radiophysics, Brian Cooper joined the project at Harvard:

> 'To learn about the maser at first hand I arranged to go to the Harvard Observatory in 1958 where Tom Gold and Nicolaas Bloembergen were keen to have a 21 cm maser built for hydrogen-line observations on their 60 ft dish. I soon began to feel that, for the next few years at any rate, the parametric amplifier would be a more generally useful device for radio astronomy, but I went ahead and built the 21 cm maser. The maser worked nicely and enabled the Harvard radio astronomers to observe the hydrogen line with much greater sensitivity than had been available to them previously.' [16]

While Cooper kept Radiophysics posted on the latest American developments in maser receivers, Brian Robinson one of the hydrogen-line pioneers played a similar role with the new parametric amplifiers:

> 'I had just completed my doctorate at Cambridge and was back on the Radiophysics payroll. Pawsey asked me to go straight to the Netherlands and learn of the Dutch progress in receivers. I took up a position at the

Kamerlingh Onnes Laboratory which had long been famous for its work on low-temperature physics. The radio astronomers at the nearby Leiden Observatory had been investigating maser receivers but the recent developments in parametric amplifiers in the United States made them swing their effort in this direction. Philips were very interested in the commercial side of things and were backing the research. Parametric amplifiers—or paramps as they were known for short—had a clear advantage of working at manageable temperatures, whereas masers had to be cooled down to just a few degrees above absolute zero with liquid helium.' [17]

Robinson stayed until the completion of the new paramp. Partly in return for Robinson's services, Pawsey had arranged with the leader of the Leiden group Jan Oort for duplicate parts of the receiver to be manufactured and shipped to Sydney, ready for assembly and testing once Robinson returned. The new paramp was installed at Parkes in 1962 and used in a number of important observations on the hydrogen content of distant galaxies and clusters of galaxies. Despite this major effort, the operation of the paramp was never entirely satisfactory.

Meanwhile the new receiver group at Radiophysics had begun developing a range of other devices. One was the first Australian-made paramp, built by Frank Gardner and Doug Milne to operate at 20 cm and used in the first Parkes survey of radio sources. Another project involved upgrading a homegrown device built several years earlier and which had been installed on a small dish situated at John Murray's orchard property at Rosebank. Known as the Murraybank receiver, this device consisted of 48 channels tuned to cover a range of frequencies each side of the natural hydrogen line at 1420 MHz and could be used to measure the radio emission from hydrogen gas clouds Doppler shifted away from the natural frequency. This provided information on the velocities of these galactic gas clouds and, overall, on the large-scale rotation of the Milky Way itself.

Work also proceeded on maser receivers in collaboration with the group led by Gib Bogle at CSIRO's National Standards Lab. In 1961 the group built a 21 cm maser receiver which used a small synthetic ruby crystal doped with the 'masing' agent chromium oxide. The receiver had performance characteristics as good as the American paramps then under development. The main drawback with the maser, however, was that its operation required cooling the core of the device to near absolute zero with liquid helium, whereas the paramp only required cooling to more 'moderate' liquid nitrogen temperatures. The bulk and complexity of the equipment for the maser, plus its greater operating costs (liquid helium costs about one hundred times as much per litre as nitrogen), cast serious doubts about its suitability for a remote site such as Parkes. For these reasons the development work on maser receivers for the new telescope came to an end—a contributing factor too was the setback to the maser program following the mysterious death of Bogle early in 1962.[18]

Much of this development work on receivers took place in parallel with the construction and commissioning of the telescope at Parkes. While Bowen was the driving force behind the telescope, much of the credit for developing the receivers belongs to Pawsey. It was his foresight that led to this state of the art technology, a factor as important as any other in making the Parkes telescope the most advanced instrument of its type in the world.

MAN at Work

By September 1959 the construction of the telescope had got off to a flying start—not only at Parkes but also at three different centres halfway round the globe. Under the watchful eyes of the Freeman Fox engineers, Associated Electrical Industries began on the servo-control systems in Manchester, Askania-Werke started manufacture of the master equatorial unit in West Berlin, and MAN began casting some of the massive steel components of the telescope at its Gustavsburg plant. By May 1960 the fabrication of the turret, hub and counterweight structures had been completed. A trial assembly of these components was then made on a section of the steel roller track so that various tests and measurements could be made on the driving gears and the level of friction forces. After the successful completion of these trials the assembly was dismantled in August and the components packed into crates ready for shipment to Australia.

In the meantime the Sydney firm, Concrete Constructions, had completed the excavations for the telescope's foundations and in November 1959 began erection of the reinforced concrete tower. Each of the three floors in the tower had been assigned a particular function. The top floor would be the control centre, fitted out with a control desk for operating the telescope and an array of data-handling equipment for processing and storing the incoming stream of signals. The intermediate level would house the bulk of the receiver equipment, connected by cables to the aerial cabin, while the ground floor would be a general purpose storage and workshop area stocked with equipment required for repairs and maintenance. By March 1960 the last concrete had been poured. The tower was then allowed to stand for several months so that any settlement of the foundations would occur well before cranes began lifting the telescope's 1000 tonnes of steel superstructure into place.

The 170 hectare site also underwent a transformation, beginning with the construction by the local council of an access road from the main Newell highway and an internal network of roads. The grass was slashed and lines of white markers laid out to form a landing strip for small aircraft. (The sheep usually had to be cleared before an aircraft could land.) Plans were also made for the construction of a cluster of buildings

Right MAN carried out a trial assembly of the telescope's turret and hub at its Gustavsburg plant near Frankfurt. *below* Casting the steel roller track which would support the massive superstructure of the telescope.

situated a kilometre north of the telescope. This 'village' area would consist of two small houses for engineering staff responsible for the maintenance and safe operation of the telescope, as well as a larger dormitory building for visiting research and technical staff carrying out observations. A combined garage and fuel store was also erected for housing site vehicles, a tractor and other essential items such as firefighting equipment. Efforts were also made to improve the layout of the flat surrounding paddocks, as station manager George Day recalls:

> 'A program of tree planting was started in conjunction with the Forestry Commission, and some 4000 seedling trees were planted along the perimeter and internal roads. A nearby farmer did an excellent job of ground preparation for the tree plantings. The lack of water was a major problem. During the construction period, and for a considerable time thereafter, water had to be carted daily by road from Parkes. Numerous test bores at and near the site were put down in an endeavour to locate water. Later after successful negotiation with the owner, an underground stream about one mile south was tapped. Eventually a supply from the Parkes–Peak Hill pipeline became available.' [19]

In September 1960 the MAN construction crew arrived on site, together with some of the steel components cast in West Germany and brought to the site from Sydney by rail and road. The first task involved fitting the circular steel roller track to the top of the tower, upon which the dish is rotated in the horizontal (azimuth) direction. Once the track had been bolted into place precision measurements showed that over the complete 35 metre circumference of track, no part deviated from a perfectly flat surface by more than 0.25 mm (less than half the design specification). Then the MAN crew retraced the steps followed at the Gustavsburg workshop and began to assemble the steelwork to support and steer the dish structure. First came the turret, a compact and very rigid structure that rotates around the azimuth roller track on four sets of bogies. The turret contains several rooms, the two main ones designed to house some of the servo-drive machinery and receiver equipment.

Next, the cylindrical hub was mounted on two strong A-shaped steel frames of the turret. The hub also contains two rooms, the upper one housing a small optical telescope, an observer's chair and various controls, as well as indicators used when the telescope is put through calibration tests or special tracking procedures. The lower room houses additional receiver equipment. The hub supports the dish structure and movement in the vertical (altitude) direction is carried out by a rack and pinion drive on each side of the hub. These racks are in turn attached to heavy counterweight boxes filled with 400 tonnes of concrete ballast mixed with steel punchings, this ballast counterbalancing the weight of the dish itself. Slightly more counterweight was added than actually needed to balance the dish, so that the dish would always have the tendency to return to its upright or zenith direction. This had the

Above Laying the foundations for the three-storey tower. The road leads north to the site of the village area. *below* At times up to fifty construction workers were on site. MAN imported a specialist group from Germany, while most of the carpenters, fitters, electricians, painters and labourers were recruited from the Parkes area or brought from Sydney.

advantage of removing backlash between the teeth of the rack and pinion drive, and also making it easier to stow the telescope in its upright position in an emergency.

With the last of the counterweight concrete poured by March 1961, work began on the erection of the dish structure. Fabrication of the thirty radial ribs to form the dish had already been under way at ground level since the previous November. By day the tubular steel was cut to length and placed in jigs set in massive blocks of concrete, and then at night welding of the joints took place when all the steel components were at a uniform temperature, minimising the possibility of locking thermal stresses into the structure. After sandblasting and painting, these rib structures were then connected in twos or threes. In April 1961, each of these segments or 'slices of pie' was lifted into place, bolted to the hub and then fixed to its neighbouring segments by a series of ring girders. Next came a period of comparative lull as each rib and ring girder was carefully surveyed and adjusted. Five weeks later, work began attaching two sets of steel cross members or purlins, each shaped in the form of an extended spiral, but opening out in opposite directions. The two sets of spiral purlins cross at right angles at each intersection and, together with a further set of radial purlins, provide the underlying steel web sup-

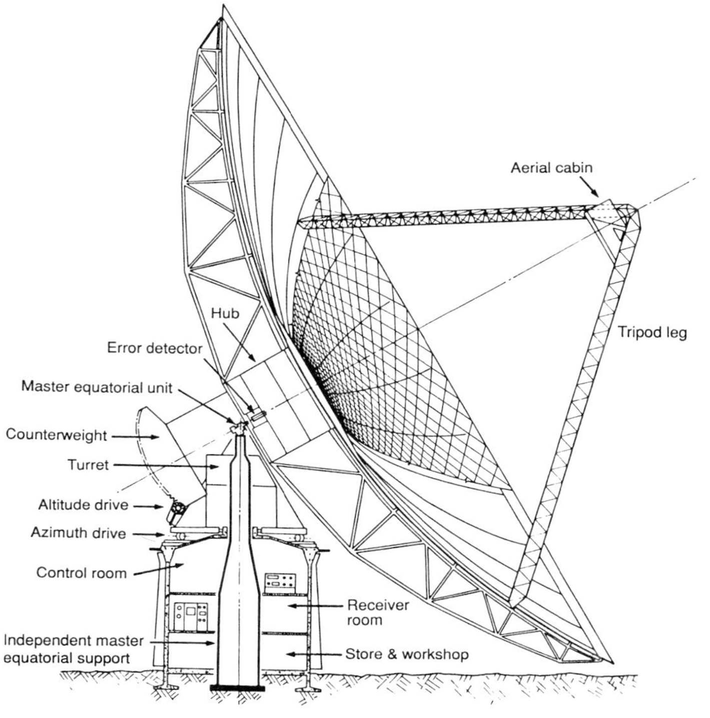

The Parkes telescope illustrating some of its principal components.

In September 1960 MAN erected a derrick and jib for lifting the steelwork. Here the turret is in place on top of the tower and, within the turret, the central pillar on which to mount the master equatorial unit.

porting the dish's reflecting surface. The geodetic nature of the spiral purlins adds the right amount of stiffness to the structure.

In the meantime the aerial cabin had been lifted into position. Supported by three legs and set slightly above the focal point of the parabolic dish, the cabin provides space for front-end receiver equipment and enough room for two people to work in relative comfort, even when the dish is tilted. Access to the cabin is provided by a small elevator in one of the tripod legs or by ladders in the other two. Perched almost 60 metres above ground level the cabin offers an impressive view of the surrounding countryside—at least for those who can control their vertigo!

With the framework of the dish largely complete early in August 1961, work began laying the reflecting surface. For the task a group of labourers was especially recruited, most of them members of the Parkes rugby league team. First, segments of thick steel plate were used to form the central section of the surface (17 m in diameter), the plates welded together to provide the dish with extra stiffness. The remaining part of the surface was then covered with over one thousand mesh panels, each woven from a special high-tensile steel wire. Problems had arisen in finding an effective technique to galvanise the mesh. The shipment of panels from West Germany only arrived at the last minute, almost bringing the construction to a complete standstill. The panels were laid out ring by ring, working outwards, with each panel fastened to its purlin support and to its neighbours with self-tapping screws.

Above The hub and pre-assembled rib sections. A thick layer of sand litters the site, used to sandblast the ribs prior to painting. *left* Lifting the first rib section into place. Each section was attached to the hub with high tensile bolts and then to its neighbouring sections with a series of ring girders.

The jewel in the crown—positioning the aerial cabin. The structure of the dish shows the network of intersecting spiral and radial purlins, one of the design features proposed by Barnes Wallis.

Over one thousand mesh panels were fitted to form the reflecting surface of the dish. The aerial cabin above can be reached via the small lift within the tripod leg.

Next came the task of fine-tuning the surface so that it would conform as closely as possible to a perfect parabolic shape, a painstaking process which would prove crucial in determining the performance of the telescope. With the dish parked in the zenith the precise positions of 800 special adjustment points over the surface were surveyed using a combination of steel measuring tape and surveyor's theodolite. Any section found to deviate from the ideal shape was then moved up or down by screwing the adjustment points in or out of the supporting purlins. John Bolton describes the work:

> 'It turned out that the surface of the dish supported by the radial purlins was slightly floppy. When the position of one panel was adjusted to give the desired level, the surrounding panels would sag. The spiral purlins were not quite stiff enough to provide rigid support to the radial purlins. Jim Roberts who had just returned from Owens Valley did the analysis of the surface with Norm Broten who had come from the National Research Council in Canada. We had such a tight schedule to keep before the opening ceremony. Roberts and Broten did the analysis during the day, while Mike Jeffery from Freeman Fox and I did the surveying and resetting at night when the temperature of the dish was nice and even. The final setting of the surface became a matter of inspired guesswork as much as anything else.' [20]

As the adjustment of the surface proceeded, the construction crew carried out a number of finishing touches on the rest of the telescope. By the end of August 1961 the structural work was virtually complete, just on two years after construction began and only slightly over the ambitious schedule MAN had promised at the project's outset. Askania was the only contractor unable to keep up the pace, having fallen three months behind early in the manufacture of the master equatorial unit and, despite constant pressure applied by the prime contractor MAN, unable to make up the time. Yet, by any standards, the construction was proceeding remarkably smoothly with few problems or delays, as Bowen could cheerfully report to the Rockefeller Foundation: 'We have high hopes that the instrument will come into use about 24 to 25 months after placing the contract, which you will recall was in July 1959. This is not quite as good as we once hoped but, compared with other projects of this magnitude, we have reason to be well satisfied with the progress which has been made.' [21]

News of the rapid progress also filtered through to other radio astronomy centres. Reflecting on the problems encountered with the 25 metre Dwingeloo dish, completed in 1957, Gart Westerhout wrote from Leiden: 'It is certainly unique in the history of radio astronomy that so large a dish has been built in such a short time and produces such an excellent beam at so short a wavelength. Having observed from close quarters the struggles people have had to arrive at the specified surface accuracy, it is enlightening to see that it is possible, after all, to build a dish of this size which not only meets the specifications, but considerably surpasses them.' [22]

The weeks leading up to the inauguration in October 1961 became a period of feverish activity for the teams of technicians, electricians and painters. To avoid congestion, the day was divided into an early shift from 7 am to 5 pm, followed by the night shift lasting until midnight. There was an abundance of wildlife at the site, some of it unwelcome. Spiders and snakes were always a danger, while at night the floodlights attracted vast swarms of insects.

The main task ahead consisted of completing the installation and testing of the drive and control systems. Early in September Askania completed the long-awaited master equatorial unit, the key element in the telescope's control system. The central feature of the unit is a small precision-built optical device accurately positioned at the intersection of the telescope's two axes of rotation (alt and az). This device can be pointed with great accuracy in any chosen direction of the sky; the dish itself is made to follow it by means of the master–slave system devised by Barnes Wallis. Within the hub is an error detector which operates with a narrow beam of light and an elaborate arrangement of mirrors, lenses and photoelectric cells. The detector senses whenever the pointing direction of the dish is out of alignment with the small optical device. This generates an error signal which is fed to the altitude and azimuth motors, which then bring the dish back into near perfect alignment. The massive superstructure is in effect 'slaved' to follow the master unit, avoiding the need for a computer to continually calculate the telescope's actual pointing direction.

By the first week of October 1961 the master equatorial had been installed and wired to the control desk in the telescope's tower. Unexpectedly a major setback then occurred with the error detector. An attempt by Askania to make up lost time by air-freighting the detector to Sydney backfired when rough handling of the crate in transit badly damaged several of its components. Meanwhile, despite other problems with the control desk, test drives of the telescope in azimuth had gone ahead. With only a subdued hum from the motors the motion of the telescope's superstructure proved impressively smooth. By the middle of October, the Parkes telescope was finally ready to tip in elevation. This was to be the most important test drive of all, the occasion that really marked the culmination of six years of planning and construction. New recruit to the Parkes staff John Shimmins witnessed the moment, a majestic if not nerve-racking event:

> 'Despite a thorough clean up, the first time we tilted the dish a shower of metal—nuts, bolts, rivet heads—rained down from above. We immediately found ourselves faced with some serious teething troubles. The two motors used to tilt the dish were found to be working in opposition to each other and this was effectively twisting the dish. We fixed that problem easily enough, but then discovered that the dish was 'walking' down tooth by tooth on the two big rack and pinion gears. This caused a sort of jerking motion instead of a nice smooth movement. After a while Jack Rothwell from

The grand moment—the dish tilts for the first time in October 1961, only a few days before the inauguration.

> Associated Electrical Industries and Harry Minnett came up with the solution in the form of a damping motor. Actually, in some ways it was a relief when the gale blew up on the big day of the opening. It meant we didn't have to try and drive the telescope in front of all those people. At that stage we knew we could tip it down—but not too sure we could bring it back up!' [23]

Test Driving a Telescope

Not exactly a case of 'back to the drawing boards', but much work remained to be done before the telescope could be made available for full-time astronomy. Freeman Fox still controlled the project and, in fact, for several months after the inauguration it was the project manager Mike Jeffery who took responsibility for putting the telescope through its paces and for carrying out thorough tests on the drive and control systems. The most common of the three operational modes would be the observation of distant radio sources where the dish moves at the sidereal rate, an imperceptible motion which cancels the earth's rotation and leaves the telescope pointing in the same fixed direction in space. The second mode of operation is automatic scanning, where a particular area of sky can be selected and then systematically searched for astronomical objects. This was used to compile the Parkes Catalogue of radio sources in the 1960s. The third mode is simply manual control where the driver at the control desk is able to select the rate at which the telescope moves about the alt and az axes. Manual control can be used for tracking fast moving local objects such as spacecraft or, more routinely, for preliminary selection of a target. In this case the telescope can be moved into rough alignment with the direction of the astronomical object of interest, the servo system switched on so that the dish becomes slaved to the master equatorial unit, and then the object followed across the sky at the usual sidereal rate. With maximum drive rates of 15° and 24° per minute in alt and az respectively, the telescope can be quickly slewed into position ready for observations.

A number of elaborate safety devices were also built in to protect the telescope from human error and certain 'acts of God'. Limit switches prevent the telescope from being twisted too far in the azimuth direction. If this occurs the switches automatically trigger the drive motors to decelerate the motion and brakes are applied to bring the structure to a complete halt. Protection from lightning strikes is also provided by heavy copper cables running from the hub down through the tower to earthing plates buried next to the tower's foundations. In winds higher than the safe operational limit, the telescope can be stowed in the zenith where a massive locking pin of high tensile steel firmly clamps the dish against any movement in altitude. In exceptionally strong winds, the load can be completely taken off the azimuth roller track by using hydraulic jacks to lift the whole superstructure a few millimetres clear of the track, and so avoid any damage to either the track or the drive bogies.

A variety of other features required attention, as John Bolton recalls:

> 'There were numerous details to sort out. We had to adapt the cabin to allow the feed at the telescope's focus to be rotated. This feature had not been included in the original design, so we set about converting the cabin by running up additional cables from below and installing the control system downstairs. This modification was certainly worth while as it led directly to the discovery that radiation from sources is polarised—the first major discovery made with the telescope. Basically anywhere you looked there was something which needed changing. The lift up to the cabin turned out to be unsafe and had to be replaced with another using a system of cables to haul the thing up and down. I was the guinea pig that tested the safety devices on the new lift! Anyway, gradually all these problems were sorted out.' [24]

One crucial task remaining was to calibrate the telescope's pointing direction. The major research program ahead at Parkes would be to locate and study new radio sources. The success of this program would depend critically on measuring source positions with enough precision to allow optical telescopes to positively identify these sources. In short, whatever the pointing direction, the dials on the telescope's control panel had to give a true reading of the celestial coordinates of right

Three of the key players—John Bolton (left), Mike Jeffery and Taffy Bowen. As project manager, Mike Jeffery played a leading role in the design, construction and commissioning of the telescope.

ascension and declination. The task of reducing the errors in the pointing direction involved several steps, beginning with a series of delicate adjustments to ensure that the polar axis of the master equatorial unit was aligned exactly parallel to the earth's axis of rotation. Then came further adjustments to make sure that the radio axis of the dish itself was slaved exactly parallel to the master unit's axis. The trickiest part then involved making estimates of a range of other possible errors—the effect of gusting winds, gravity distorting the shape of the tripod supporting the aerial cabin, and refraction bending the path of radio signals in the earth's atmosphere. The size of these effects could in some cases be worked out by observing special 'calibrator' sources whose positions were already known to great accuracy by astronomers in the northern hemisphere (none of these benchmark sources were available in the southern sky at this time). Once the size of all these errors had been estimated, some amount of automatic compensation could then be built into the telescope's drive and control system.

This painstaking work carried out by Jim Roberts, Don Mathewson, John Whiteoak and others continued well after the start of the astronomical program in April 1962. By the end of 1963, the pointing accuracy over the full range of sky had been determined to within 20 arcseconds. For radio sources roughly overhead in the sky, positions could be measured to an accuracy approaching 6 arcseconds. To put this in perspective, a pointing accuracy of 6 arcseconds corresponds to a marksman hitting a small coin at a distance of one kilometre. This pointing accuracy was far better than that achieved by any other large dish at this time and, for that matter, better than all but a few of the world's optical telescopes. Taking into account that optical telescopes are much lighter structures, protected from the elements in domed enclosures, this high pointing accuracy illustrates how Parkes turned out to be an extraordinary achievement in precision engineering.

CHAPTER EIGHT

Radio Astronomy in Transition

After the Sydney Harbour Bridge and Opera House, the Parkes telescope is probably the best publicised structure in Australia.

Chris Christiansen, 1988

The inauguration of the Parkes telescope in October 1961 marked an important stage in the development of science in Australia. On a purely scientific level the telescope provided Australian astronomers with the most powerful and versatile instrument of its type in the world, one which immediately produced a stream of significant and, at times, fundamental discoveries. A review of these achievements forms most of Part 3 of this book. On a different level the Parkes telescope also had a major impact in shaping the way astronomy developed in Australia. In contrast to the 1950s, when small teams of radiophysicists-turned-astronomers built and had exclusive use of their own telescopes, Parkes would operate as an observatory and have more in common with the great optical telescopes of the world. A full-time specialist group of technicians would look after the maintenance and routine operation of the telescope and radio astronomers would now have to compete with their peers for observing time on the new instrument. A new breed of radio astronomer emerged and also a new way of doing radio astronomy (see Chapter 11). In this respect the completion of the Parkes telescope marked the arrival of radio astronomy in Australia as a mature scientific discipline. In some respects too, it marked the end of the most innovative and interesting period.

The Parkes telescope also had major repercussions on the broader development of astronomy in Australia. With the massive investment of staff and funds in Parkes, only very limited resources remained to support a number of other research programs which had flourished at Radiophysics in the 1950s. In the present chapter, we see how this heavy commitment to Parkes sparked off a fierce struggle for these remaining resources which, in turn, precipitated the departure of several leading members of the Radiophysics group to continue careers in astronomy elsewhere.

SuperCross or Radioheliograph?

Even before the first blueprint had been drafted for the Parkes dish, the Radiophysics group realised it faced a major reorganisation of its resources. At a series of meetings in mid 1959 to plan its long-term future the group—consisting of about thirty research scientists backed by a similar number of technical and laboratory staff—had reached agreement that its effort would be divided roughly equally between Parkes and other projects. Initially this was to be a flexible arrangement, so that promising leads could be followed in any field as they occurred. For the Parkes group the future was clearcut. With a superlative instrument on the way the main tasks ahead were to ensure that the telescope was brought into operation as quickly as possible, to establish some sort of priority among the wide variety of possible observing programs, and then to start working through these programs in what promised to be richly rewarding years of astronomy. The Parkes group possessed all the optimism, enthusiasm and cohesion that comes with a genuine sense of purpose.

In sharp contrast, for the radio astronomers who would not be directly involved in the Parkes project, the future seemed far less promising. As mentioned earlier, Radiophysics operated a number of instruments at this time—the radio spectrograph built by the solar group at Dapto, two small dishes for 21 cm work at Potts Hill and Murraybank, and the Mills Cross, the Shain Cross and the Chris Cross at the Fleurs field station. Each of these instruments had been brought into operation during the period 1952–57 and, though none were ready for the scrap-heap, by 1959 serious thought had to be given to a new generation of telescopes for the years ahead. Aside from the 21 cm group which eagerly awaited the completion of the Parkes dish, the future of the solar group and those committed to research programs with the cross instruments became an issue of major importance. Any new instrument would inevitably be larger, more complex and more expensive than its predecessor, and clearly not all of the research programs of the 1950s could continue to flourish. The Parkes telescope now meant that only about half the staff and financial resources were available to stretch across all these programs. Unavoidably, severe cutbacks were the order of the day for the remainder of the radio astronomy group.

Two proposals for new instruments emerged as the major contenders for the 1960s. One was a metre-wavelength Radioheliograph, devised by Paul Wild to supersede the Dapto spectrograph as the principal instrument for studying bursts of radio emission from the sun. The other proposal by Bernard Mills called for a SuperCross, a larger and more elaborate version of the instrument built in 1956 to continue the highly successful program on the detection and cataloguing of remote radio sources. Joe Pawsey outlined the relative merits of the two proposals in a memorandum, dated June 1959:

> 'In the first case, the field looks like a thoroughly rewarding one for study, the Radiophysics Laboratory is at present out in front and Wild is very keen to continue. The observations are beyond the capabilities of the Parkes telescope and will be quite independent of the latter. This is the only serious solar investigation proposed. In the second (Mills) case, the field of study—galaxies at extreme distances—is perhaps the most intriguing problem of all astronomy, our position is currently right in the front rank and Mills is keen to continue at longer wavelengths as well as at the shorter wavelengths for which the Parkes telescope is suited. A cross can be constructed at a wavelength of about a metre which, for the restricted purpose of a survey at one wavelength, should be both more sensitive and more directive than the Parkes telescope at its optimum wavelength. Observations with this cross would be exceedingly valuable in their own right.' [1]

Both proposals undeniably had outstanding scientific merit. And yet, because of the limited resources available there was little possibility of proceeding with both at the same time. The issue would be one resolved on non-scientific grounds and here, on several counts, Wild's proposal gained the upper hand. The Dapto spectrograph had been built in 1952 and since then the cosmic group had constructed both the Mills and the Shain versions of the cross. For what it was worth, fair play demanded that next it was the turn of the solar group. Another consideration was the question of diversification. As Pawsey's memorandum spelt out, the SuperCross would carry out an important research program separate and complementary to the Parkes telescope, but nevertheless it would mean a total commitment by the Radiophysics group to galactic and extragalactic radio astronomy. Inevitably the solar group, by then widely recognised as world leaders, would run down and eventually be disbanded. A vote for Wild's proposal would at least be a vote for diversification, rather than specialisation of the group's activities. Further, there was also the sensitive issue of productivity. While concentrating solely on observations of the sun, the heliograph at least promised greater versatility in that a wider variety of research programs could be undertaken and so 'feed' a larger number of radio astronomers. Because a cross spends long periods of time set in the same observing mode, only a small number of astronomers are required to exploit the astronomical potential of the instrument. For all these reasons, at a meeting in August 1959 the radio astronomy group decided in favour of Wild's proposal.[2]

When the Radioheliograph received the blessing of the group the instrument itself was no more than a vague concept. Wild had initially proposed an instrument in the form of a cross and, only later, devised the plan for an array consisting of 96 identical aerials positioned at equal distances around a large circle. The plan had then been to build the instrument at Parkes but the sheer size of the circular array, 3 km in diameter, proved too large for the 170 hectare site. Negotiations then began with the local farmers in the Goobang Valley to lease land surrounding the site. Sheep would simply graze around each of the 96 aerials. However, this plan was blocked because of regulations which decreed that CSIRO facilities had to be sited on land owned by the government. A second plan to purchase a large new site south of the Parkes township also came to nothing when two of the landholders refused to sell. This led to the decision to head north to less fertile and less expensive farming country and to the eventual selection of the site at Culgoora, near Narrabri in northern New South Wales.

Funding for the Radioheliograph was obtained from the Ford Foundation in two instalments totalling $US630,000—approximately the same amount as the combined Carnegie and Rockefeller grants made earlier towards the Parkes telescope. In contrast to the Parkes dish and the engineering challenges it presented, the simple form for each of the 96 aerials enabled the entire heliograph to be built by Radiophysics staff. After eight years of planning and construction, the heliograph recorded its first signals in August 1967 (details are given later in Tangent 7).

Beyond the Doors of Radiophysics

The decision at Radiophysics, in August 1959, to proceed with the Radioheliograph at the expense of the SuperCross meant that Bernard Mills had to quickly assess his future:

> 'The result was expected and there was certainly no opposition to it, even from me. Making an all-out attack on galactic and extragalactic problems using a sufficiently large general purpose reflector, together with an even larger special purpose cross, would have effectively absorbed all the available funds and facilities and led to the eventual demise of the solar program … . It was a choice of specialisation or diversity. News of the decision spread quickly and I was soon offered a position at the US National Radio Astronomy Observatory with the necessary funds to construct a cross there. However, my interests were in the southern hemisphere and I was not particularly interested in building a cross *per se*. Therefore I naively approached several Australian universities to see if there was any interest in forming a radio astronomy group having the eventual aim of constructing a large instrument to complement the deficiencies of the CSIRO reflector.' [3]

Mills did not have to look far. In May 1960 the offer of a position and resources to build the SuperCross came from Harry Messel, head of physics at the University of Sydney.

Earlier in Chapter 5 we saw how, shortly after his arrival in Sydney in 1952, Messel had begun a vigorous campaign to recruit new staff and to attract financial backing. His aim was to transform a scientifically moribund department into a research centre of world class. A recent history of the department records:

'In those days the School of Physics was in a low state and to Harry it seemed more like a museum than a modern physics laboratory. The building was fine but dilapidated, the budget was exiguous, the permanent staff were few and there had been no permanent Head for six years. Nevertheless Harry realised that, given the position and history of the University of Sydney, its School of Physics ought to be making much more of a contribution to the development of science in Australia than it was. He could see that, given more staff, more money and better direction, it would be possible to transform it into a major centre for teaching and research which, as he had aimed to do in Adelaide, would retain promising young scientists in Australia and attract good staff and students from overseas.' [4]

After making a solid start in nuclear physics and cosmic ray research Messel decided to broaden the scope of the School and, in 1958, announced plans to form a small astronomy group. Messel's star recruit to the new group was Robert Hanbury Brown, Lovell's right-hand man at Manchester and a scientist with an outstanding record. Since joining the Jodrell Bank staff in 1949 Hanbury Brown had become a leading exponent of intensity interferometry, a technique which allows the structure of individual radio sources to be measured with very high angular resolution. In the early 1950s Hanbury Brown realised that the same idea might work at optical wavelengths and he developed the design of a stellar intensity interferometer for measuring the sizes of visible stars. Ignoring vigorous objections that the idea violated basic principles of physics, Hanbury Brown, John Davis and Richard Twiss went ahead with a pilot model of the interferometer. In 1955, they confounded their critics by measuring the diameter of Sirius, the brightest star in the sky. It was the first time a technique developed originally for radio astronomy had been successfully transplanted to the traditional field of optical astronomy.

For a full-scale version of the instrument Hanbury Brown decided to leave Cheshire and head for Australia with its much clearer skies. The new interferometer was to consist of two large parabolic mirrors on movable mounts which could be separated over large distances on a circular railway track. Unlike the pilot model, which had used the mirrors from two ex-Army searchlights, the reflecting surfaces would be comprised of a large honeycomb array of small hexagonal mirrors to focus light from a star onto a photomultiplier tube. An analysis of the correlation between the signals from the two mirrors would then yield the diameter of a star. The intensity interferometer came into operation in 1962 at Narrabri, not far from the site chosen for the Radioheliograph. Hanbury Brown later wrote:

'Before it started in 1962 the angular diameters of only six bright stars were known: all of these stars were giants or supergiants and none of them represented the population of common stars which is called the main sequence. In a program which took about ten years to complete the angular diameters of 32 single stars were measured at Narrabri, many of them being main sequence stars. This was the first time in the history of astronomy that

> the angular size of a main sequence star had ever been measured, and the data from Narrabri made it possible to establish the first wholly empirical scale of effective temperature for stars in the spectral range O–F. The work of this instrument represents a classical contribution to stellar astronomy.'[5]

Following the appointment of Hanbury Brown to the new chair of astronomy, together with Richard Twiss as senior lecturer, Messel decided to strengthen the new group in the School of Physics by recruiting another astronomer to establish an optical observatory. His choice was Colin Gum who had impressive credentials in both optical and radio astronomy. Trained at Mt Stromlo and, in 1955, awarded one of the first doctorates at the new Australian National University, Gum had then spent several years on the staff at Radiophysics. The famous Gum Nebula was one of his discoveries, a spectacular complex of ionised hydrogen surrounding the Vela constellation. After accepting Messel's offer, Gum set off to Europe to order a 40 inch telescope for the new observatory. Tragically, these plans came to an end when Gum was killed in a skiing accident in Switzerland. This forced Messel to decide on a new plan of action. Aware of the mounting pressures at Radiophysics, Messel decided to abandon plans for the optical observatory. Instead, he would establish a new radio astronomy centre.

It would have been wishful thinking to expect harmonious relations between Radiophysics and the new centre. With Radiophysics now facing its first real rival in Australia, a power play involving several strong personalities would inevitably lead to friction. The controversy that erupted centred on the attempts by Bernard Mills to obtain funds for the SuperCross, shortly after joining Messel's department. Even though a substantial sum of £80,000 previously earmarked for the optical observatory had been made available through the Science Foundation for Physics, Mills estimated that the SuperCross would cost four times that amount. Meanwhile Messel had learnt that the National Science Foundation in Washington was, for the first time, ready to consider making grants to scientists outside the United States. With great speed Mills submitted a detailed funding application. Relations between Bowen and Messel, poor in the past, deteriorated sharply over the issue. The matter came to a head during a visit to Sydney by a senior NSF official to assess the merits of Mills' application. Bowen took the opportunity to criticise the scientific value of the SuperCross, which swiftly ended any hope of a working relationship between Radiophysics and the new centre. The feud then went public with an article, misleadingly entitled 'Mills Cross versus Parkes Dish', in the influential magazine *The Bulletin*:

> 'The split between the University group on the one hand, and the CSIRO group on the other, is so marked that there have indeed been efforts to influence the US National Science Foundation against financing the Mills Cross project and rather devote funds to the CSIRO group. Scientifically this split is most unfortunate as there are many aspects in which the two sides

Robert Hanbury Brown and the Narrabri intensity interferometer, used for measuring the size of stars. The first stellar diameter measured was for the blue-white subgiant known as Beta Crucis, the second brightest star in the Southern Cross.

> should complement each other, and it is to be hoped that eventually a measure of cooperation and collaboration will develop between them.' [6]

Later in 1962 the NSF announced that it would back the SuperCross with a sum of $US746,000 (£336,000) spread over five years. This grant is still the largest made to any project outside the United States by the NSF, a body which focuses almost exclusively on funding research in American universities and laboratories. For Mills: 'It was celebrated as a great victory!' [7] For more on the SuperCross, see Tangent 7.

While the feud over the funding of the SuperCross broke out largely because of personal rivalries among some of the key figures involved, it also highlighted a deeper problem which afflicted Australian science at this time. The fact that the proponents of the cross and the heliograph had turned to the United States for funds—as Bowen had done earlier for the Parkes telescope—reflected the lack of funding avenues open in Australia. While large projects of this nature were rare within CSIRO,

TANGENT 7

SuperCross, SuperCircle

While the Parkes telescope paved the way for big science in Australia, two other giant instruments followed shortly after. Next came the SuperCross inaugurated in November 1965, and then the Radioheliograph completed in September 1967.

The 96 aerials comprising the Radioheliograph were spaced at equal intervals through the thin scrub at the Culgoora site, tracing out a circle almost 10 km in circumference. The aerials were identical parabolic reflectors 13 m in diameter, giving a total collecting area for the instrument four times larger than the single dish at Parkes. In comparison, the SuperCross at Molonglo is made up of two arms, each 1.6 km long and 12 m wide, providing a total collecting area more than ten times larger than Parkes.

More important than sheer size was the type of astronomy suited to these specialised instruments. The Radioheliograph observed the sun at a single frequency, with the signals from each of the 96 aerials fed to a central receiving station for analysis. A complete radio map could be recorded once every second and then, frame by frame, a moving picture formed to show the evolution of violent disturbances occurring across the sun's disc. These radio 'movies' led to a detailed understanding of the rapidly varying bursts of radiation first identified and classified in the 1950s with the radio spectrographs developed by Paul Wild's group at Penrith and Dapto. In addition to studies of these violent processes, the sun provides a natural laboratory for observing the behaviour of plasmas in conditions far too extreme to reproduce on earth. These studies may eventually provide the key to the peaceful harnessing of fusion power.

The Radioheliograph enabled the solar group at Radiophysics to maintain its position at the forefront of solar research for well over a decade. In 1984 the instrument was closed down to make way for the construction of the Australia Telescope at Culgoora, bringing to an end almost forty years of leadership by Australia in solar radio astronomy. The site has since been named the Paul Wild Observatory.

After the controversy surrounding its beginning, the construction of the SuperCross proceeded smoothly at a site near the Molonglo River south–east of Canberra. Ironically, this was the same site selected earlier when plans had been made to locate the Parkes dish near Canberra.

Similar to the earlier version at Fleurs, the Molonglo Cross (as it became known) operates as a transit instrument recording a thin strip, 40 arcminutes wide, across the sky as the earth rotates. By tilting the east–west arm and electronically phasing the other, the cross could observe successive strips and build up a radio map of the southern sky.

The collecting area and angular resolution of the cross made it a survey instrument of unrivalled power. During its first ten years, the cross was used to compile a comprehensive catalogue of 12,000 strong sources, along with several other major catalogues of weaker sources. The cross also proved ideal for detecting pulsars. During the 1970s over one-half of these strange new objects known were discovered at Molonglo (see Chapter 12).

Some of this research was carried out in association with the Parkes telescope, although it took a number of years before the feuding ended and regular collaboration began between the Radiophysics and Sydney University groups.[8] In 1978 the north–south arm of the cross was closed down and the east–west arm underwent a major refit to form a new instrument known as the MOST—the Molonglo Observatory Synthesis Telescope.

Paul Wild in 1969, shortly after the completion of the Radioheliograph at Culgoora.

they were virtually unheard of in Australian universities. Messel recognised this when he reported in *Nature* that the SuperCross 'is without parallel in the University of Sydney, or indeed any Australian university, and may help substantially in setting a pattern for the more effective introduction of advanced research into the universities of Australia.' [9]

Messel's own solution to the problem had been to go out and raise funds off his own bat, but the overwhelming majority of scientists in Australian universities had no access to funds on this scale. The core of the problem was that the universities, except for the Australian National University, were funded by state governments that traditionally saw the primary role of universities to be educational institutions. In general the states recognised the importance of scientific research, but saw its funding to be a federal responsibility. However, the government in Canberra did not itself have the power to directly fund scientific research in state institutions such as universities—all of which added up to a catch-22 situation for university scientists. The small amount of federal money that did filter through to these scientists came in most cases via the federally funded CSIRO. Each year CSIRO doled out a modest proportion of its budget to universities in the form of fellowships, travel grants and funds for specific items of equipment. Thus, when Messel approached CSIRO and asked the organisation to chip in £10,000 towards the SuperCross he was quite within his rights to do so. (Given the controversial circumstances Messel was of course turned down, but he can at least be admired for trying.) While most research projects stood a reasonable chance of receiving CSIRO support, nevertheless there was mounting anger among university scientists at what they saw to be a virtual monopoly by CSIRO in shaping the priorities and overall direction of scientific research in Australia. Pressure by the universities led to small amounts of direct federal funds being made available in 1957. However, conditions for university research did not markedly improve until the formation of the Australian Research Grants Committee in 1965, under the direction of the distinguished biologist Rutherford Robertson. Beginning with just $4 million annually, the new body did much to promote the rapid growth of university research from the mid 1960s onwards.

Aside from giving state universities their first taste of big science, the SuperCross is notable for being the project that finally broke the monopoly held by Radiophysics on radio astronomy in Australia. Some radio astronomy and some closely related research areas had been initiated in Australian universities, but these programs hardly measured up as genuine competitors to Radiophysics. As one example, in 1949 a small group at the University of Adelaide under Leonard Huxley began a study of meteor showers using radar, but this research had more in common with similar work at Jodrell Bank than with any of the Radiophysics programs. In Hobart another group led by Bill Ellis carried out observations of radio emission from the galaxy at very low frequencies (20 MHz), a program boosted by the arrival in 1955 of Grote Reber, the American pioneer of radio astronomy. Reber had been attracted by an unusual hole or 'window' in the ionosphere in this part of the world which makes Tasmania a particularly good location for radio astronomy at very low frequencies. Characteristically, using his own resources, Reber constructed a radio telescope at Dennistoun, north of Hobart, a vast array of wooden poles and wire. In terms of its physical dimensions, it was the world's largest telescope. Working alone, Reber carried out a survey of radio emission from the Milky Way at 2 MHz, a frequency much lower than anything attempted at Radiophysics.

However, whether it was Hobart, Adelaide or elsewhere, the resources available to these other groups remained small, the research programs either separate or complementary, and in no sense could these centres be looked on as rivals. As related before, the only real competition had come from the Australian National University in the early 1950s with its plan to start a major centre at Mt Stromlo. Nothing came of the plan and the Radiophysics dominance of radio astronomy had been left intact. To this extent, Messel succeeded where Mark Oliphant had earlier failed.

Some of the heat generated by the SuperCross arose because of the way it challenged several comfortable assumptions CSIRO held about itself. In addition to Mills, a number of other Radiophysics staff joined the Sydney University group, a significant brain-drain from the organisation. CSIRO was not accustomed to having its best researchers poached from outside. Until then the flow had almost exclusively been the other way, with Australia's premier research organisation regularly recruiting the most talented young scientists trained by its poor cousin institutions,

Bernard Mills (left) and Grote Reber at the Fleurs field station in 1955, shortly after Reber's arrival in Australia.

the universities. Radio astronomy had been nurtured within CSIRO and had grown to be its brilliant success story of the postwar years. Now radio astronomy was outgrowing CSIRO. Perhaps it could not stop the process, but neither did CSIRO feel obliged to go out of its way to help. Fred White who succeeded Clunies Ross as CSIRO chairman in 1959 put it this way: 'Our attitude was good luck to anyone else who wanted to get into the game, but they could not expect funds or the help of our own staff into the bargain. Nor could anyone expect to take it away from us.' [10]

☆ ☆ ☆

A significant part of the radio astronomy program at Radiophysics in the 1950s had used instruments based on variations of the same principle. As explained earlier, the Mills Cross had been used for surveys of galactic and extragalactic sources, the Shain Cross for a survey of the galaxy at low frequency and the Chris Cross for studies of the sun. In 1960, in a period of just a few months, each of these three programs came to an abrupt halt. In February Alex Shain died, aged only 38, and the radio astronomy group lost one of its foundation members. Three months later, both Mills and Chris Christiansen joined the University of Sydney. In contrast to Mills, however, who left because of much better research prospects, Christiansen was forced to leave because of his fundamental objection to the direction radio astronomy had taken at Radiophysics.

Paradoxically, Christiansen had contributed solidly to the planning of the Parkes telescope, serving on various committees and carrying out searches for a suitable site. Whereas Major Thomas Mitchell had been the first European to spot the Goobang Valley near Parkes, Christiansen had actually been the first radio astronomer to do so. And yet, throughout the project he was the telescope's most vocal critic. Christiansen had already had outstanding success in developing new radio telescopes, beginning with the interferometers built at Potts Hill in the early 1950s. The grating interferometer had been one of the first arrays to use the technique of earth-rotation synthesis, a technique which later became the central principle in most high-resolution telescopes built around the world, including the Australia Telescope. Given the brilliant record by Radiophysics in devising and building new types of telescopes, Christiansen believed the new Parkes dish to be too conventional, too mediocre, indeed, too old fashioned. The 'last of the Windjammers' he liked to call it.[11]

Christiansen's outspokenness put him on a collision course with his chief:

> 'My leaving CSIRO to the chair of electrical engineering at Sydney was not of my own choosing. The circumstances are as follows. Pawsey, Mills and I were extremely dissatisfied with the way things were developing at Radiophysics. Pawsey was negotiating with US scientists and Mills was doing

> the same with Messel. Pawsey told me that Bart Bok was thinking of setting up a radio astronomy group at the ANU in Canberra and suggested that I should write to him telling him what I felt was happening at Radiophysics. I did this and included some comments on Bowen's 'interference' in our radio astronomical policy and also my opinion of Bowen as a scientist (very uncomplimentary). One morning about a fortnight later I was called down to see Bowen who was angrily flourishing the letter that I had sent to Bok. It had been posted to him by someone at the ANU to whom Bok had apparently passed on my letter. There was an angry scene and I had no option but to hand in my resignation.' [12]

Earlier in 1960 Christiansen had been invited to apply for the chair in electrical engineering at the University of Sydney, but had turned down the offer to be a candidate. He had no formal training in engineering, the position carried a heavy teaching and administrative load, and like physics in the pre-Messel era research in the department was virtually non-existent. But in view of the row with Bowen and also the fact that the department would now be closely associated with the new radio astronomy centre, Christiansen put in a late application for the position which was accepted immediately.[13]

Before commencing at Sydney, Christiansen decided to take up an offer from Jan Oort in Leiden to spend time in charge of a design team working on a new type of cross. The Australian success with the cross had led a number of groups in the northern hemisphere to emulate the idea. New versions were already under construction at Serpukhov near Moscow and at Bologna in Italy. Christiansen's new team planned one to top them all, consisting of 100 parabolic dishes—each half the size of the Parkes dish—and with a sensitivity and resolving power greater than any instrument built or on the drawing board. To launch this ambitious project, Oort had negotiated the formation of a consortium of scientific organisations from the Benelux group of countries, Belgium, the Netherlands and Luxembourg. The Benelux Cross reached an advanced stage of planning but, predictably enough, the sheer complexity and cost of the project stirred up too many political problems. In 1963 the Belgian government withdrew its funding and Oort's consortium fell apart. The Leiden group decided to continue alone and came up with a far simpler plan to build twelve 25 m dishes set out along a straight east–west line. The new array eventually came into operation at Westerbork in the Netherlands in 1970.

Returning to Sydney in 1962, Christiansen faced the daunting task of starting his new university position with no telescope and no ready source of funds. He then learnt of a plan to demolish his abandoned Chris Cross at the Fleurs field station. Following some swift negotiations, CSIRO agreed to transfer ownership of the Fleurs site to the University of Sydney. Christiansen revamped the old cross by adding several medium-size parabolic dishes to the array which increased its resolving power by a factor of ten. The new instrument, known as the Fleurs

Synthesis Telescope, was used throughout the 1970s to study the structure of objects such as large radio galaxies, supernova remnants and emission nebulae. Although smaller in scale than the new generation of northern hemisphere instruments such as the Westerbork array, the Fleurs Synthesis Telescope had the highest resolving power of any southern radio telescope (20 arcseconds) until its operation was shut down in 1988. The last observing session at Fleurs took place on the same day as the inauguration of the Australia Telescope.

Changing of the Guard

Both Christiansen and Mills, aged in their forties at the time of their controversial departure from Radiophysics, were able to reestablish themselves and carve out successful new careers in radio astronomy. For Joe Pawsey, the future did not hold the same good fortune. Since the beginning of the radio astronomy group in 1945, Pawsey had provided leadership and cohesiveness to a loose federation of half-a-dozen small research teams; his wisdom, intellect, modesty and international stature as a scientist had won him the admiration and, in most cases, fierce loyalty of his staff. By 1960 circumstances had changed radically. With the formation of the two largely independent Parkes and Culgoora groups, Pawsey's position had suddenly become redundant.

The changing of the guard had been signalled by the appointment of John Bolton to take charge of the Parkes telescope, a post which later was given the title of Director of the Australian National Radio Observatory (ANRAO). Bolton had outstanding credentials for the position. After an illustrious start in the late 1940s with the discovery and identification of the first radio sources, Bolton had spent two years with the cloud physics group. Then, in 1955, he had been appointed joint Caltech professor and director of the new Owens Valley Observatory near Bishop in California. Assisted by Bruce Rule and Gordon Stanley, Bolton organised the construction of the twin-dish interferometer and soon established Owens Valley at the forefront of radio astronomy. His talents both as an astronomer and as a radio engineer with plenty of hands-on experience won him many admirers, one of them Lee DuBridge, later to become chief science adviser to the White House. In offering to use his influence to obtain further funds for the Parkes telescope, DuBridge told Bowen in 1959: 'As you know, we are only repaying a tiny portion of our debt to you, as it was you who gave us John Bolton in the first place, without whom we would not have had the finest radio observatory which will be in existence until [October] 1961.' [14]

One further point in Bolton's favour was that, from the outset, he had been a strong supporter of building the giant Australian dish. While at Caltech he had done much to interest American firms in the project and on several occasions had offered advice on particular design problems. He wrote at the time:

> 'There has been for many years a conflict of opinion on whether the correct way to do galactic and radio source work is with partially filled arrays [crosses] at metre wavelengths or steerable dishes at the centimetre and short decimetre wavelengths. Those who have left Radiophysics are of the former school and Taffy and I of the latter. In my mind there is absolutely no question that the steerable dish with the ability to choose the time of observation, repeat it immediately afterwards, and the knowledge that the beam is the same anywhere in the sky has enormous advantages.' [15]

Bolton's support for the project, combined with a special rapport he had developed with Bowen, made him the leading candidate for the director's position. In June 1960 Bolton informed Caltech that he would be returning to Australia and, by the time the position was advertised later in October, the outcome had become a foregone conclusion.

Bolton's responsibilities were to be wide ranging. Aside from pursuing his own research interests, he would oversee the allotment of observing time on the telescope, supervise a procession of postgraduate students from the ANU and elsewhere, and vet all research papers before they went out for publication—duties which in practice made him leader of the cosmic radio astronomy group. In terms of professional classifications within CSIRO, Bolton had been appointed on his return to Australia at the level of chief research scientist. Earlier, Paul Wild had also been elevated to the same classification after being offered a new chair in astronomy at Cornell University in New York State, at more than three times his CSIRO salary. Although CSIRO had not been able to match the American offer in monetary terms, the promise of promotion and support to build the Radioheliograph had been enough to persuade Wild to stay in Australia. In practice, Wild was now leader of the solar radio astronomy group. Pawsey too was classified as a chief research scientist. Although he was not outranked by his two former charges, his position had basically become redundant.[16]

Pawsey was not to be stranded. In December 1961 he accepted an offer to succeed Otto Struve as director of the National Radio Astronomy Observatory, recognised as the top-ranking post in radio astronomy in the United States. Since its foundation four years earlier, the NRAO had worked towards its goal of establishing a national facility with state of the art equipment, shared by radio astronomy groups across the country as well as its own permanent research staff. An 85 ft telescope had been completed in October 1958 at the Green Bank site in West Virginia and a 300 ft dish, steerable in elevation only, was scheduled for late 1962. As mentioned earlier, the main problem at the NRAO had been a series of setbacks in the construction of the 140 ft telescope where the shaft and yoke of the polar-mounted dish had been plagued by stress fractures. Aside from these engineering difficulties, there had been one other problem at NRAO as Isidor Rabi confided late in 1961: 'Pawsey is saving my life by coming as director of NRAO next year. The present incumbent although a great optical astronomer has no administrative talent and no knowledge of radio astronomy. We hope now we are off to a better start.' [17]

In March 1962 Pawsey arrived at Green Bank for an initial six-week visit, but a few days later suffered paralysis of the left arm and leg. Bernard Lovell wrote later:

> 'The full extent of his illness was not immediately recognised and after partial recovery following some weeks in hospital in Washington, he proceeded to Princeton to convalesce. Unfortunately, the recovery was short-lived, and within a few days he had to return to hospital in Boston where in the middle of May he was operated on for the removal of a malignant brain tumour. He recovered sufficiently to return to Sydney at the end of July and for the next two and half months he was able to visit the Laboratory several times a week and again take up the editing of the special radio astronomy issue of the *Proceedings of the Institution of Radio Engineers* (Australia), much of which was done at home. Early in October it became evident that his condition was deteriorating, but under great difficulties he continued with his work. He went into hospital for the last time on 26 October 1962.' [18]

One of Pawsey's last visitors was Fred Hoyle, who presented him with the Hughes Medal on behalf of the Royal Society.

Pawsey's death marked the end of perhaps the most exciting era in Australian radio astronomy. It can also be said to have marked the end of the period of upheaval begun several years earlier, triggered by the decision to wind down support for the research programs of the 1950s and to direct the resources into the two giant instruments at Parkes and Culgoora. Several factors had come into play which intensified this turbulent period of transition. Paradoxically, one factor was the very success of the group. Although the size of the group had undergone a steady growth since the early postwar years, its key personnel had changed remarkably little. With the group's growing international prestige and, until 1960, its comfortable level of funding, there had been no incentive for its members to leave. Until the end of the 1950s the success of the group had given it a certain stability and cohesiveness. One measure of this success is that eight members of the group (Bolton, Christiansen, Mills, Minnett, Pawsey, Piddington, Robinson, Wild) were elected Fellows of the Australian Academy of Science, recognised as the highest honour Australia can bestow on its scientists, in most cases for their research carried out prior to 1960. This success in many cases led to accelerated promotion through the ranks faster than most scientists in CSIRO and for some, at a relatively young age, to an eventual promotion barrier. The best prospects for advancement were then to be found beyond the doors of Radiophysics.[19]

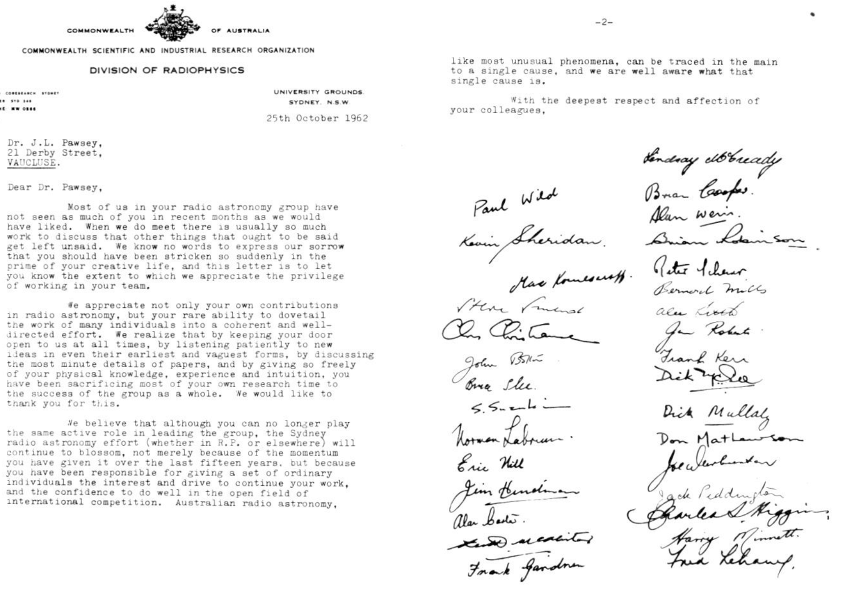

COMMONWEALTH OF AUSTRALIA
COMMONWEALTH SCIENTIFIC AND INDUSTRIAL RESEARCH ORGANIZATION
DIVISION OF RADIOPHYSICS

UNIVERSITY GROUNDS.
SYDNEY. N.S.W.

25th October 1962

Dr. J.L. Pawsey,
21 Derby Street,
VAUCLUSE.

Dear Dr. Pawsey,

Most of us in your radio astronomy group have not seen as much of you in recent months as we would have liked. When we do meet there is usually so much work to discuss that other things that ought to be said get left unsaid. We know no words to express our sorrow that you should have been stricken so suddenly in the prime of your creative life, and this letter is to let you know the extent to which we appreciate the privilege of working in your team.

We appreciate not only your own contributions in radio astronomy, but your rare ability to dovetail the work of many individuals into a coherent and well-directed effort. We realize that by keeping your door open to us at all times, by listening patiently to new ideas in even their earliest and vaguest forms, by discussing the most minute details of papers, and by giving so freely of your physical knowledge, experience and intuition, you have been sacrificing most of your own research time to the success of the group as a whole. We would like to thank you for this.

We believe that although you can no longer play the same active role in leading the group, the Sydney radio astronomy effort (whether in R.P. or elsewhere) will continue to blossom, not merely because of the momentum you have given it over the last fifteen years, but because you have been responsible for giving a set of ordinary individuals the interest and drive to continue your work, and the confidence to do well in the open field of international competition. Australian radio astronomy,

-2-

like most unusual phenomena, can be traced in the main to a single cause, and we are well aware what that single cause is.

With the deepest respect and affection of your colleagues,

A final letter to Joe Pawsey, from the Australian group he had founded.

Early in the 1960s the great exodus began: Jack Piddington transferred to the CSIRO Division of Physics and joined its solar physics program; Richard Mullaly and Arthur Watkinson followed Christiansen to electrical engineering at Sydney University; Alec Little joined Mills and rose to the position of associate professor in the Chatterton Astronomy Department; Jim Hindman transferred to the ANU Siding Spring Observatory; Don Mathewson moved to and later became head of the ANU Astronomy Department; and Kevin Westfold went on to become professor of mathematics and later deputy vice-chancellor of Monash University in Melbourne. This migration by the Radiophysics group into mainly academic posts eventually proved beneficial to the long-term development of astronomy in Australia. From the comparatively insular environment of the Radiophysics Lab these radio astronomers were placed in positions where they could train and inspire a new generation of astronomers. As Pawsey wrote in his introduction to the special radio astronomy issue, referred to by Lovell above:

> 'In recent years, research in radio astronomy in Australia has overflowed into the universities and there are now active radio astronomy programs at the University of Tasmania and at Sydney (Mills, Christiansen). This

> development into the universities has the major advantage from the Australia-wide point of view of making direct contact between some of the most successful research scientists in the country and students, so that successful research efforts should have a much more profound influence on Australian science in the future.' [20]

This overflow of Radiophysics staff extended beyond Australian universities and into the international arena as well. The rapid growth of radio astronomy elsewhere in the 1950s meant that, when it came to choosing a career path, radio astronomers had the world at their feet. The well-established groups at Jodrell Bank, Cambridge and Leiden continued to expand, new centres sprang up elsewhere in Europe, and particularly in the United States radio astronomy experienced a runaway growth. In addition to the Caltech group at Owens Valley and the NRAO at Green Bank, a number of other university centres underwent rapid development in the late 1950s including Cornell, Harvard, Illinois, Michigan and Ohio State. Most of these groups were in advanced stages of planning a major new instrument for the 1960s. In fact the main problem in the United States was not a shortage of new instruments (or of course funds), but the lack of suitably qualified radio astronomers. In their book *Astronomy Transformed*, David Edge and Michael Mulkay estimate that, around 1960, for every American-born radio astronomer there was at least one other from either Radiophysics or Jodrell Bank active in the United States.[21]

In view of the fact that the Radiophysics group was by far the world's largest at this time, Australian radio astronomers naturally enough were in great demand in the United States. As one illustration, following Pawsey's death a concerted effort had been made to persuade Paul Wild to take up the directorship of NRAO. Some such as Ron Bracewell, Gordon Stanley and Frank Kerr took up permanent academic posts, but most went for extended visits on fellowships and helped to accelerate the growth of radio astronomy in the United States: Brian Cooper (1959 Harvard), Alec Little (1963 Stanford), Dick McGee (1963 Michigan), Jim Roberts (1959–61 Owens Valley), Kevin Westfold (1959 Owens Valley) and Don Yabsley (1960–2 Cornell).

It would be wrong to label this migration of expertise to the United States as another alarming case of the 'brain-drain' which has adversely affected many other branches of Australian science. The exodus over decades of many of Australia's best young physicists to overseas centres such as the Cavendish Laboratory is one good example. On the contrary, here was the first and probably only occasion when a major new science pioneered in Australia played a leading part in promoting the growth of its counterpart in the United States. Rather than a brain-drain, Australian radio astronomy provided a fountainhead.

Was Taffy Bowen Right?

Part 2 of this book has described in some detail the story behind the funding, the design and the construction of the Parkes telescope while, in the present chapter, we have looked at the repercussions the new telescope had for the Radiophysics group as a whole. In view of the substantial resources committed to Parkes and the major transformation it triggered in Australian astronomy, it is time to pose the question: Should the Parkes telescope have been built in the first place?

In Part 3 we turn our attention to the research carried out at Parkes over the years and how this fitted in with the development of astronomy elsewhere—perhaps this broad overview will provide the answer to our question. Some years ago I put the question to a number of astronomers who had been closely involved in the events described in this chapter. The responses showed forcefully that the question is still one which can produce a strong reaction. To close this chapter I quote the views of just three—first, Bernard Mills:

> 'There was no doubt expressed in the early 1950s that Australia needed a large general purpose reflector capable of working at what were then considered to be high frequencies. Initially there were some differences at the scientific level between the Bowen and Pawsey camps as to how this might best be achieved. However, it was soon apparent that for any progress at all, it was necessary to go along with the Bowen vision and these differences were put aside. The degree of enthusiasm was obviously dependent on how this would further one's own scientific interests. Kerr and the hydrogen-line people were extremely enthusiastic. I personally supported the concept quite strongly but, in part, for the possibility of programs which were never implemented ... . The real troubles arising at this time were political rather than scientific and related to the progressive undermining of Pawsey as leader of the radio astronomy group, which was much resented by his supporters.' [22]

Next, Chris Christiansen, also responded:

> 'Joe Pawsey decided to leave CSIRO and go to the USA and Bernard Mills and I also left *because the Parkes telescope was to be built.* We three felt that this was a complete break with the past activities of the group, in which we had obtained a considerable reputation overseas by our inventions in radio telescope techniques and designs that put us in the lead in high resolution radio astronomy. (High resolution was the main problem in radio astronomy at that time.) We felt that with all the money going into an instrument which involved nothing new except in mechanical features (provided in the main by overseas engineers) that the era of great innovations in the Radiophysics Lab would end. It certainly did. All work on high resolution moved to the University of Sydney and CSIRO made no significant contribution to the technique until much later when plans emerged for the Australia Telescope.' [23]

The third response, short and sweet, comes from Taffy Bowen:

> 'Over twenty years of successful operation and over one thousand research papers from the Parkes telescope says yes.' [24]

We return briefly to this question at the end of Chapter 12.

Notes to Part Two

CHAPTER 5

Epigraphs: Plato's Republic, see A. L. Mackay, *Harvest of a Quiet Eye*, 120.5 (Inst. of Physics, Bristol, 1977); Lovell, *Scient. Am.* **188**(1) (1953), p. 21.

1 See R. R. Burman and P. M. Jeffery, 'Solar radio astronomy at the University of Western Australia 1946–48' *Proc. Astron. Soc. Aust.* (in press).
2 A. R. Hyland and D. J. Faulkner, 'From the sun to the universe—the Woolley and Bok directorships at Mt Stromlo' *Proc. Astron Soc. Aust.* **8** (1989) 216–28; see also S. C. B. Gascoigne, 'Astrophysics at Mt Stromlo: The Woolley era' *Proc. Astron. Soc. Aust.* **5** (1984) 597–605.
3 CSA file KE12/11, Woolley to White 12/2/51.
4 Note 3, White to Woolley 23/2/51.
5 R. v. d. R. Woolley, Presidential address to Section A of the Australia New Zealand Association for the Advancement of Science (ANZAAS), Perth, 1947. See also J. P. Wild, 'The beginnings of radio astronomy in Australia' (Inaugural Harley Wood Memorial Lecture) *Proc. Astron. Soc. Aust.* **7** (1987) 95–102. Ironically, one of Woolley's papers appears to have been the first Australian contribution to *cosmic* radio astronomy. It was written before, though not actually published until after, the Bolton–Stanley paper on Cygnus A [see Woolley, 'Galactic noise' *Mon. Not. R. Astron. Soc.* **109** (1948) 123].
6 Note 3, Martyn to White 26/2/51.
7 Note 3, Oliphant to W. A. McLaren (Dept of the Interior) 22/3/51.
8 Note 3, White to McLaren 10/4/51.
9 Note 3, Bowen to White 17/4/51.
10 RPA file A1/3/11/3, Part I, 2 letters Oliphant to Bowen 12/6/52 and 15/8/52.
11 B. Lovell, *The Story of Jodrell Bank.*
12 See Chapter 4, note 23. See also Sullivan in *Australian Science in the Making* (Ed. R.W. Home), pp. 336–8.
13 Interview with author, February 1986. Woody Sullivan makes the interesting observation: 'If Radiophysics had had access to US-level funds in the late 1940s, one can speculate that its high-energy accelerator work might have thrived to the point of dominating the entire laboratory's program. In such a case it would seem unlikely that Radiophysics staff would have achieved such a distinguished international stature as they in fact did with radio astronomy' (see Sullivan in *Australian Science in the Making*, p. 344). Rutherford would have undoubtedly applauded the observation.
14 RPA file Z1/14A, DuBridge to Bowen 21/2/52.
15 Note 14, Bowen to DuBridge 22/5/52. See also RPA file A1/3/11/1 Part I.
16 Bowen interview with author, August 1983. Quite apart from the Caltech offer, Bowen had also been approached to head a proposed research centre for cloud and rain physics in the south-western United States (see CSA file PH/BOW/9, Bowen to White 7/11/52).
17 RPA file Z1/14A, Bowen to DuBridge 11/7/52. See also RPA file A1/3/11/3, Bowen to Tizard 15/7/52; Tizard to Bowen 11/8/52 and 20/8/52. Fred White also wrote to Edward Appleton at this time (27/6/52) asking him to use his influence to lobby Australian government officials for a giant radio telescope during Appleton's visit to the URSI Congress in August 1952.
18 RPA file A1/3/11/3, Bowen to Bush 22/8/52.
19 CSA file KE2/14, Dollard to Bowen 21/5/54.
20 For both statements of support see RPA file A1/3/11/3 Part II.
21 CSA file TZ797/31/3, Clunies Ross to Casey 4/4/55.

22 Note 21, Menzies to Casey 20/4/55. Menzies position had not been unexpected. In May 1954 Bowen had discussed the funding issue with H. C. 'Nugget' Coombs, governor of the Commonwealth Bank, who indicated that the government would probably only go as far as matching all the funds raised privately on a £ for £ basis. See e.g. CSA file KE2/14, White to Roland Wilson (Treasury Secretary) 2/6/54.

23 Note 18, White to Bowen 6/4/54.

24 The four trustees were Walter E. Bassett, a Melbourne based consulting engineer; Sir Edward Lefroy, a West Australian pastoralist and member of the CSIRO Council; Donald Mackinnon, another pastoralist and a former president of the Victorian Liberal Party; and Andrew Tyne Reid, a chartered engineer in New South Wales. Mark Oliphant initially accepted the invitation to become a trustee and then withdrew. With Richard Casey in the chair, the Trust held its first meeting in February 1956 and then met each year, usually in Casey's ministerial office at Treasury Place in Melbourne, until it wound up its affairs in 1963. Another purpose in establishing the trust was to maintain control over the funds raised. At the time any donations or earnings intended for CSIRO were absorbed into Treasury consolidated revenue and could for one reason or another be diverted from their original target.

25 See P. Collier and D. Horowitz, *The Fords*, p. 258 (Collins, London, 1988). Shortly afterwards the Ford Foundation altered its policy. In the late 1950s it donated $1 million to help found the European Southern Observatory (a consortium of European countries which constructed several optical telescopes at La Silla in Chile) and then in the 1960s it funded the next large radio telescope built by Radiophysics, the Radioheliograph at Culgoora (see Chapter 8).

26 See e.g. P. Robertson, *The Early Years: The Niels Bohr Institute 1921–30*, p. 91 (Akademisk Forlag, Copenhagen, 1979); F. Aaserud, *Redirecting Science: Niels Bohr, Philanthropy, and the Rise of Nuclear Physics* (Cambridge Univ. Press, 1990).

27 See note 18, Part III. The Rockefeller grant was made at the expense of a competing proposal to fund the Caltech interferometer at Owens Valley. Instead, the instrument was built with funds from the US Office of Naval Research.

28 Note 19, Bush to Bowen 17/5/54.

29 CSA file KE38, Report to Radio Astronomy Trust, 31/1/60.

30 F. W. G. White, 'Biographical Memoir: Casey of Berwick and Westminster, Baron' *Rec. Aust. Acad. Sci.* **3** (1977) 54–83. The Phytotron Trust had originally planned to raise at least £150,000 from private sources.

31 S. Cockburn and D. Ellyard, *Oliphant*, p. 231.

32 J. B. T. McCaughan in *The Messel Era* (Ed. D. D. Millar), Chapters 2–4.

33 *Christian Science Monitor* 19/12/61. Along the same lines, a favourite among radio astronomers in the 1950s concerns the story about the British scientist Michael Faraday. During a public lecture explaining the fundamental discovery he had just made of the connection between electricity and magnetism a woman in the audience interrupted him with the question: 'But Mr Faraday, what is the use of this knowledge?' Faraday responded with the counter-question: 'Madam, of what use is a new born baby?'

34 The National Standards Laboratory in fact consisted of three divisions—Metrology, Physics and Electrotechnology. The other divisions in the secondary industry group were Building Research, Mathematical Statistics, Meteorological Physics and Tribophysics.

35 See e.g. J. F. H. Wright, *Measurement in Australia 1938–88: A History of Australia's National Standards Laboratory*, p. 84 (CSIRO, Sydney, 1988).

36 CSA file KE8/3, Report on Future Activities of Radio Astronomy Group. See also White to Bowen 23/8/56.

37 Note 36, Bowen to White 20/8/56.

38 The meeting with the CSIRO Executive took place in Melbourne with Bowen, Pawsey, Christiansen and Arthur J. Higgs in attendance. (Higgs joined Radiophysics

in 1940 as a research officer and from 1945 had been technical secretary, effectively Bowen's administrative right hand.) After this meeting in 1956 the mix between radio astronomy and cloud physics at Radiophysics remained essentially unaltered until the cloud physics group was set up as a separate CSIRO division in 1971.

CHAPTER 6

Epigraph: Hanbury Brown, see D. D. Millar (Ed.), The Messel Era, p. 143.

1 J. G. Bolton, 'Radio astronomy at Dover Heights' *Proc. Astron. Soc. Aust.* **4** (1982) 349–58.
2 See RPA file A1/3/11/2. Bowen developed the proposal for the rolling parabolic cylinder late in 1952 after he had already sent his proposal for a giant parabolic dish to Caltech (see Chapter 5). Bowen did not consider the rolling cylinder a substitute for a large dish, either in the United States or Australia, but rather as a very large instrument which would serve in the interim and be affordable with the Radiophysics funds available: Interview with author, August 1983.
3 RPA file A1/3/11/1, J. L. Greenstein to Bowen 27/5/54.
4 L. H. Mitchell, 'Preliminary design of a radio telescope', Aeronautical Research Labs, Structures and Materials Note 209. 29 pp. (Dept of Supply, Melbourne, 1953).
5 The Radio Telescope Planning Committee, consisting initially of Bowen, Pawsey, Kerr, Mills, McCready and Bolton, met 14 times between May 1954 and November 1955 until its role was overtaken by a Technical Advisory Committee (see note 26). See RPA files A1/3/11/2 and A1/3/11/6.
6 A. K. Head, 'A new form for a giant radio telescope' *Nature* **179** (1957) 692–3. Although Head's telescope was never built it formed the basis of the giant 1000 ft diameter fixed reflector constructed by Cornell University in a natural hollow near Arecibo, Puerto Rico in the 1960s. Actually, it is not clear whether a Head telescope as large as the Parkes dish would have turned out that much cheaper to build. An ARL report by L. H. Mitchell and F. S. Shaw in February 1956 costed a 250 ft Head telescope at £463,000, only slightly less than the projected cost of the Parkes instrument. The report took no account of gradual earth movements which can amount to several centimetres over a period of years, even in relatively stable geological formations, movements which would distort the shape of the overlying reflector surface. The Arecibo telescope overcame the problem by allowing for periodic adjustment of the mesh surface which returns it to a true hemispherical shape, a design feature that added substantially to the cost. The Arecibo telescope has a diameter almost five times that of Parkes, and it cost about six times as much to build (see the table in Chapter 11).
7 RPA file A1/3/11/2, Part B, 8/3/55.
8 RPA file Z1/14A, Bowen to Bush 23/10/52.
9 RPA file A1/3/10, Pawsey to Lovell 13/6/51.
10 RPA file A1/3/11/3, Lovell to Bowen 23/6/54.
11 AUI was formed immediately after the war mainly through the initiative of the physicist Isidor Rabi. The nine eastern universities were Columbia, Cornell, Harvard, Johns Hopkins, MIT, Princeton, Pennsylvania, Rochester and Yale.
12 Note 10, Tuve to Bowen 3/3/55. Although Tuve had played a major part in the formation of the NRAO, he had grave doubts about the wisdom of promoting large-scale government support of basic scientific research. It was others such as Berkner who pushed for the construction of large dishes: see A. A. Needell, 'Lloyd Berkner, Merle Tuve and the federal role in radio astronomy' *Osiris* **3** (1987) 261–88.
13 Note 10, Bowen to Clunies Ross 10/3/55. Clunies Ross replied a week later agreeing that Bowen had made the right decision in opting for a smaller dish.
14 Note 10, Bowen to Tuve 10/3/55.

15 Lovell, *The Story of Jodrell Bank*, p. 47.
16 Note 3, Bowen to Pawsey 10/6/55.
17 CSA file KE22, Bowen to Pawsey 4/7/55.
18 Note 10, Lovell to Bowen 23/6/54. Lovell maintains that the project would not have necessarily benefited by having a main contractor and that this might have added to the total cost. The construction of other large radio telescopes at Jodrell Bank in later years proceeded with a similar management structure and without serious design problems or delays: Lovell to author 8/3/84.
19 Note 15, p. 158.
20 D. Saward, *Bernard Lovell: A Biography*, p. 201.
21 Note 3, Lovell to Bowen 10/8/54.
22 See J. E. Morpurgo, *Barnes Wallis—A Biography*, Chapter 13; see also A. Pugsby, 'Barnes Neville Wallis 1887–1979' *Biog. Mem. R. Soc. London* **27** (1981) 603–28. Wallis' development of deep penetration bombs was an achievement of comparable importance. Numerous aircraft and crews had been lost in futile attempts to destroy German submarine pens by conventional bombing. Deep penetration bombs exploded near or below foundation level, attacking the pens at their weakest place.
23 Note 3, Wallis to Bowen 14/4/55.
24 RPA file A1/3/11/32, proposal by Barnes Wallis September 1955.
25 Bowen in *The Early Years of Radio Astronomy*, p. 103 (Ed. W. T. Sullivan). Similar to many 'Eureka' stories in science Bowen's recollection is probably apocryphal. Wallis had already given the problem some thought in connection with the Isaac Newton Telescope. The 25 metre dish then under construction at Dwingeloo in Holland used a type of master equatorial unit but located in a hut next to the telescope. Pawsey had visited Dwingeloo immediately before his initial discussions with Wallis in September 1954. Nevertheless, Wallis' suggestion to place the master equatorial unit at the intersection of the telescope's two axes of rotation was a stroke of genius. (For more on how the master equatorial operates see Chapter 7.)
26 The Technical Advisory Committee consisted of White, Pawsey and Higgs (Radiophysics), H. A. Wills (Aeronautical Research Labs) and three academics from the University of Sydney—Professor Jack Roderick (Civil Engineering), Professor William H. Wittrick (Aeronautical Engineering) and Dr Hugo K. Messerle (Electrical Engineering). The committee met eight times between July 1955 and June 1959 (see RPA file A1/3/11/4).
27 The two other firms approached for a design study were Sir William Halcrow and Partners and Head, Wrightson and Company, both based in London.
28 CSA file KE10/3/2, 'Specification of a giant radio telescope for which a design study is required'. 15 pp. 23/11/55. The second part of the document contained five appendices, written by Harry Minnett.
29 There is a third way to counteract the effect of gravity distortions. This is done by allowing the dish to deform when it tilts, but requiring that the new shape also be a parabola, with a different focus. This concept was not understood at the time of the Parkes design but, as it turned out, the Parkes dish like many others does deform in this way. In later years it was possible to achieve good performance well below the original design wavelength by moving the feed to keep it at the focal point, which changes as the dish tilts. The concept, known as the homology principle, was used with brilliant success by the German designers of the 100 m Effelsberg telescope (see Chapter 11).
30 See RPA file A1/3/11/1, Part IX. The Freeman Fox report was sent to a number of people for comment including Lloyd Berkner (AUI), John Bolton and Lee DuBridge (Caltech), Bruce Rule (Mt Wilson–Palomar) and Merle Tuve (Washington). A copy sent to the Mt Stromlo Observatory led the design engineer Kurt Gottlieb to suggest the addition of a small sighting telescope to assist in the calibration of the dish, a feature later incorporated into the design.
31 The Giant Radio Telescope User's Group, consisting of Kerr, Christiansen, Higgs,

McCready, Mills and Wild, held the first of its nine meetings in September 1958. One proposal made by the group was to use a digital display system for the celestial coordinates, a feature that later became standard in all radio telescopes. See RPA file A1/3/11/15.

32 RPA file A1/3/11/10, Minnett to Bowen 7/2/57.

33 Note 3, Roberts to Bowen 14/9/56.

34 CSA file TZ797/31/2, Bowen to White 7/11/58. Bowen's complaint led to pressure being applied at a higher level. In January 1959 Richard Casey informed Freeman Fox 'that we are not satisfied' with the progress made and set out in some detail the reasons why. In reply Freeman Fox regretted 'the note of recrimination' and that any implication 'that we have not been doing our best for you is completely unjustified' (see note 3, Casey to Freeman Fox 27/1/59 and, vice versa, 13/2/59).

35 Note 3, Bowen to White 7/3/59. Much of Bowen's anger was directed at Gilbert Roberts whom he held personally responsible for delaying the project by twelve months. Bowen later tried unsuccessfully to persuade Ralph Freeman to have Roberts taken off the project (see RPA file Z1/14A, Freeman to Bowen 20/11/59). This friction was generated in part by their different personalities. Minnett knew both: 'Taffy and Gilbert were poles apart in temperament and approach to problems. Taffy was dynamic, impatient of delay and tough when opposed or frustrated. He had a sense of urgency for the project driven by the quite understandable fear that some other telescope might come into operation first and skim off the major discoveries. This energetic drive greatly helped the progress of the project, but inevitably left some casualties. Gilbert, on the other hand, was a perfectionist and liked to 'chew over' problems. He was sensitive, idiosyncratic and often exasperating even to his closest associates. Stories of his eccentric deeds were legion and, after four years, I could contribute a considerable number myself! He was not a good communicator under pressure and at times did not do justice to the merits of his case and the quality of his thinking. I respected his great design abilities and extensive engineering experience' (Minnett to author 19/6/90 and 26/9/91).

CHAPTER 7

Epigraph: Lovell, *Out of the Zenith*, p. 228.

1 Lovell, *The Story of Jodrell Bank*, pp. 170–4.

2 CSA file TZ/797/2, Bowen to White 12/2/57. Others such as Gilbert Roberts had reservations about the Cliffvale area, in particular the possibility of flooding.

3 Note 2, Martyn to S. H. Bastow (CSIRO's new chief executive officer) 10/8/57 and 14/8/57. The Camden research plans were known at Radiophysics but thought not to be a problem. The frequencies for the radar work were understood to be well below the operating range of the new telescope (see RPA file A1/3/11/5A Higgs to G. Munro 15/8/57).

4 Note 2, Martyn to Bastow 10/8/57.

5 RPA file A1/3/11/3, White to Oliphant 30/6/54. Oliphant and Martyn were two of the principal figures involved in the foundation of the Australian Academy of Science in 1954. Over the following years membership of the Academy often became controversial because of the 'personal incompatibilities' referred to by White. Originally composed of 11 Australian scientists who were Fellows of the Royal Society of London, the Academy invited a further 13 scientists to join later in 1954, one of them being Joe Pawsey. According to Bowen, after he had received an invitation to be part of this second intake, Martyn had then contacted him to say that the invitation had been a mistake. Bowen later became a Fellow in 1957, after Martyn had completed his term in the influential position of Secretary, Physical

Sciences, and at the same time that the Cliffvale site was under discussion. Fred White was made a Fellow in 1960, after his elevation to chairman of CSIRO.

6 R. T. Tindall (Ed.), *Parkes: One Hundred Years of Local Government,* Chapter 1.

7 A. W. Martin in *Australian Dictionary of Biography*, Vol. 5 (Melbourne Univ. Press, 1974).

8 *Champion Post,* 14/8/58. The description 'acre of ear' is only a tiny exaggeration—the actual surface area of the 64 m dish is 0.9 acre (or 0.35 hectare).

9 Wild to author 21/6/90. The matter was finalised at a meeting in Melbourne on 17 March 1958 between five senior Radiophysics staff and the CSIRO Executive. Again Bowen decided not to attend. Although conceding the technical superiority of Parkes, he still favoured Cliffvale because of the 'costs, convenience and the speed with which we can get cracking'. Facing defeat on the issue, he explained to White: 'I seem to spend all my time pushing people into things they don't particularly want to do, or steering them up paths they are reluctant to tread—sometimes with a particular lack of success' (see CSA file TZ/797/2). Radio astronomy came later to the Canberra area, first with the SuperCross built near Hoskinstown (see Chapter 8) and then the 64 m NASA dish at Tidbinbilla (Chapter 10).

10 MAN initially quoted £A593,000 but then adjusted the figure to cover the extra costs of the trial assembly of the turret, hub and drive systems at Gustavsburg. MAN in fact submitted two final bids, the second being slightly more (4%) expensive and based on the assumption that the British firm of Grubb Parsons would continue as subcontractors developing the master equatorial unit, rather than its own choice Askania-Werke. Only one of the seven firms had any prior experience in telescope construction—the lowest of the American bidders, Loewy Hydropress, which was engaged in fabricating parts for the disastrous US Navy project to construct a 600 ft instrument (see Chapter 10). It is worth noting that no Australian firms were invited to bid on the project, reflecting the grim truth that no local engineering firm could compete with the best in the northern hemisphere. In sharp contrast, at the start of the Australia Telescope project in the early 1980s, Australian expertise had improved so much that it became possible to maximise the involvement of local companies. With the completion of the project in the bicentennial year 1988, in straight dollar terms 80% Australian content had been achieved.

11 Note 2, Bowen to White 20/4/59.

12 *New Scient.* **6** (1959) 142. Gilbert Roberts reacted with an angry letter to the editor, objecting to the article's title and pointing out that no mention had been made of Freeman Fox or of how much of the development work on the control and drive mechanisms had already been carried out *ab initio* by British firms (MetroVick and Grubb Parsons). Roberts' attack was justified but nevertheless only one British firm (John Brown) showed a competitive interest in becoming prime contractor for the job. Informing Henry Tizard of the MAN contract, Bowen added almost apologetically: 'We tried hard to have it built by a British firm, but obtained a very disappointing response … . The Germans and the Americans were wide awake, keen and anxious to do business, but this was conspicuously lacking among British firms' (RPA file Z1/14A, 20/7/59).

13 Rockefeller had already demonstrated its willingness to give further assistance to the project. Under its rules the first grant made in December 1955 would have been revoked if unspent after a mandatory three year period, but at the end of 1958 the Foundation agreed to an extension until 1962. Lee DuBridge at Caltech again lobbied Rockefeller on behalf of the project (see RPA file A1/3/11/3, Parts IV and V).

14 For the physical principle of the maser see Chapter 12. Townes liked to point out that when it came to government agencies funding his research the acronym maser had a second, hidden meaning—*m*eans for *a*cquiring *s*upport for *e*xpensive *r*esearch. In 1964 Townes shared the Nobel Prize in physics for his work on the maser/laser principle.

15 G. S. Bogle, 'Masers—new amplifiers for microwaves' *Aust. Scientist* **1** (1961) 128–32.

16 Cooper to author 28/3/84. Gold was one of Hoyle's collaborators on the steady-state theory and later achieved fame for his theoretical pulsar model (see Chapter 12). Bloembergen was another maser/laser pioneer to receive a Nobel Prize (1981) for his work.

17 Interview with author, May 1985. Robinson's reports from Leiden and Cooper's from Harvard were supplemented by two other Radiophysics staff—Alec Little studying for a master's degree on paramps at Stanford and Lou Davies at the Bell Labs in New Jersey. Initially it was hoped that the Leiden group would build two identical paramps and, in return for Robinson's input, supply one gratis to Radiophysics. However, Oort estimated the development costs for one paramp to be £20,000, far in excess of Robinson's salary. Pawsey balked at the prospect of finding this amount, knowing that funds raised for the Parkes telescope were earmarked for the structure itself and could not be diverted to receivers. The figure of £20,000 was approximately the same as the annual budget allocation for equipment at Radiophysics which had to stretch over the rest of the radio astronomy program, as well as equipment used in other research areas such as cloud physics. Pawsey compromised by deciding to carry out the labour-intensive task of assembling and testing the paramp in Sydney. This eventually reduced the amount owed to the Leiden group for the paramp to £6000. Pawsey had good reason to complain: 'The finance for receivers for the big dish is going to drive us to drink' (RPA file A12/1, Pawsey to Robinson 12/2/60).

18 The deaths of Bogle and Margaret Chandler early on New Year's Day 1962 began one of Australia's most sensational homicide cases. One theory speculated that Bogle's involvement in maser research might have been the reason behind the double homicide. Admittedly the maser, or rather the laser, was seen by some as potentially the most significant new weapon since the atomic bomb. The US military immediately began investigating the idea of a 'death ray' for shooting down Russian missiles (the idea found even stronger support later in the 1980s with the Star Wars project). Although the maser/laser was clearly to become of great military and industrial importance, there is no evidence to suggest that Bogle's research was in advance of similar work underway in the United States. The Bogle–Chandler case remains unsolved.

19 Day to author 9/4/84.

20 Interview with author, April 1984. For more on the construction and commissioning of the telescope see: 'Radio Telescope' *Current Affairs Bulletin* **28** (no. 10) (1961) 146–60; E. G. Bowen and H. C. Minnett, 'The Australian 210-ft radio telescope' *Proc. Inst. Radio Eng. Aust.* **24** (1963) 98–105; H. C. Minnett, 'The Australian 210-foot radio telescope' *Sky & Telescope* **24** (1962) 184–9; and M. H. Jeffery, 'Construction and operation of the 210-ft radio telescope at Parkes, Australia' *Ann. New York Acad. Sci.* **116** (1964) 62–88.

21 CSA file TZ797/31/2, Bowen to R. S. Morison 19/4/61.

22 RPA file A1/3/11/1, Westerhout to Bowen 28/11/61.

23 Interview with author, March 1984. Shimmins was recruited from the Weapons Research Establishment in South Australia after ten years experience on missile guidance systems. Problems with some of the electronics in the master equatorial (apart from the damage caused in transit) led to the decision to proceed without Askania's further involvement, with Shimmins placed in charge of the final commissioning of the electronics.

24 Note 20.

CHAPTER 8

Epigraph: Christiansen, *Proc. Astron. Soc. Aust.* **8** (1989), p. 100.

1 RPA file A1/1/7, draft memorandum 29/6/59.
2 In a note written for the meeting Pawsey stated: 'We cannot take on both [the Wild and Mills proposals] simultaneously with current resources (both men and finance). Both are outstanding proposals in my opinion, but Wild's retains a balance, solar and cosmic, in the laboratory' (see note 1). Bowen also considered the SuperCross to be a first-rate proposal but favoured the Radioheliograph because it would maintain a diversification in the research program as well as benefit a larger number of the group: 'It was a victory for the solar group—Wild, Steve Smerd, Kevin Sheridan and others—many with 8 to 10 years experience. They were already well in front of the field, unlike the stiff competition Mills faced from overseas. Mills and Christiansen, and also Shain with his own version of the cross, were essentially one-man bands, so the decision for a heliograph was one which would benefit more members of the group' (interview with author, August 1983). This same line of reasoning in Bowen's view also provided solid justification for the Parkes telescope—an instrument of versatility and flexibility that could accommodate the competing interests of a large number of users, not just from within the Radiophysics group but from the astronomical community at large.
3 Mills to author 8/3/84. A month after the offer to Mills by the NRAO, Bowen wrote to Lloyd Berkner at Associated Universities Inc. (the NRAO's funding body) and suggested a collaborative effort on the SuperCross. Bowen proposed a two-way agreement between Radiophysics and AUI, or possibly even a tripartite one including the Netherlands, with a pooling of resources leading to the construction of a SuperCross in the northern hemisphere, followed by a second one in the southern hemisphere (see CSA file KE5/5/10). The formula was similar to the one Bowen had proposed in 1952 in an attempt to get a giant dish built in Australia (see Chapter 5). The interest by the Dutch in large arrays led to the abortive Benelux Cross and then to eventual success with the Westerbork array (see later in this chapter).
4 D. D. Millar (Ed.), *The Messel Era*, p. 137. It is interesting to note that the chair filled by Messel in 1952 had been vacant since the retirement of O. U. Vonwiller in 1945. The position, first advertised in 1946, apparently failed to attract a candidate of sufficient stature and so several direct approaches were made. One in fact was made to Bernard Lovell, via a letter from Oliphant (then at Birmingham): 'The University is a large one but physics therein has not been in the forefront for some time. The place needs a man who is prepared to go into it in rather a bull-headed fashion and drive through the apathy which has gripped it .... I think a man of your ability and drive could have a great deal of fun out there and could play a vital part in the scientific work of Australia ...'. Even with a salary of £2000 a year, double the amount he earnt at Manchester, Lovell turned down the offer. Lovell also refused the offer of a chair in physics at Adelaide the following year (see Saward, *Bernard Lovell: A Biography*, p. 135). Radio astronomy in Australia might have taken a radically different course had Lovell accepted either offer.
5 Note 4, p. 142.
6 *The Bulletin*, 31/3/62, p. 33. Bowen viewed the anonymously authored article as a 'beat up' engineered by Messel. As related earlier, one of the major financial backers of the revitalised School of Physics had been Sir Frank Packer, owner of the *Bulletin* and Sydney's *Daily Telegraph*. Bowen had been particularly incensed when, on the day following the inauguration of Parkes, the *Daily Telegraph* had run a front-page article on the event accompanied by a large photograph of Messel and Mills standing in front of the telescope. That day of all days had belonged to Bowen! The title of the *Bulletin* article 'Mills Cross versus Parkes Dish' was itself inflammatory, implying as it did that Mills had lost out to the Parkes telescope and been forced to pursue the SuperCross at the University of Sydney. The misconception became quite widespread outside Radiophysics and has still not been properly laid to rest. Bowen had attempted to set the record straight two years before the *Bulletin* article:

'There has not been a decision in this Lab to go ahead with the giant radio telescope in preference to a SuperCross. The giant radio telescope was a going concern long before the SuperCross came into the picture. The decision which had to be made was this: Now that we are well along with the construction of the giant radio telescope, what other large radio astronomy project may we tackle? The choice lay between Paul Wild's new Radioheliograph and Mills' SuperCross. It is going to be difficult to support just one of them, and quite impossible to tackle both. It was a decision of the radio astronomy group—not mine—that they would go ahead with the Radioheliograph and not the Cross' (CSA file KE23, Bowen to White 10/2/60). Irrespective of the personal feuding between some of the principal players, the record shows that the issue at stake had not been SuperCross versus Parkes telescope; rather, the contest had been SuperCross versus Radioheliograph.

7 Mills to author 4/4/84.

8 The first collaborative paper, published in 1969, consisted of a study of galactic sources at 5000 and 408 MHz using, respectively, the Parkes and Molonglo instruments. Mills' own attitude towards Radiophysics had hardened after his application for observing time on the Parkes telescope, shortly after its opening, had been rejected. Mills recounts being informed that he 'had no rights of use and, moreover, my project had no value. Application refused! I guess this had a salutary effect and, although I have visited Parkes on a few occasions, never again have I requested observing time there—although, of course, these problems disappeared with Bowen's retirement' (see note 7). Recently, the collaboration has been strengthened with a successful program of observations using the MOST and the Parkes telescope as a long baseline interferometer system at 843 MHz.

9 H. Messel, 'The new radio astronomy centre of the University of Sydney' *Nature* **188** (1960) 528–9.

10 Interview with author, March 1984. White's stance at the time was much more conciliatory than this quote suggests. Writing to Alan T. Waterman (director of the National Science Foundation) on the SuperCross in February 1962, White stated: 'We in CSIRO are determined to maintain our position in this science, but have, of course, every intention of encouraging others in Australia to come into the field as well. I am personally very happy indeed that the University of Sydney has chosen to embark on research in this field and that two of our best men—Mills and Christiansen—have seen fit to leave us to join the staff of the University. This can do nothing to hinder radio astronomy in Australia; in fact the opposite may well be the case, that through them younger men in physics will enter this very exciting field of research' (see CSA file KE20/14).

11 Christiansen to author 27/2 and 9/4/90. The work on earth-rotation synthesis was described in W. N. Christiansen and J. A. Warburton, 'The distribution of radio brightness over the solar disk at a wavelength of 21 centimetres, Part III' *Aust. J. Phys.* **8** (1955) 474–86. See also W. N. Christiansen, 'An omission in radio astronomy histories' *Q. J. R. Astron. Soc.* **30** (1989) 357–8. The technique of earth-rotation or aperture synthesis was also under development at this time at Cambridge principally by Martin Ryle, for which he was awarded the Nobel Prize for physics in 1974.

12 Christiansen to author 16/4/84. It was rumoured that Christiansen's letter had been returned from the ANU to Bowen by Mark Oliphant.

13 Mills and Ron Bracewell (the former Radiophysics member who had gone to Stanford) were two original applicants for the engineering chair. Bracewell had been offered the chair but then turned it down.

14 RPA file A1/3/11/3, DuBridge to Bowen 19/8/59.

15 CSA file PH/BOL/5, Bolton to White 22/6/60.

16 At this time, it was rare within CSIRO for a scientist to be promoted to the level of chief research scientist without actually being chief of one of its divisions. Wild was quoted in the Melbourne *Herald* (10/4/58) as saying one reason for turning down Cornell's offer was that he preferred to do research fulltime 'rather than become

bogged down with the red tape and paper work which goes with every professorial position'. We can mention here that the Parkes site had been given the title Australian National Radio Astronomy Observatory in expectation that the Radioheliograph would be situated there and share the same facilities. However, as noted above, the economics of the area of land required for the instrument had forced the shift north to Culgoora. If, instead, Parkes had developed with more than one major instrument (such as occurred with the Manchester group at Jodrell Bank and with the NRAO at Green Bank), there would have been some logic in appointing an 'overlord' such as Pawsey as director of ANRAO itself, with Bolton and Wild in charge of the two facilities. However, with the physical separation of the two telescopes, ANRAO came to be an unnecessary and somewhat grandiose title for the Parkes site alone.

17 RPA file Z1/14A, Rabi to Bowen 2/12/61. Rabi had just succeeded Lloyd Berkner as president of Associated Universities Inc.

18 B. Lovell, 'Memoir of J. L. Pawsey' *Biog. Mem. R. Soc. London* **10** (1964) 229–43. Pawsey followed Mark Oliphant (1943) and Harrie Massey (1955) as the third Australian to receive the prestigious Hughes Medal. In Pawsey's place, Dave Heeschen was appointed director of NRAO, a position he held until 1978.

19 An interesting feature was the academic progression of a number of the radio astronomy group. Several started in humble clerical or technical grades, working their way up to the status of 'radio astronomer' by part-time study. Most of those who at some stage added 'doctor' to their names, earnt the title by a Doctor of Science. A D.Sc. is awarded on the basis of published research, as distinct from the Ph.D. awarded for postgraduate study. This type of 'bootstrap' advancement is most unlikely to occur in the scientific institutions of today.

20 J. L. Pawsey, *Proc. IRE (Aust.)* **24** (1963) 95–7.

21 Edge and Mulkay, p. 53.

22 Mills to author 8/3/84.

23 Christiansen to author 31/1/84.

24 Interview with author, August 1983.

PART THREE

JOHN BOLTON

THE ASTRONOMER

CHAPTER NINE

Starting with a Splash

> It has now become clear that the exploration of the universe, as conducted by physicists, astronomers and cosmologists, is one of the greatest intellectual adventures of the mid-twentieth century. It is no exaggeration to say that their achievements, especially in the last few years, constitute a revolution in our knowledge and understanding of the universe without parallel in the whole recorded history of mankind.
>
> Dennis Sciama, 1971

In time the period 1946 to 1960 will be remembered as the golden age of Australian radio astronomy. Within Australia and indeed for the whole southern hemisphere, the Radiophysics group had the southern sky to itself. In the northern hemisphere major groups had been established at Jodrell Bank, Cambridge, Leiden and in the United States but, in comparison with Radiophysics, each group was smaller and each had concentrated its resources into certain areas of radio astronomy. In contrast Radiophysics had made fundamental contributions throughout the whole of radio astronomy. After 1960 the proliferation of new groups and telescopes worldwide meant that it would be much harder for Radiophysics to maintain its leadership. During the 1960s radio astronomy became progressively international in the way it was practised, with astronomers travelling more frequently, visiting each other's institutions, attending conferences, sharing ideas and techniques and, to an increasing degree, seeking observing time on the telescope best suited to their own research interests. The internationalisation of radio astronomy began to smooth out the differences between these groups which, until 1960, had worked in relative isolation. The completion of the Parkes telescope undoubtedly gave the Radiophysics group a head start but, as the 1960s progressed, the group's ability to lead the way in radio astronomy would depend more on intellectual resources than on the technical superiority of the new instrument.

Part 3 of this book describes in broad outline the development of radio astronomy since 1960, though without attempting to give a comprehensive account of a scientific field which has undergone tremendous growth since then. Similarly, it is beyond the scope of this book to attempt a systematic review of the numerous research papers from Parkes, produced at the rate of about one per week since the telescope opened. Instead the main aim will be to describe some of the important developments in radio astronomy in which the Parkes telescope has played a significant role. The present chapter looks at some of the highlights that occurred mainly in the 1960s, while later Chapter 12 examines several major developments in radio astronomy since the early 1970s. Chapter 10 discusses the occasional but interesting work carried out by the Parkes telescope on the tracking of spacecraft, especially its central role in the manned Apollo missions. Chapter 11 reviews the growth of radio astronomy internationally and also gives an impression of how astronomy is carried out at Parkes.

☆ ☆ ☆

The Parkes telescope recorded its first signals in October 1961, a trial observation on the radio galaxy Fornax A which passed within a few degrees of the zenith.[1] Over the next few months other trial observations were made to explore the general features of the radio sky, all part of the commissioning work required to calibrate, test and fine tune the telescope ready for operation. Full-time observations did not begin until the following March. By then plans for the research program had already received a great deal of thought. As early as June 1959 Joe Pawsey had outlined a research strategy for the new instrument in a report entitled 'Future program for radio astronomy'. Pawsey cautioned that even though it would be inappropriate to plan rigidly too far in advance, nevertheless the telescope 'implies a considerable reorientation of the Laboratory program and it is desirable to set down in general terms a

13/14 Dec 1961

EST.	δ	Az	Zd	21 det	Gain	Tc	75 det	Gain	Tc	Observations
1950	-37°20'	108 38	24 49	100	2	3	50	2	4	Fornax - A
2010	-37°25	109 23	20 51	100	3	3	25	1	4	"
2030	-37 15	109 51	16 54		"			"		"
2045	-37 10	112 09	12 59	100	3	3	25	1	4	"
2105	-37 30	121 46	09 21		"	"	"	"	"	"
2125	-37 35	138 33	06 16		"			"		"
2145	-37 20'	180 03	04 20		"			"		"
2205	-37 05'	222 46	05 44		"			"		"
2225	-37 40'	237 07	09 25		"			"		"
2245	-37 50	247 47	12 55		"			"		
2255	Power failed temporarily in nearby lightning – Rain very heavy.									
2308	Telescope stowed.									

The first observing session at Parkes, entered in the logbook by John Bolton. The first four columns give the time (far left), the declination, the azimuth and zenith angles of the telescope, while the next six columns give the settings of two receivers being tested. Observations of the source Fornax A were in progress, until a storm shut down operations late in the evening.

plan for experiment and observation for the period ending a year or so after its construction.' [2]

In response to Pawsey's call for research proposals, the radio astronomy group compiled a list of 24 topics worthy of investigation as soon as the opportunity arose. The issue of assigning priority to these proposals was then examined by Gart Westerhout from Leiden, who spent several months during 1960 discussing the future research program with the Radiophysics group. At the end of his visit Westerhout advised:

> 'As soon as the telescope can be operated, some spectacular program should be performed, quickly yielding results which will show the excellent performance of the dish and also be of great importance for astronomy; in other words, the observations should start with a splash.' [3]

Two of the 24 research topics proposed by the group in mid 1959 created just the type of splash called for by Westerhout. Both will now be explored in depth. One proposal led to the discovery of quasars, the other, as we see next, to the detection of magnetic fields in interstellar space.

Compass on the Cosmos

> But even if radio astronomy has not so much destroyed our older astronomical viewpoint, it has enormously enlarged and enriched it. It is like that magical moment in the old Cinerama, when the curtains suddenly opened still further, unveiling the grandeur of the wide screen. Optical astronomy in the 1950s, on that narrow, central screen, offered a quiescent view of a slowly burning universe, the visible radiations from thermal disorder. But then the curtains abruptly parted, adding a grand and breathtaking vista, a panorama of swift and orderly motions that revealed themselves through the synchrotron radiation they generated—the so-called violent universe.
>
> Owen Gingerich

By 1960 an impressive collection of astronomical objects had been shown to emit radio waves. Radio sources ranged from nearby, well-known objects such as the sun, the moon and several planets in our solar system, to galactic objects such as the debris of exploded stars known as supernova remnants, to more distant, poorly understood extragalactic sources. It seemed likely that many of the fainter radio sources might be at distances beyond the most distant optical objects known and that radio astronomy might possibly provide a new and more penetrating view into the furthest reaches of space. Even though sky surveys of the type carried out by the Sydney and Cambridge groups in the 1950s had been unable to provide decisive evidence in favour of either of the two main cosmologies—the big-bang theory and the steady-state theory—few doubted the potential of radio astronomy to enlarge our understanding of the universe, of how it began, and of how it evolved.

Despite the detection of radio signals from a growing number of astronomical objects by 1960, the physical process responsible for the emission from most sources remained a mystery. A number of possible emission mechanisms were already known and could be divided into two main categories. The first category is where an object emits thermal radiation as a direct result of its intrinsic heat or temperature. The emission from our near neighbours such as the moon or the planets Venus, Mars and Saturn is of this type. Thermal emission also occurs from vast clouds of hot ionised gas known as HII regions, of which the Orion Nebula is a particularly good example.

The second category contains several types of radiation. One example is line radiation from hydrogen where, as mentioned earlier, the spin of the electron undergoes a flip in direction which sees the emission of a radio photon at the wavelength of 21 cm. The detection and study of the 21 cm hydrogen line had been one of the great triumphs of radio astronomy during the 1950s. Another type of radiation arises from plasma oscillations where charged particles such as electrons, protons or ionised atoms move or oscillate in bulk. Much of the radiation from the sun stems from plasma oscillations caused by streams of high-energy electrons bursting outwards through the solar atmosphere.

A third type of emission is known as synchrotron radiation, named after a type of particle accelerator first constructed by American nuclear physicists in the late 1940s. The early operators had noticed a new and unusually intense radiation emitted by the beams of electrons, confined by a strong magnetic field and accelerated to high energies after numerous circuits of the synchrotron's chamber. In 1950 the Scandinavian physicists Hannes Alfvén and Nicolai Herlofson suggested that the newly discovered emission from radio 'stars' might also result from this same acceleration of electrons in magnetic fields associated with the source. Shortly afterwards the Soviet astrophysicists Iosif Shklovsky and Vitaly Ginzburg extended the theory of synchrotron emission to distant radio galaxies, sources which were far too strong to be explained by any of the other known emission mechanisms. The theory was convincing and soon the synchrotron mechanism had become accepted as the principal way in which emission occurs from radio sources.

The key feature which distinguishes synchrotron radiation from other types is its strong polarisation.* When electrons move in a magnetic field they trace out long spiral paths, steadily emitting a narrow cone of

* Television antennas provide a familiar illustration of the polarisation of radio waves. The signals transmitted by TV stations are polarised and so the TV antenna at home must be correctly aligned to pick up the strongest signal. Polaroid sunglasses also illustrate the principle of polarisation for visible light.

radiation. More precisely, the electric field vector of the radiation emitted by the electron is perpendicular to the magnetic field lines. But then came the mystery. Even after the synchrotron theory had been widely accepted, very little observational evidence could be found in favour of its distinguishing feature, polarisation of the radiation. One piece of evidence had been found in 1954 when, following directly from Shklovsky's prediction, visible light from the Crab Nebula was shown to be polarised. Three years later Cornell Mayer and a group at the Naval Research Labs in Washington also detected a small amount of polarisation from the Crab at the radio wavelength of 3 cm. However, aside from the curious case of Jupiter (see below), no other evidence for polarisation could be found from any of the hundreds of radio sources known at the time. Later, Fred Hoyle explained the reason for the mystery:

> 'Following the acceptance of the synchrotron theory, polarisation was also looked for at radio wavelengths, but was not found for several years due to an unfortunate circumstance. It has turned out that there is indeed polarisation at radio wavelengths, but that the polarisation tends to be lost for sources that lie along the plane of the Galaxy. Now the famous Cygnus source, the most intense of all the extragalactic sources, lies smack along the galactic plane. The natural thing to do was to look first at this source, simply because it is so intense. When no polarisation was found, the observers became naturally discouraged; if no polarisation could be found in the very brightest source there seemed little chance of finding any in fainter sources. It was not until the big precision radio telescope at Parkes, New South Wales, came into operation that it became readily possible to conduct systematic polarisation measures.' [4]

The importance of polarisation to the synchrotron theory had long been recognised at Radiophysics. Several unsuccessful attempts to detect polarisation had been made, dating back to the early 1950s. Joe Pawsey had taken a particular interest in the subject. Among the 24 research topics drawn up for the new telescope in mid 1959, Pawsey had written the proposal for polarisation himself:

> 'Magnetic fields in interstellar space play a major part in astrophysical theory but evidence for their existence is indirect and sparse ... . It is currently believed that the cosmic radio continuum is due to synchrotron type emission and hence should be plane polarised at the source. But the observed emission is, at least approximately, randomly polarised ... . Observations on the giant radio telescope will be well worth while.' [5]

The breakthrough finally occurred early in 1962. First, the Naval Research group in Washington detected 8% plane polarisation from Cygnus A at a wavelength of 3 cm. At 10 cm however the polarisation had all but disappeared which largely accounted for the earlier failures, described by Hoyle above, to observe polarisation from this source at longer wavelengths. When the news from Washington reached Parkes,

The underside of the aerial cabin at Parkes shows the rotatable feed at the dish's focus. This simple feature led to the discovery that the emission from most radio sources is polarised.

observations began immediately on Centaurus A, the intense source consisting of two extended components discovered by Bolton and Stanley in 1947. The Parkes observations by Ron Bracewell, Brian Cooper and Tom Cousins detected 13% polarisation at 10 cm from one of the two compact central sources of Centaurus A. Further work by Frank Gardner and John Whiteoak showed that the extended components were polarised by as much as 40%. A quick survey then revealed a substantial amount of polarisation in seven out of the first nine galaxies examined and also in the galactic supernova remnant Vela X. Solid evidence for the synchrotron emission theory, which had eluded astronomers for twelve years, was now produced in abundance at Parkes. There could no longer be any doubt that the synchrotron theory was correct.[6]

The discovery of polarisation from radio sources was quickly followed at Parkes by an even more important result—the discovery of a phenomenon known as the Faraday effect. Early last century the British scientist Michael Faraday had found that certain types of rock crystal have the ability to rotate the plane of polarisation of light. A light beam polarised in one particular direction would pass through the crystal and emerge with its plane of polarisation rotated to another direction. Faraday also discovered that the amount of rotation depended on the wavelength of light, that is, its colour. Much later, it was realised that the same phenomenon should also occur for polarised radio waves passing through a magnetic field. Shortly after the first observations of plane polarisation at Parkes, Brian Cooper was joined by Marc Price, a young Fulbright fellow from Colorado:

> 'We installed a 20 cm receiver on the telescope and Marc had a look at Centaurus A to see if there was substantial plane polarisation at that wavelength. There was indeed, but he also noted that there was a substantial difference in the position angle of the polarisation compared with that at 10 cm. The difference was about 40°. From his reading of the literature Marc wondered if the difference could be due to Faraday rotation, but as a conservative engineer I was inclined to suspect an instrumental effect. The way to find out would be to see if the position angle had a wavelength-squared dependence. Over the next couple of nights we improvised receivers and feeds for as many wavelengths as possible up to 30 cm, and found that the feed position angle for maximum source intensity did indeed have a wavelength-squared dependence. The Faraday rotation was surprisingly large. That was one of the nice plums waiting to be picked from the new telescope and it was a privilege to be in on the discovery.' [7]

This surprising result immediately raised the question of what caused the Faraday rotation. The polarisation observations had established that radio sources have magnetic fields which force high-energy electrons to follow spiral trajectories and emit synchrotron radiation. But were these magnetic fields associated with radio sources the cause of the Faraday rotation? One theory put forward by Dennis Sciama at Cambridge proposed that a weak magnetic field belonging to our local cluster of galaxies was responsible for the effect. Further observations of a larger sample of radio sources carried out by Gardner and Whiteoak at Parkes then revealed the true cause. Sources lying along the direction of the galactic plane displayed the greatest amount of Faraday rotation, while sources in the direction of the galactic poles displayed virtually none. This immediately ruled out the sources themselves or some intergalactic magnetic field as the cause of Faraday rotation. Instead, the effect could only arise from the existence of a magnetic field within our own galaxy, a field apparently of remarkable uniformity aligned over vast distances of space.[8]

The existence of a magnetic field within the Milky Way had in fact already been predicted by astronomers studying the theory of galaxy formation and structure. Theories based on gravitational and rotational dynamics alone had been unable to explain why galaxies vary so greatly in size and shape, so astronomers had been forced to 'invent' magnetic fields to explain the wide variety of galaxies observed. The detection of Faraday rotation at Parkes not only revealed the existence of a galactic magnetic field, but also provided a ready-made compass to measure the features of the field in our local arm of the Milky Way. This was done by observing a large number of radio sources over all directions of the sky. The amount of Faraday rotation then provided information on the strength and orientation of the magnetic field along the line of sight to each source. These observations showed that the magnetic field is aligned along the spiral arms of the Milky Way. The field is extremely weak—only about one-millionth the strength of the earth's magnetic field—but the vast distances of space have enabled the field to exert a major influence on the evolution of the galaxy. Over a period of several billion years the field has gently shepherded billions of stars and gas clouds into the gigantic alleyways we now observe as spiral arms. In the 1950s, the discovery of the 21 cm hydrogen line provided a brilliant method for mapping the spiral structure of the Milky Way; in the 1960s the discovery of Faraday rotation and the galactic magnetic field provided a convincing explanation for the evolution and stability of these spiral arms.

The twin discoveries of polarisation and Faraday rotation were the first triumph for the astronomers at Parkes, made within months of the new telescope becoming operational. Further studies building on this initial work were to provide a substantial part of the total research program over the next decade. As John Bolton recounts:

> 'It was easy enough to detect these polarisation effects initially but then we followed this up by looking at hundreds of radio sources in some detail. To show that the amount of Faraday rotation varies over the sky and that it gives the magnetic field of the galaxy involved months of observations spread over quite a few years. Our techniques were always on the improve, but these large projects still took a great deal of time. Our enthusiasm for this work was maintained for a long time.' [9]

A major landmark in these polarisation studies was the completion in 1965 by Cooper, Price and Doug Cole of the first magnetic field map of an external galaxy, Centaurus A. The map revealed that the orientation of the magnetic field is fairly uniform along the two large components of this source. However, within the two bright core regions the field shows abrupt changes in direction and intensity, an important clue to understanding the nature of radio galaxies.[10] By 1970 an extensive polarisation survey had been completed of close to 400 sources, which included a wide range of galactic objects, normal spiral galaxies, radio galaxies, and the strange new objects known as quasars.[11] These polarisation studies, a research program originally proposed by Joe Pawsey and begun shortly before his death, created for Pawsey a fine scientific memorial.

☆ ☆ ☆

The curious case of Jupiter formed another research program at Parkes, carried out in parallel with the polarisation studies of galactic and extragalactic sources. As described earlier, unusually strong radio emission from Jupiter, the largest and most massive planet in the solar system, had been detected in 1955 by Burke and Franklin at the Carnegie Institution. The discovery had been confirmed immediately by Alex Shain at Radiophysics after checking through old records made several years

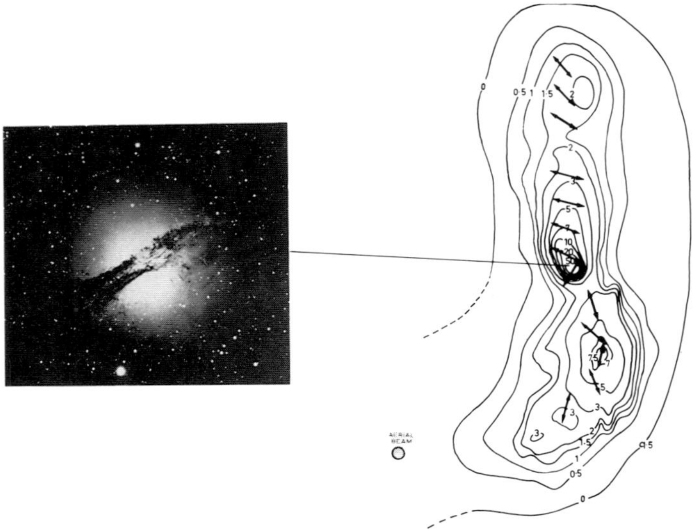

Right Radio contours measured at Parkes for Centaurus A illustrate the large size of the source. The photograph of the nebula NGC 5128 located at the core of the source has been enlarged 25 times relative to the radio map. Double-headed arrows indicate the orientation of the source's magnetic field. *below* Polarisation studies of distant radio sources were carried out at Parkes by John Whiteoak (left), a recent graduate from the Mt Stromlo Observatory, and Frank Gardner, one of the leading receiver specialists at Radiophysics.

earlier. Further observations by the Naval Research group in Washington had then shown that, at a wavelength of 10 cm, the Jupiter radiation is several times stronger than expected for normal thermal emission from the planet. The puzzle was solved shortly afterwards at Owens Valley by Jim Roberts (visiting from Radiophysics) and the Indian astrophysicist Radhakrishnan. Not only is Jupiter's emission polarised, indicating synchrotron radiation, but the emission comes from an area about three times the size of its visible disc. This indicated that Jupiter has a large radiation belt surrounding the planet, apparently similar to the Van Allen belts around the earth.

On returning to Australia, Roberts in collaboration with Max Komesaroff began a series of observations at Parkes which extended the work by the American groups. The first involved observing Jupiter during its eclipse by the moon, which confirmed that the synchrotron radiation is emitted from an invisible halo extending over a much wider area than the visible planet. As the planet rotates on its axis, the direction of the plane of polarisation rocks back and forth through an angle of ±10°. The rocking shows that the magnetic axis of Jupiter, and thus its belts of radiation, are inclined at 10° to its axis of rotation. This rocking motion, however, is not completely regular which indicates that these radiation belts are not distributed symmetrically around the planet.[12]

The Jupiter watch at Parkes continued intermittently into the 1970s leading to new information on properties of the planet such as the strength of its magnetic field and the density of charged particles trapped in its radiation belts. Its rotation period—the length of the Jovian day—was also measured to a remarkable accuracy of 0.02 seconds. Another interesting finding was that Jupiter's magnetic field is 'anti-parallel' to its direction of rotation; in other words, in comparison with the earth, Jupiter's field is upside down. A number of these properties were subsequently confirmed during the encounters with Jupiter by the spaceprobes Pioneer 10 (December 1973), Pioneer 11 (December 1974) and the later Voyager missions. Although not as extensive as the polarisation studies of galactic and extragalactic sources, the Jupiter program carried out by Roberts, Komesaroff and others has brought about substantial progress in unravelling the mysteries of this unusual planet. Often, small research programs turn out to be among the most productive.

Star Like, Star Bright

Twinkle, twinkle, quasi-star
Biggest puzzle from afar
How unlike the other ones
Brighter than a trillion suns
Twinkle, twinkle, quasi-star
How I wonder what you are.

Joseph Silk

The discovery of quasars with their fantastic energy output in the early 1960s took the astronomical world by storm. Thirty years later there is still much to learn about the origin and nature of these celestial objects, but already astronomers believe that quasars will play a central role in our understanding of the creation and evolution of the universe. Since the birth of radio astronomy sixty years ago, the discovery of quasars is perhaps the finest single achievement in radio astronomy. It was one in which the Parkes telescope played a major part.

One strand in the detective work leading to the discovery of quasars began late in 1960 when Caltech astronomer Tom Matthews measured an accurate position for 3C 48, one of the sources listed in the third Cambridge catalogue. The first indication that there was something unusual about this source came from observations at Jodrell Bank using the 76 m dish* in conjunction with a small antenna located 100 km away to form a long baseline interferometer. The Jodrell Bank observations by Henry Palmer had shown that the angular diameter of 3C 48 is less than 4 arcsec, making it an exceptionally compact source in comparison with the usual 30 to 60 arcsec for most radio galaxies known at this time.

The challenge then became to identify the source with an optical object. As a first step, John Bolton, then at Owens Valley, examined the plates in the Palomar Sky Survey. The position measured by Matthews for 3C 48 seemed to coincide with a sixteenth magnitude star. The agreement between the two positions was close enough to stir the interest of optical astronomers. Observations with the Hale telescope by Allan Sandage, Jesse Greenstein and Guido Münch showed the star to have an unusual blue colour made up of a combination of strong emission and absorption lines unlike that of any star known. The most puzzling feature was that the object's spectrum apparently contained no evidence for the presence of hydrogen, the major constituent of all normal stars. In December 1960, Sandage gave a brief report on this bizarre new object to a meeting of astronomers in New York:

> 'Since the distance of 3C 48 is unknown, there is a remote possibility that it may be a very distant galaxy of stars; but there is general agreement among the astronomers concerned that it is a relatively nearby star with most peculiar properties. It could be a supernova remnant. The radio output may be intrinsically ten million times stronger than the sun's.' [13]

All the evidence seemed to point to the detection of the first true radio star somewhere in our galaxy.

While the riddle of 3C 48 preoccupied the Caltech astronomers, another strand in the quasar story had begun elsewhere. One way to measure radio positions is a technique known as lunar occultation which uses the moon to pinpoint the exact position of a source. The technique was well known in traditional astronomy having been used to study the detailed shape and motion of the moon against the backdrop

* From here on we use metric units for the world's telescopes. For example, the diameter of the 250 ft Mark I at Jodrell Bank is 76 m and the 210 ft Parkes telescope is 64 m. Similarly, the 200 inch Palomar telescope becomes the 5 m Hale telescope, named in 1969 after the American astronomer George Hale.

Caltech astronomer Tom Matthews measured an accurate position for 3C 48 late in 1960, but for the next two years the nature of the source remained a complete mystery.

of fixed stars. In radio astronomy the technique can be used in reverse. Because astronomers know the exact position of the moon's edge or 'limb' at any given time, a careful measurement of the times when a radio source disappears behind the moon (immersion) and then later reappears (emersion) enables the position of the source to be calculated to great precision. The accuracy of the technique is limited basically by uncertainties in the detailed shape of mountains on the moon, some of which rise as high as 6000 metres. In December 1960, shortly before the puzzling results for 3C 48 had been announced, Cyril Hazard at Jodrell Bank used a lunar occultation to measure the position of 3C 212 to an accuracy of 3 arcsec, the first time the technique had been used for a radio source.[14]

In 1962 a second opportunity arose for Hazard to use the technique to measure the position of another source, 3C 273. By sheer coincidence 3C 273 would undergo a series of occultations by the moon at southern latitudes during 1962, an event that occurs only about once every twenty years. By then, Hazard had joined the new astronomy group at the University of Sydney to work with Hanbury Brown on the intensity interferometer, but had also applied successfully for observing time on the new Parkes dish. The first occultation in April 1962 proved inconclusive. Only the immersion of 3C 273 was recorded before the source passed outside the field of view at Parkes, but the occultation curve did at least indicate the source to be of very small angular size. Preparations for the next occultation were made with utmost care. All roads leading to the telescope were closed, unnecessary electrical equipment on the site shut down, and all the electronic equipment made ready in duplicate. Yet, as John Shimmins recalls, there was a problem:

> 'It seemed likely that at emersion, as 3C 273 reappeared from behind the moon, the zenith angle of the telescope would be slightly greater than 60°. The telescope will only go to 60° and then there is a solid stop. When you reach it bells ring and the brakes are applied. So, we simply removed the safety stops from operation for that observation. Just to be certain, John Bolton took a grinder and cut away some of the housings of the zenith angle bearings. This meant that we could tip the dish by almost another degree. During the occultation the rim of the dish was practically touching the ground.' [15]

The observation of the occultation carried out by Hazard, Shimmins and Brian Mackey during the early evening on 5 August 1962 went according to plan. Immersion of the source occurred at 5.46 pm and 80 minutes later 3C 273 reappeared from behind the moon. After the elaborate preparations nothing was left to chance. As one final, macabre precaution Hazard and Bolton carried duplicate records of the observation back to Sydney on separate planes.

An analysis of these records combined with the results of a third occultation in October revealed that 3C 273 is not a simple point source, but a double with each component about 6 by 2 arcsec in area with a separation of about 20 arcsec between the two centres. In the meantime, the immersion and emersion times had been sent to the Nautical Almanac Office in Sussex which then supplied Hazard with data on the shape and motion of the moon's limb. After a painstaking analysis, the position of 3C 273 was calculated to an accuracy of 1 arcsec—the most accurate position ever determined for a radio source. By coincidence, Rudolph Minkowski happened to be visiting Radiophysics at this time, following his recent retirement from Mt Wilson–Palomar. Minkowski brought a number of plates from the Hale telescope, including one for the region of sky containing 3C 273. The radio position calculated by Hazard coincided exactly with a double source on the plate, the compact core of one radio component matching with what appeared to be a blue star, the other with a peculiar jet extending out from the star in one direction. At Jodrell Bank, Bernard Lovell learnt of the news from Hazard:

> 'On 9 February 1963, I received a handwritten letter from him which I treasure as a classic of British understatement and phlegm in the face of great discoveries. Most of the letter complained about the temperature of 104 degrees at Narrabri where he was carrying out some work with Hanbury Brown, and about the problem of his house and cost of moving his family. Then came the following paragraph about the work he had been doing with the steerable telescope at Parkes using the technique of lunar occultation:

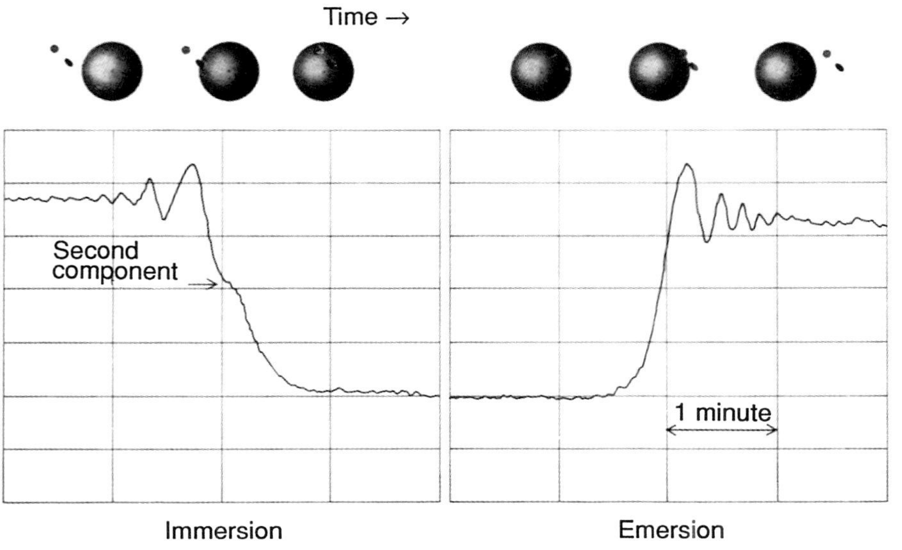

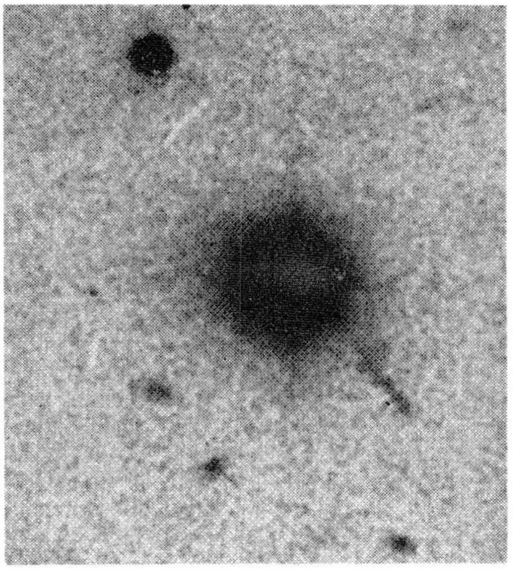

Above Lunar occultation of the quasar 3C 273 recorded at Parkes in August 1962. From left to right, the signal from the two-component source is blacked out, with the curve showing where the second component starts to disappear. Because of their orientation with respect to the moon's limb, both components emerge together. The waviness is caused by diffraction of the signal around the moon's limb. *left* Palomar photograph of 3C 273 showing the unusual jet extending at lower right.

> 'The work I have done at Parkes has turned out very well. 3C 273 has now been identified with a thirteenth magnitude star! We noticed this star some months ago on a 200 inch plate, it is a peculiar object with a jet protruding at an angle of 45° which is the inclination of the radio source. The source is double and the separation of the components approximately equal to the length of the jet. A very intriguing identification'.[16]

Next came the turn of Maarten Schmidt, a Dutch astronomer who had succeeded Minkowski at Palomar. In December 1962 Schmidt measured the optical spectrum of 3C 273 and found six strong emission lines. Four of these lines formed a simple harmonic series with a separation and strength similar to the type of series which characterises hydrogen and certain highly ionised atoms. However, a search of the extensive tables of spectroscopic data failed to identify the atom responsible. After two months of work, Schmidt finally found the solution. Rather than belonging to an unusually rare atom, the four emission lines in the spectrum belonged to none other than hydrogen. The confusing part had been that these lines did not have their expected wavelengths but were shifted to the red or long wavelength end of the spectrum. The amount of redshift measured by Schmidt was 16%, much higher than the values for most of the distant galaxies then known. If this redshift arose because of the expansion of the universe, then according to the Hubble law of redshifts 3C 273 would be about 3 billion light-years distant. And yet, in view of the brightness measured for the object (thirteenth magnitude) this had a startling implication: 3C 273 is radiating at least 100 times more energy than the most luminous galaxy.

Schmidt's success in identifying the spectrum of 3C 273 prompted Jesse Greenstein and Tom Matthews to return to the spectrum of 3C 48, which had remained a mystery since Sandage had announced the peculiar features of this starlike object at the New York meeting two years earlier. To their surprise Greenstein and Matthews were now able to recognise a number of lines in the spectrum belonging to magnesium, oxygen and neon, but shifted to the red by a staggering 37%. 3C 48 was

Two panels from the feature 'The Universe of Hoyle' which appeared in Australian newspapers in 1966.

apparently at a distance of 7 billion light-years, over twice the distance of 3C 273, and with a speed of recession close to one-third the speed of light. By this stage most astronomers favoured the view that 3C 48 and 3C 273, along with several other 'quasi-stellar' objects under suspicion, were compact radio stars in our galaxy. Greenstein had already sent a theoretical paper to the *Astrophysical Journal* arguing that these objects were nearby, ultra-dense stars formed from the collapsed cores of supernovas, with emission spectra arising from intense radioactive decay of material on the object's surface. The redshifts measured for 3C 273 and 3C 48 dealt the theory a fatal blow, leading Greenstein to quickly withdraw his paper from publication. Rather than nearby stars within the galaxy, these were among the most distant objects known in the universe.

The paper by Hazard, Mackey and Shimmins on 'Investigations of the radio source 3C 273 by the method of lunar occultations' appeared in *Nature* in March 1963, accompanied by three others. One by Maarten Schmidt reported the redshift of 3C 273, noting that the object is 'about 100 times brighter optically than the most luminous galaxies which have been identified with radio sources thus far.' The second paper by J. Beverly Oke at Mt Wilson–Palomar reported on the energy characteristics of the 3C 273 optical spectrum, while the third by Greenstein and Matthews gave the results for 3C 48, concluding with the understatement that 'so large a redshift, second only to that of the intense radio source 3C 295, will have important implications in cosmological speculation.' [17]

Jesse Greenstein, head of the Caltech astrophysics group at the time of the discovery of quasars. Like many astronomers, Greenstein at first believed quasars to be ultra-dense radio stars closeby in our galaxy.

The publication of this quartet of papers marked the start of a major new field of astronomy. The name quasi-stellar radio source for these objects became shortened to QSR and then later changed to quasar, a word coined by Hong-Yee Chiu at the Goddard Institute for Space Studies in New York. More recently, the name quasar has itself fallen from favour, following the discovery that the great majority of quasars currently known are not strong radio emitters. Many astronomers now prefer the term QSO (quasi-stellar object) to quasar.

The discovery of these extraordinary objects generated a considerable amount of interest beyond the world of astronomy. The Australian media focused its attention on Parkes, the first of many occasions since, and the radio astronomers were applauded for making such an important breakthrough. Parkes had played a key role, but it would be an exaggeration to claim quasars as an Australian discovery. If anything, the quasar story provides a good example of how difficult it can be to clearly assign priority for a major discovery. Was 3C 48 or 3C 273 the first quasar discovered? As mentioned, 3C 48 was the first quasar observed optically, but the fact that it was wrongly believed to be a nearby radio star meant that quasars as a new class of exceptionally distant objects had gone unrecognised. 3C 273 was the first to be correctly identified, the crucial step in recognising that these objects were not nearby galactic radio stars, but distant extragalactic objects. Thus, 3C 273 was the first quasar to be *discovered*, though paradoxically it was not the first to be *observed*. Opinion among astronomers remains divided on whether 3C 48 or 3C 273 deserves priority as the first quasar. On one side John Bolton, who took part in the 3C 48 observations at Owens Valley and then coordinated the 3C 273 operation at Parkes, favours 3C 48:

> 'The important point is that these objects were already known and had been named quasi-stellar sources at Caltech in 1960. They looked like stars photographically, although their spectrum was not understood. The significance of 3C 273 was that it established the spectrum and enabled the redshift—and thus the distances—of these objects to be worked out. But the importance of this should not be over-emphasised. For example, we know of more than 1500 quasars now, but we only know the distances of about a half. Quasars were known as an astronomical entity before their vast distances were appreciated. For me, 3C 48 was the first quasar, not 3C 273.' [18]

On the other side, Cyril Hazard argues that the publication of the first 3C 273 paper marks the true discovery date:

> 'There can be no question that quasars were recognised as a new class of extragalactic object directly as a result of the position measurement of 3C 273, its subsequent identification and the determination of its redshift. This work owed absolutely nothing to previous work on 3C 48 and other compact radio sources which had previously been identified with stellar-like objects. It is true that 3C 48 turned out to be a quasar, once quasars had been

> recognised as a new class of object, but this is precisely what the work on 3C 273 established. Up to then, 3C 48 and the few apparently similar objects known, were considered to be stars in our galaxy. It was the publication of the 3C 273 result which created the big upheaval. The 3C 48 study was a parallel investigation and, like it or not, it was the occultation work that succeeded.' [19]

Aside from the question of priority, the discovery of quasars highlighted the increasingly international character of astronomy. On the radio side, the 3C sources had first been detected at Cambridge, their small diameters measured at Jodrell Bank, and their precise positions located at Owens Valley and Parkes. On the optical side, the giant telescopes at Mt Wilson–Palomar had finally revealed the extraordinary nature of the new objects. Astronomers were also becoming more mobile. Cyril Hazard had made the first occultation measurement of a radio source at Jodrell Bank, transferred to the optical astronomy group at Sydney University, and then made the crucial 3C 273 observation on a radio telescope operated by a different institution. The discovery of quasars illustrates the increasing collaboration which took place among astronomers on an international level, both optical and radio.

Looking Out, Looking Back

> The history of astronomy is a history of receding horizons.
>
> Edwin Hubble

The publication of the first papers on 3C 273 and 3C 48 in March 1963 triggered one of the most intensive hunts ever witnessed in astronomy. Shortly afterwards, Schmidt and Matthews announced that 3C 47 is a quasar with a redshift of 0.43 and also that 3C 147, another of the objects which had been under investigation since 1960, showed a redshift of 0.55, making it the most distant astronomical object known.

In April 1964 Hazard and Mackey in collaboration with Bill Nicholson at the Nautical Almanac Office in Sussex announced the discovery of a further three quasars detected by the lunar occultation technique.[20] By the end of the year a total of forty quasars had been reported, leading some astronomers to confidently predict that these extraordinary objects make up a substantial fraction of the thousands of extragalactic radio sources already catalogued but unidentified. Several of these quasars were found to have redshifts approaching 2, corresponding to distances around 10 billion light-years. In less than two years the discovery and study of quasars had led to a doubling of the distance scale of the known universe. One astronomer triumphantly stated: 'Man was on the march to the edge of the universe.' [21]

As the hunt to find more quasars intensified, the nature of these objects remained a mystery. A puzzling feature discovered by Sandage and Matthews was that the intensity of light emitted by 3C 48 varies by as much as 30% within a single year. Similarly, a study of photographic plates dating back to the 1880s at the Harvard College Observatory revealed that 3C 273 fluctuates in brightness over a time scale of about ten years. On several occasions the object appeared to have experienced a fourfold increase in brightness in a matter of months. Such a large variability in brightness was well known for certain types of stars, but for quasars there was an astonishing implication. According to the theory of relativity, when an astronomical object shows a marked change in brightness, the physical mechanism causing this change cannot be transmitted across the breadth of the object at a rate faster than the speed of light. Thus, if an object varies in brightness over one year, then this implies that its physical dimensions cannot be more than one light-year. The variability in brightness of 3C 48 and 3C 273 suggested that they might be even less than a light-year in diameter, and yet these objects were emitting vastly more energy than all the stars combined in a large galaxy like our own. How could any object not much larger than the size of our solar system generate such an enormous output of energy? What caused the extreme variability in the energy output of quasars, equivalent to switching on and off over short periods of time the output from about 100 billion suns? These questions deepened the mystery surrounding quasars. Not even the most extreme processes of nuclear burning known to occur in stars came anywhere near accounting for the energy produced by these objects. Quasars were obviously much more than an unusual addition to the astronomical catalogue; they revealed the presence of new and unknown physical processes occurring in the universe.

In addition to the prodigious energy output from these objects astronomers were equally intrigued by their extreme distances. A powerful motivation in the search for more quasars has in fact been the desire to find ever more distant objects. The most distant objects not only provide a yardstick for determining the dimensions of the known universe, but also its age. The light from a quasar such as 3C 48 — measured from its redshift to be 7 billion light-years distant—has taken 7 billion years to reach observers on earth. Quasars therefore provide long 'look-back' times, the opportunity to study the early history and evolution of the universe. Any increase in the time scale of the universe will have an important bearing on other fundamental issues such as the theory of galaxy formation and the evolution of stars, including our own sun and solar system. Even the age of the earth and the time available for the evolution of life on earth depend on the time scale of the universe. Because of its central importance to astronomy, the search for the most distant objects has become for astronomers a blue ribbon event (its early history is given in Tangent 8).

The rapid discovery of new quasars saw the distance record tumble with almost monotonous regularity during the 1960s. Several record-breaking quasars were detected at Parkes, the first being the source* PKS 0106+01 found in 1966 to have a redshift of 2.1. A year later John Bolton in collaboration with Chip Arp at Palomar and Tom Kinman at the Lick Observatory reported a redshift of 2.22 for the quasar PKS 0237–23. For a time this quasar was also the most luminous object known in the universe, with an unusual emission spectrum indicating the presence of a much higher abundance of heavy elements than any other known quasar.[25] By 1973 the redshift record had passed to OH 471 and then, three weeks later, to OQ 172, two sources first catalogued by John Kraus and his group at Ohio State University. These 'Ohio Specials' with redshifts of 3.40 and 3.53 were the two most distant objects known until 1982, the year the record returned to Australia.

"It's the telescope of the future. It can look back millions of years."

Molnar in the *Sydney Morning Herald*, November 1961.

The discovery and identification of PKS 2000–330 by Bruce Peterson, Ann Savage, David Jauncey and Alan Wright involved a series of observations with four telescopes, two radio and two optical. The source itself had been detected at Parkes as early as 1970, but there had been nothing about the observations to suggest anything unusual. In 1978 the radio position of the source was measured to greater accuracy with the new Tidbinbilla interferometer operated by NASA. Next it was possible to

* Most new radio sources were given a designation after the observatory to first detect them. PKS 0106+01 indicates a Parkes source with the numerals specifying its position: the first two digits give the hour of right ascension, the next two the minutes of right ascension; these are followed by the sign and the degrees of declination. In 1973 a designation similar to this was adopted by the International Astronomical Union to standardise the names of sources from all observatories.

TANGENT 8

Professor Eddington's Trophy

The search for the most distant objects in the universe dates from a survey of nebulae begun by the American astronomer Vesto Slipher in 1912. Slipher's survey showed that most nebulae have emission lines shifted to the red end of the visible spectrum, indicating that these objects are rushing away from our local region at very high velocities. By the end of the 1920s Edwin Hubble realised that the redshift of an object provides a direct way of measuring its distance: the larger the redshift, the higher its velocity of recession and the greater the object's distance. The hunt for the most distant object in the universe then began in earnest.

In 1933 Arthur Eddington wrote in his classic book *The Expanding Universe*: 'Slipher's first determination of the velocities of 40 nebulae included a dozen with velocities from 800–1800 km per sec. The survey has since been extended to fainter and more distant nebulae by Milton Humason at the Mt Wilson Observatory, and much higher velocities have been found. The speed record is continually being broken. The present holder of the trophy is a nebula forming one of a faint cluster in the constellation Gemini, which is receding with a velocity of 25,000 km per sec. This is about the speed of an alpha particle. Its distance is estimated at 150 million light-years.' [22]

Later, Humason found an object with a redshift of 0.1, corresponding to a distance about four times further than the nebula in Gemini referred to by Eddington. Success in the search for distant objects of course depended critically on the power of the instruments available. As expected, the completion of the 5 m Hale telescope at Palomar, which superseded the 2.5 m Hooker instrument at Mt Wilson, led almost immediately to an extension of the record redshift to just over 0.3 (about 2 billion light-years).

Objects at such large distances are usually extremely faint. In some cases, up to ten full nights of observation with the Hale telescope were necessary to obtain a single spectrum suitable for analysis. Because of limitations such as obscuring interstellar dust and the effects of the earth's atmosphere, many astronomers began to believe that the discovery of objects with redshifts much larger than 0.3 would be most unlikely. In 1950 Fred Hoyle stated that 'even with a perfect telescope, we could penetrate only about twice as far into space as the new telescope at Palomar. This means that there are no new fields to be opened up by the telescopes of the future, and this is a point of no small importance in our cosmology.' [23]

Hoyle had failed to foresee the potential of radio astronomy. In 1952 Walter Baade and Rudolph Minkowski identified the radio source Cygnus A with what appeared to be a pair of colliding galaxies (see Chapter 3). The redshift measured corresponded to a distance of about one billion light-years. Though less than the distances known for a number of faint optical objects, the fact that Cygnus A was one of the strongest radio sources in the sky, even at this huge distance, was an omen of things to come.

In 1960 radio astronomy lived up to its promise with a radio galaxy in the Boötes constellation known as 3C 295. From a measurement of the radio position made at Owens Valley, John Bolton identified the source with a faint object visible on one of the Palomar survey plates. Next, Minkowski examined the radio galaxy with the Hale telescope and, following hours of careful observation, eventually identified an emission line in its spectrum that had been redshifted by 0.46. This smashed the barrier for values around 0.3 which astronomers had been struggling to better by using optical telescopes alone. Milton Humason congratulated Bolton with the comment: 'You have succeeded in doing what I spent the last ten years of my life trying—and failed.' [24]

The achievement of 3C 295 was completely overshadowed by the discovery of quasars shortly afterwards. The redshift for 3C 295 of 0.46 remained, however, the largest known for a galaxy for fifteen years. Eventually the record fell to a group led by Hyron Spinrad who measured, in quick succession, three record-breaking radio galaxies, with 3C 467 having the largest redshift of 0.63. More recently, a new class of radio galaxies with redshifts greater than 3 has been reported, but they too continue to be out-distanced by quasars. Since 1963, Eddington's trophy for the most distant object in the universe has been held by a succession of quasars, each with redshifts undreamt of by early astronomers.

Edwin Hubble at the Mt Wilson Observatory in 1924. Hubble's greatest achievement was to show that the universe is expanding, with the most distant galaxies receding at the highest velocities.

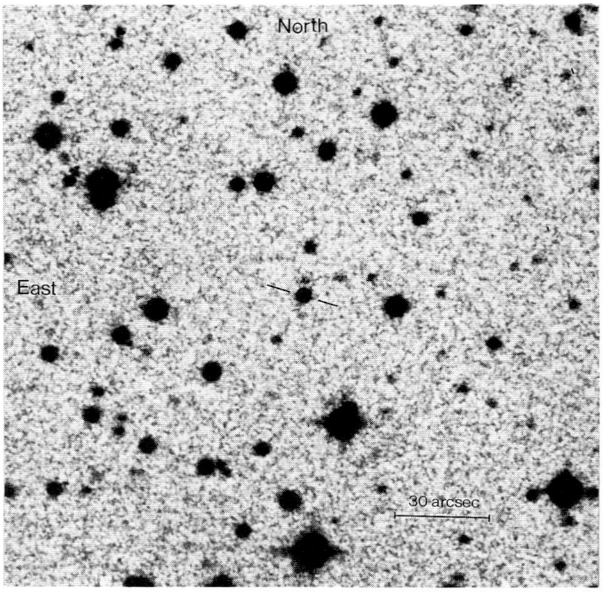

The source indicated by the bars is the quasar, PKS 2000–330, identified in 1982 as the most distant object known in the universe. Most of the other objects shown in this negative print of an Anglo-Australian Observatory photograph are stars in our galaxy.

reliably identify PKS 2000–330 on a photographic plate taken with the UK Schmidt telescope at Siding Spring. Observations with the larger Anglo-Australian telescope in March 1982 then revealed a record redshift of 3.78 for PKS 2000–330, corresponding to a look-back time of 90% of the age of the universe.[26]

Predictably, the record did not remain in Australia for long. Several quasars with redshifts greater than 4 have recently been reported. Theoretical astrophysicists routinely talk of redshifts in the range 5 to 10, corresponding to times earlier and earlier in the universe, closer and closer to the big bang. The search will go on but most astronomers now agree that the age of the universe is 14 billion years, give or take a mere 2 billion!

After thirty years of quasar research, not a great deal is understood about these objects or the mechanism responsible for their tremendous energy output. One early theory by Geoffrey Burbidge suggested that a sequence of supernova explosions at the crowded core of a galaxy would yield this huge release of energy. The shockwave from one explosion would trigger an irregular chain reaction of explosions in nearby stars and so explain the rapid variation in brightness observed for some quasars. Another early theory by Fred Hoyle and William Fowler proposed that a quasar is formed by the gravitational collapse of a single 'mega-star' with a mass at least 100 billion times as great as the sun. Currently, the theory generally favoured pictures quasars as arising from the accretion of matter onto a supermassive black hole. Matter in the whirling accretion disc is accelerated close to the speed of light, radiating

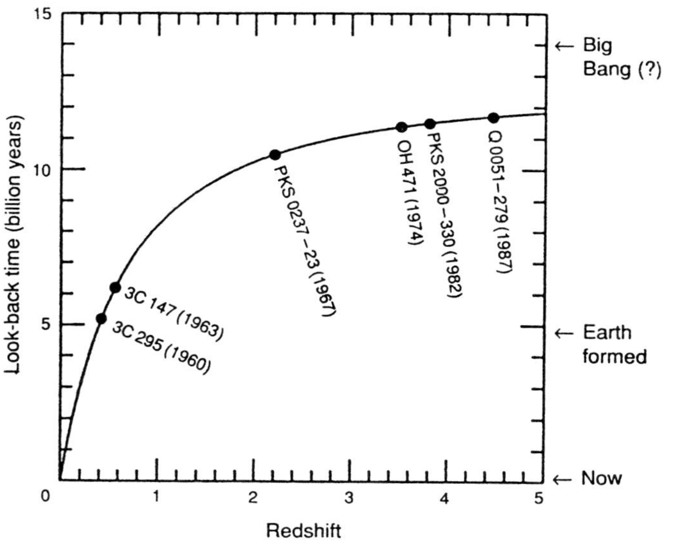

Look-back time for distant objects, plotted against their redshifts. Six record-breaking objects are shown, together with their year of discovery. The detection of these increasingly remote objects has pushed back the known age of the universe.

huge amounts of energy, before spiralling into the hole. The in-falling matter causes the black hole to rotate in the same direction at extremely high speeds, leading to the formation of a doughnut-shaped magnetic field similar to that surrounding the earth. The quasar's magnetic field is, however, extremely strong and capable of focusing the accreting matter into powerful jets ejected outward along the black hole's axis of rotation. Jets of this type are clearly observed in 3C 273 and in many other quasars.

There are other even more imaginative theories of quasars. One proposal is that they are white holes, the topological 'other side' of black holes where all the radiation and matter devoured by black holes gushes back into the universe. For all this, the words of Jesse Greenstein in 1963 still ring true: 'In any event, we have encountered a most baffling group of astronomical objects. Whether fundamental new processes lie behind their brilliant but ephemeral appearance, or whether our imaginations are still too limited, remains for the future to determine.' [27]

Charting the Southern Skies

> The radio sky is no carbon copy of the visible sky; it is a new and different firmament, one where the edge of the universe stands in full view and one which bears the telltale marks of a violent past.
>
> John Kraus, 1976

The Parkes telescope played a major part in the search for quasars following the discovery of this new type of object in the early 1960s. By

1970 approximately half the several hundred known quasars had been first detected at Parkes, many of which had then been studied in detail with the large optical telescopes in the northern hemisphere. The detection of these quasars came about as a direct result of a series of systematic surveys of southern radio sources which began shortly after the commissioning of the telescope and which continued for the next two decades. These sky surveys have undoubtedly been the most important of all the research programs at Parkes, both in terms of the amount of telescope time and their overall contribution to astronomy.

Before the construction of the Parkes telescope a number of major surveys of radio sources had been completed by various groups around the world. These surveys had been carried out at a variety of wavelengths, ranging from as long as 16 m to as short as 10 cm, but in most cases each survey covered only a limited region of sky. The two most comprehensive surveys had been those by Bernard Mills, Bruce Slee and Eric Hill at Radiophysics (the MSH catalogue) and by Martin Ryle and his group at Cambridge (the 2C catalogue). As described in some detail in Chapter 4, the almost total lack of agreement between these two surveys led to what Francis Graham-Smith tactfully called 'a ding-dong competition between Sydney and Cambridge'.[28] The Cambridge survey seemed to indicate the existence of a relatively high proportion of weak sources at large cosmological distances. In turn, this apparently provided evidence in favour of the big-bang theory at the expense of the steady-state theory.

The publication of the new Cambridge 3C catalogue in 1959 and its impressive agreement with the Sydney survey at last gave astronomers confidence that a sizable number of radio sources had been reliably catalogued. But while the 3C and MSH surveys could be counted among the major achievements of radio astronomy, the actual detection and cataloguing of radio sources would never be of profound importance in itself. What mattered more was to identify the types of astronomical object responsible for this radio emission, which meant studying the object with an optical telescope. By 1960 this had been achieved for only a few per cent of the 3000 or so radio sources catalogued. A handful had been identified with objects in our own galaxy, such as hot gas clouds or supernova remnants. The remainder turned out to be galaxies, often with unusual distinguishing features such as excessive brightness, irregular shape, or in the case of Cygnus A apparently consisting of two galaxies in collision. The reason so few sources had been identified was that the position of very few had been measured with anywhere near the required accuracy. Even within an area of sky as small as one minute of arc square—and the sky divides into a grid with 500 million areas this size—a number of stars or galaxies are usually visible through a large telescope, any one of which could be the candidate radio source.

Early in the 1960s the main challenge facing radio astronomy was to substantially increase the number of optical identifications. The importance of the program can be illustrated by an analogy—for example, with the type of taxonomic work carried out in a science such as botany. Imagine a young botanist on his first field trip chancing upon an unexplored forest which harbours a varied and exotic collection of trees. On returning home, the inexperienced botanist reports that he has counted the total number of trees, estimated the height of each, and also its approximate position in the forest. His colleagues would hardly be impressed. On his return to the forest, the botanist would record the information of real interest. This would include details of the various types or species of trees, especially those unknown to botanists, the types of terrain favoured by each species, which species clustered with other species, their relative populations, plus a host of other information.

Radio astronomers in 1960 found themselves in much the same position as the young botanist after his first visit to the forest. A large number of sources had been counted, their positions measured as best possible, their emission intensities recorded, and in some cases their angular sizes and structures determined, but not much more systematic information was known. An extensive taxonomy of these radio sources would only be achieved by optically identifying a much larger number. Identifications as such would be the key to integrating radio astronomy into optical astronomy with its accumulated wealth of data. Early in 1962, Frank Kerr outlined the task ahead at Parkes:

> 'While it is interesting to count observable radio objects, the main point of [this] work is to increase the number of objects that can be identified optically (or definitely shown to have no observable optical counterpart). We believe that the majority of discrete radio sources are very distant galaxies, but the number optically identified is still very small—only a few dozen out of the thousand known. One of the main difficulties has been that positions could be determined well only for the strongest sources. Our new dish should yield accurate positions for numerous fainter sources, permitting many more identifications, perhaps even reaching the stage where the properties of identified objects can be treated statistically.' [29]

The first survey of radio sources began at Parkes in 1962, observing at a wavelength of 75 cm (408 MHz). The strategy chosen was to divide the sky accessible from Parkes into four zones and then to make a preliminary scan of each zone in a square-search pattern. Even with a rapid telescope drive rate, over twenty full nights of observation were required to cover each zone. Once this preliminary 'finding' survey had been made at 75 cm, a selection of stronger sources were then examined in more detail at the shorter wavelengths of 21 and 11 cm. These more detailed observations of each zone took a further forty nights to complete.

The survey of the first zone from declination –20° to –60° by Bolton, Gardner and Mackey was published in April 1964 and listed 297 sources.[30] Surveys of the other three zones by a group which included Margaret Clarke, Doug Cole, George Day, Ron Ekers, Doug Milne, Marc

Price and John Shimmins were completed by the end of 1965. The combined survey, covering the sky from the south celestial pole to near the telescope's northern limit of declination, brought the number of sources to a grand total of 2133. The final catalogue recorded the position of each source, its flux density (i.e. its emission intensity), its degree of polarisation, its angular diameter, and whether or not the source had been detected in the earlier Mills–Slee–Hill survey. Another parameter listed was the spectral index, describing how the intensity of the source varies with wavelength, which could be determined from separate observations at 75, 21 and 11 cm. Often the value of this index provides a strong clue to the nature of the source, independent of whether it is optically identified.

The outstanding feature of the new catalogue was the greatly improved accuracy in the position measurements. Observations on about 700 of the strongest sources yielded values for the celestial coordinates accurate to within 10 to 15 arcsec, an 'error window' around each source small enough to attempt an identification. Most of the subsequent optical identifications were made by John Bolton, during several extended visits to the United States. Bolton's resource was the Palomar Sky Survey, a comprehensive collection of 2000 photographic plates covering the northern sky and a good portion of the southern sky down to declination –33°. To make an identification, a transparent overlay was first prepared for each radio source showing its position in relation to a handful of nearby bright stars, each marked with a cross and positioned so that the scale would correspond exactly to that of the sky survey plates. When the transparency was overlayed on the plate, and the crosses representing the bright stars made to match exactly with those on the photograph, the radio source could then be identified with its optical counterpart. One of Bolton's regular bulletins captures the excitement of the hunt:

> 'Since arriving here almost two weeks ago, I have not been out of the 48 inch plate vault in daylight except for meals. I have just added up my score: From 87 positions of the Bolton–Gardner–Mackey catalogue between –33° and –44°, where John Whiteoak has taken special 48 inch plates, I have:
> 16 definite identifications (galaxies brighter than 17th magnitude)
> 8 probable identifications (galaxies between 17th and 19th)
> 4 objects which have a good chance of being quasi-stellar.
> From about 130 positions of the weaker sources in the Day–Shimmins–Cole catalogue between 0° and +20° I have:
> 11 definite identifications with galaxies (brighter than 17th magnitude)
> 23 probable identifications with galaxies (between 17–19th magnitude)
> 15 quasi-stellar candidates.
> If you add up the galaxies, 58, this now exceeds the total northern identifications with galaxies.' [31]

In this way nearly 250 optical identifications were eventually made directly from the sky survey plates. For a large number of sources, however, no object appeared on the plate, indicating that the object was faint and below the sensitivity of the Schmidt telescope used to compile the survey. Further searches were then made with the Hale telescope, the 3 m telescope at the Lick Observatory in California and, for a small number of southern sources, the 2 m (74 inch) telescope at Mt Stromlo. By the end of 1967 the tally of Parkes sources identified with visible objects had risen to over 400. By any measure, this was a staggering increase on the several dozen or so identifications astronomers had struggled to make before the Parkes survey.

Bolton's identification program showed that the majority of sources are radio galaxies, though well over 100 turned out to be quasars. Such a large sample of objects enabled just the type of statistical analysis referred to by Frank Kerr above. In many cases a quasar could, for example, be distinguished from a radio galaxy by its distinctive curved or flat spectrum over the wavelength range from 11 to 21 cm. One quasar identified in this way was PKS 0237–23. As mentioned earlier, this source with a redshift of 2.22 was for a time the most distant object known in the universe.

By the end of the 1960s the first Parkes catalogue had become a standard astronomical reference, renowned for its accuracy, detail and breadth of coverage. Its success immediately led to plans for a far more ambitious project to survey the sky at the shorter wavelength of 11 cm (2700 MHz). The project carried out principally by Bolton, Shimmins, Jet Merkerlijn, Ann Savage, Jasper Wall and Alan Wright took thirteen years to complete. The publication of the fourteenth and final part in 1979 brought the number of Parkes sources to 8000.[32]

The previous year the Molonglo Cross near Canberra had also completed its ten year survey of southern sources. With its much larger collecting area, the Molonglo Cross was an even more powerful instrument for source surveys. The Molonglo Reference Catalogue listed over 12,000 sources, the largest number for any survey in either hemisphere. To put these numbers in perspective, on a clear moonless night away in the country, the total number of stars visible to the eye is about 6000.

Many of the new sources in the second Parkes catalogue were then examined in detail at the shorter wavelengths of 3 and 6 cm in a series of further surveys that continued well into the 1980s. These surveys showed dramatically how the overall appearance of the radio sky varies according to the observing wavelength. In the first catalogue, most of the strong sources detected at 75 cm were shown to be giant elliptical galaxies, while in the second catalogue at 11 cm a higher proportion of the strong sources are quasars. The extensive data in these catalogues could also be used to draw a number of important conclusions for cosmology. An analysis of sources in the first catalogue yielded a slope for the number versus intensity relationship of 1.8, in agreement with the value derived from the earlier Mills–Slee–Hill survey (see Chapter 4).

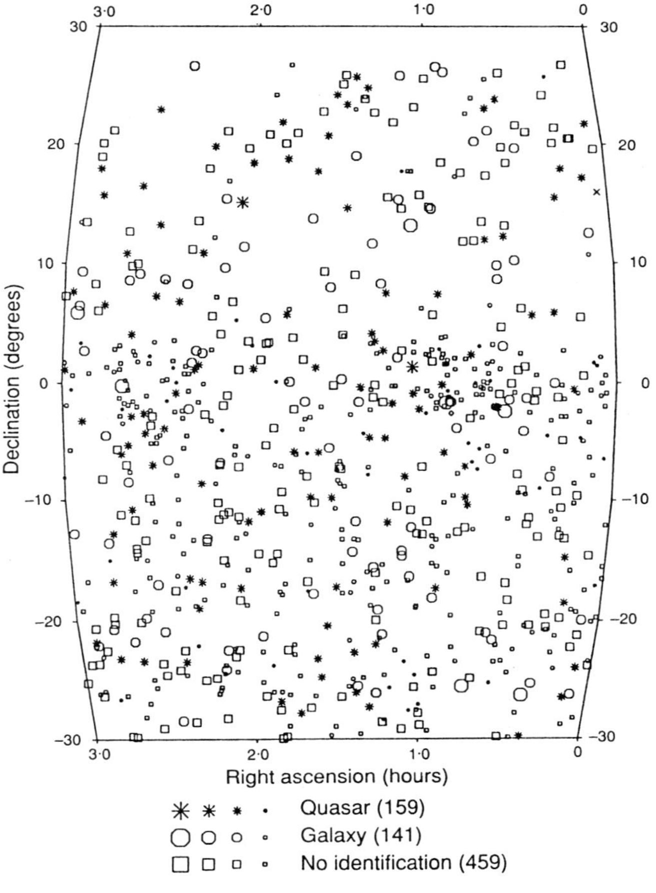

Radio sources in the Parkes catalogue. The plot is for a patch of sky looking away from the galactic plane, with an angular area of 60° by 45°. Shown are over 700 sources, some of which are quasars, some galaxies, but the majority are as yet unidentified. The size of the symbols gives a rough indication of the strength of the source.

The value for the slope had been at the heart of the controversy between the Sydney and Cambridge groups, and whether or not this lent support to the big-bang or steady-state cosmologies. Further analysis of the Parkes sources showed that the slope of the relationship for quasars is slightly greater than that for radio galaxies. As expected, this indicated that quasars are the most distant class of object in the universe and—because of the look-back effect—that quasars must have been created relatively soon after the big bang.

While the majority of sources in the Parkes catalogue are extragalactic objects at vast distances, several surveys along the galactic plane by Jim Caswell, George Day, Ray Haynes and others have yielded detailed information on sources within our own galaxy.[33] These surveys have catalogued close to 1000 galactic objects, most of which fall into one of two broad categories. The first consists of HII regions which are clouds of hot ionised hydrogen with strong thermal emission at radio wavelengths. Usually these regions are large in size, but there is a particular class of compact HII objects known as planetary nebulae which have been studied by Doug Milne and Bruce Slee at Parkes. These nebulae are formed when a red giant blows off its entire outer shell to expose the extremely hot burnt-out core of the star, which then rapidly cools to form a white dwarf. The expelled outer part forms a glowing shell of nebulosity illuminated by the radiation from the white dwarf. These shells had been named planetary nebulae by early astronomers because of their resemblance to planetary discs.

The second category of galactic sources are those that emit synchrotron radiation, the most common of which are supernova remnants. These objects are formed when a very large star has burnt all its available nuclear fuel. The star then undergoes a sudden gravitational collapse or implosion leading to the release of immense amounts of energy, far greater than in the formation of planetary nebulae. The outer shell of the star is blown far out into space where it continues to give off intense synchrotron radiation for thousands of years. Many of these supernova remnants have been observed in detail by Doug Milne at Parkes, yielding information on the structure and evolution of their expanding shells. Two of the best known remnants are the Crab Nebula, the remains of the star that exploded in 1054 AD, and Vela X, the debris from a star that exploded perhaps as long as 20,000 years ago. The collapsed cores of both these supernova remnants have been found to be pulsars (see Chapter 12).

As described earlier, many of the sources in the Parkes catalogue were identified with optical objects using the collection of photographic plates in the Palomar Sky Survey, backed up when required by deeper searches made with the large telescopes at the Palomar and Lick observatories in California. For the southern sources beyond the reach of the northern telescopes, however, the optical identification program was seriously hampered by the lack of a southern equivalent to the Palomar survey. The attempt to identify quasars was particularly frustrating.

By the end of the 1960s, the Parkes astronomers had been credited with the discovery of close to 200 quasars—more than half the total number known—but only one of these quasars was located in the southern half of the sky visible from Parkes.[34] The situation improved dramatically in 1970 with the installation of a new type of image intensifier for two-colour photography on the 2 m telescope at Mt Stromlo. Work by Bruce Peterson with the new intensifier yielded 140 identifications over the next few years, most of which were south of declination –30°. Another major advance was the construction of the UK Schmidt telescope, completed at Siding Spring in 1973. Similar in design and function to its Californian counterpart, this new telescope completed the Palomar Sky Survey from –30° declination down to the south celestial pole, enabling the optical identification of many new southern sources. The completion in 1974 of the Anglo-Australian telescope, also at Siding Spring, then provided a southern equivalent to the Hale telescope to identify very faint, distant sources (for details see Tangent 9).

TANGENT 9

High in the Warrumbungles

In comparison with their radio counterparts, Australian optical astronomers in the 1960s were at a distinct disadvantage. Australian radio astronomy was the equal of that produced by any country in the northern hemisphere, but optical astronomy had fallen well behind. The problem was easily identified. Of the twelve largest optical telescopes in the world, only two were located in the southern hemisphere—the 2 metre telescope at Mt Stromlo and its sister instrument at Pretoria in South Africa.

The need for a large new telescope in Australia became even more urgent in the mid 1960s. A group of five European countries banded together to form the European Southern Observatory and similarly, in the United States, the National Science Foundation established the Cerro Tololo Inter-American Observatory. Both observatories announced plans to build large telescopes in Chile, located at sites high in the Andes. Northern astronomers would now have direct access to the skies which had been the traditional preserve of southern observers.

In the meantime, British and Australian astronomers had been planning a large new telescope in Australia, to be a joint venture between the two countries. Negotiations at government level first began

The Anglo-Australian telescope under construction at Siding Spring in New South Wales. Its completion in 1974 led to a resurgence in optical astronomy in both Britain and Australia.

in 1959, but soon ran into a morass of political and financial problems. The estimated cost of the telescope made it the most expensive scientific project ever undertaken in Australia, which inevitably provoked sharp criticism from the rest of the scientific community.

Agreement was eventually reached in 1967, with both countries to share equally the costs of construction and operation of the new instrument. The site chosen was Siding Spring Mountain in the Warrumbungles, the location of a field station operated by the Mt Stromlo astronomers. To minimise costs, the design was modelled closely on a telescope then under construction at the Kitt Peak National Observatory in Arizona.

Although in size it ranked only fifth in the world league, the 3.9 metre (150 inch) Anglo-Australian telescope became renowned for its outstanding performance. Ben Gascoigne had the privilege of observing 'first light' through the instrument: 'The telescope was accepted almost at once as a technological tour de force. The mounting and optics were clearly of the highest standard, but what created the real impression was the computer control system, which was comprehensive, versatile and efficient to a degree beyond anything previously contemplated. For a while, indeed, visiting astronomers were nonplussed by it.'[35]

The achievements of the Anglo-Australian telescope are far too numerous to describe here, but mention can be made of its success in locating extremely remote quasars. The AAT has held the trophy for the most distant object known in the universe on several occasions, most recently in 1987 with the discovery that the source Q0051–279 has a redshift of 4.43. Fred Hoyle summed it up in his speech at the inauguration: 'A large telescope is a good example of the things which our civilisation does well.'[36]

The source surveys combined with the subsequent optical identifications have formed the single largest research program at Parkes since the early 1960s. Viewed in terms of its overall impact on astronomy, this program stands out as the most significant and successful of all the research carried out at Parkes. Much of the credit for this achievement belongs to John Bolton, who had pioneered the detection and identification of radio sources in the late 1940s. As John Whiteoak writes:

> 'Although many people took part at one stage or other, the motivator was undoubtedly Bolton. He was involved in just about every facet of the work, even from time to time roaming the world collecting photographic data, and building measuring machines for the collected plates. The excellent source catalogues that we have today must be accredited to him.'[37]

The survey of radio sources begun at Parkes in 1962 took place in parallel with a number of major surveys carried out in the northern hemisphere. The Cambridge 4C catalogue completed in 1965 listed over 5000 sources while the Ohio State survey, published in several instalments during the 1960s, also ran to several thousand sources. The excellent agreement between these surveys, followed by the optical identification of a large proportion of their sources, carried through the program which, at the beginning of the decade, had been the overriding challenge facing radio astronomy. By 1970, radio astronomy and optical astronomy had been successfully united into a powerful new science to investigate the universe beyond.

CHAPTER TEN

Sailing on a New Ocean

We have a long way to go in this space race. We started late. But this is the new ocean, and I believe the United States must sail on it and be in a position second to none.

John F. Kennedy, 1962

From the source surveys of the sky we turn now to another part of the operations at Parkes, one which has been minor in terms of the amount of observing time involved, but one which has been significant in a quite different area of science and technology. The same sensitivity and precision that enabled the Parkes telescope to detect quasars and radio galaxies, the most distant objects in the universe, also made it a powerful instrument for tracking objects as close as any natural celestial body—manned and unmanned spacecraft.

The intensely competitive space race that developed between the United States and the Soviet Union in the late 1950s led to a surge of interest in large dishes. Antennas similar in design and operation to a radio telescope were suddenly in great demand for monitoring and communicating with spacecraft. The Parkes telescope became involved in the space program in two ways. First, the telescope itself provided a model for the large tracking dishes constructed by NASA as part of its Deep Space Network. Second, the Parkes telescope became directly involved in tracking spacecraft, beginning with the Apollo missions to the moon (1969–72), followed more recently by the Giotto probe to Halley's Comet (March 1986) and the Voyager 2 encounter with Uranus (January 1986) and Neptune (August 1989). Before describing the part played by Parkes in these areas beyond the traditional realm of astronomy, we begin by briefly tracing the origins of the space race.

Launch of the First Fleet

The launch of Sputnik by the Soviet Union in October 1957 was a spectacular beginning to the space age. The Soviet success in catapulting this 83 kg aluminium sphere with its massive rocket casing into orbit stands as one of the great technological advances of the twentieth century. Sputnik orbited the earth in an elliptical path for three months before descending into the denser parts of the atmosphere, burning up like any medium-sized meteor. Less than a month later the Soviet Union sent aloft a larger satellite, Sputnik II, carrying the dog Laika and miniaturised instruments for measuring the effects of space on the dog's physiology and for studying various phenomena such as ultraviolet emission from the sun and cosmic rays from interstellar space.

In the United States, the Soviet success with Sputnik dealt a serious blow to national pride. Since World War II most major technological innovations such as atomic energy, computers and the transistor had been pioneered in the United States. The country had become accustomed to the idea, to an extent bordering on complacency, that American science and technology would lead the way and the rest of the world would follow. Sputnik was a type of technological Pearl Harbor. However, much more was at stake than a severe bruising of the national ego. With the intensification of the Cold War in the late 1950s, the Soviet Union had powerfully demonstrated its superiority in a crucial area of military technology. With the right type of instrumentation installed in Sputnik, the Soviets would have the world's first spy satellite. More potently, by replacing the satellite on top of the launch rocket with a nuclear warhead, the Soviets would be armed with the first intercontinental ballistic missile. Overnight, the United States had become more vulnerable to military aggression than at any other time in its history.

Before Sputnik, the American rocket development program had been spread across the three branches of the armed services. The Navy won a spirited interservice contest to have first attempt at developing a rocket with enough thrust to lift a satellite into orbit. In 1956 the Naval Research Laboratory began work on the Vanguard series of rockets with the aim of launching the first American satellite during 1957, the International Geophysical Year. The first Vanguard launch in December 1957 turned out disastrously when the rocket exploded in front of millions of viewers on live television. The following month, the Army won the interservice prize when one of its Jupiter series of rockets succeeded in launching the first American satellite, Explorer I. Aside from its military importance Explorer I made the major scientific discovery of the Van Allen belts, an immense doughnut-shaped region high above the atmosphere where charged particles from the solar wind are trapped within the earth's magnetosphere.

The sense of triumph experienced by the Americans with Explorer I proved shortlived. In May 1958 the Soviet Union launched Sputnik III containing instrumentation which could store information and then relay it to earth as the satellite passed over selected receiving stations, thus avoiding the possibility of military eavesdropping. Of greater significance Sputnik III weighed over 600 kg, dwarfing the 14 kg Explorer I, and once again emphasising the widening gap in rocket technology between the United States and the Soviet Union. In the United States the response to the crisis was swift. Rather than a space program fragmented among the three branches of the military and at the mercy of interservice rivalries, in July 1958 Congress established the National Aeronautics and Space Administration (NASA) to coordinate the national effort. Over 5000 personnel under the direction of Wernher von Braun were transferred to the new agency. Strong links with the military were to be maintained, but NASA would be primarily a civilian organisation with the task of propelling the country to world leadership in space science and exploration.

In addition to a number of further Explorer satellites NASA began its central mission late in 1958 with the launch of the first Pioneer lunar probe, one of a series of spacecraft designed to explore space beyond the immediate pull of the earth's gravitation. NASA soon learnt that the success of this program would depend on more than a series of increasingly advanced spacecraft, launched by increasingly powerful rockets. Early in 1959 one of the Pioneer shots missed the moon and, to its embarrassment, NASA was unable to detect the faint signals transmitted back to earth by the probe. Earth-bound satellites presented no difficulty because of their close proximity. Even a small dish like those seen on modern buildings could serve as a receiving station. But for spaceprobes at great distances, equipped with miniaturised transmitters of low power, the need for a network of sensitive tracking stations became a major priority for NASA.

In 1959 there was only one instrument in existence capable of detecting the faint signals from the Pioneer probe, the large dish at Jodrell Bank. Bernard Lovell and his staff had already become closely involved in the tracking of spacecraft, beginning with the first of the Sputniks in October 1957. Even before the new telescope had begun any radio astronomy, Sputnik's sensational launch led to the hurried installation of equipment to convert the instrument into a giant radar dish. Several days later the telescope managed to locate Sputnik's carrier rocket drifting in a separate orbit, a feat neither the United States nor the Soviet Union could achieve with their own tracking stations. The intense interest in Sputnik, combined with a blackout of information from the Soviet Union, helped to focus worldwide attention on Jodrell Bank, the instrument most likely to provide new information on the satellite. Lovell described the occasion when radar echoes were first detected from the carrier rocket: 'We were transfixed with excitement … . The entire place was seething with reporters, live TV was being transmitted and the noise and glare of light and publicity made any scientific task ten times more difficult … . At that stage the BBC alone had many more engineers on the site than I had on my whole staff at Jodrell Bank.' [1]

Sputnik launched Jodrell Bank to the front pages of the world's newspapers in a way that radio astronomy could never have done. Within days Lovell became a public celebrity in Britain. The timing could not have been better. As we see later, the fact that only the Jodrell Bank telescope had succeeded in detecting and tracking Sputnik's carrier rocket helped silence the many critics who had attacked the project as a costly failure.

Global Networking

In response to its pressing need to track spacecraft, NASA began planning a worldwide network of receiving stations early in 1959. The network was to consist of three medium-size dishes, 26 metres in diameter, stationed at three sites around the globe, each approximately 120 degrees apart in longitude. At any one time a spacecraft would be in view of at least one, and at times two, of these tracking stations. The three sites chosen for this system of dishes—known as the Deep Space Network—were Goldstone in California, Johannesburg in South Africa and Woomera in South Australia.

Woomera was a logical choice for NASA on both geographical and logistical grounds. Situated in an isolated stretch of country belonging to a close ally, Woomera already offered the infrastructure needed to support a tracking station. In 1947 a rocket range had been established at Woomera, following an agreement between the British and Australian governments to develop a site for the launching of British guided missiles and scientific rockets. Extensive laboratories and workshops were established at a former wartime explosives factory at Salisbury near Adelaide, while further north in the middle of South Australia thousands of square kilometres of gibber desert were transformed into a vast outdoor missile laboratory. Although the primary function of Woomera was to remain as a missile range, a far more deadly enterprise began later at Maralinga, situated at the northern end of the range, the main base for the series of British nuclear tests.[2]

By the mid 1950s Woomera had developed into one of the most advanced and best equipped ranges in the world, a fact that did not escape the attention of the United States. In 1957 the US established several military facilities at Woomera, one being a radio tracking system known as Minitrack for following the trajectory of long-range missiles. Another was an optical tracking system consisting of an extremely sensitive Baker–Nunn camera, capable of detecting and photographing satellites at distances well beyond the range of military radar. Formal recognition of the presence of these installations came later in February 1960 with the signing of a US–Australia agreement relating to space tracking and communications. The agreement spelt out the function and operation of these multi-million dollar facilities at Woomera, as well as areas of further cooperation on the space program. At the heart of the agreement was a plan to establish a deep space tracking station at Woomera, consisting initially of a 26 m dish, together with two other smaller tracking stations. One of the smaller stations would be near Woomera, while the other at Muchea in Western Australia would be devoted to the Mercury project, NASA's early program of manned spaceflight.

The 26 m dish was a close replica of the first NASA dish built at Goldstone in California. The instrument was constructed in the US, shipped to Australia in segments and then erected late in 1960 near a dry lake bed known as Island Lagoon, a short distance south of the Woomera township. The third dish at Johannesburg followed shortly afterwards, completing NASA's first deep space network. However, even though these dishes were relatively large by the standards of the early 1960s, the network was looked on only as an interim measure. Sending spacecraft to the moon and the planets beyond demanded a network of dishes of the maximum possible size. For several reasons NASA's interest focused on the newly completed telescope at Parkes, located 900 km to the east of Woomera. NASA was particularly impressed by the outstanding design

The 26 metre antenna built in 1960 by NASA at Island Lagoon, near Woomera in South Australia. With two similar instruments at Goldstone and Johannesburg, this dish formed NASA's first deep space network for tracking spacecraft.

of the instrument and, even for an organisation with an almost unlimited budget, impressed too by the economy of the project. The Parkes telescope had cost less than the Woomera dish, though more than double the size. One further development that sharpened NASA's interest in Parkes was the plan by the US Navy to build a 180 m (600 ft) dish, a bizarre episode worth recounting in some detail.

This extraordinarily ambitious plan to build a telescope almost three times the diameter of the Parkes dish originated at the Naval Research Laboratories, the research and development arm of the US Navy. The NRL investigated a wide range of new technologies in the 1950s, with mixed success. One was the series of Vanguard rockets mentioned above. In contrast to the success of Sputnik, only three of the eleven Vanguard launches occurred without mishap, earning the ill-fated rocket the name 'fizznik'. Radio astronomy was a much more successful venture. With a 15 m dish, built on the roof of its Washington laboratory in 1950, the NRL group made a number of fundamental discoveries. One was the detection of thermal emission from several planets, another the first evidence for the polarisation of extragalactic radio sources (details are given in Chapter 9). Up to 1960 the NRL group had contributed more to radio astronomy than any other American group. Another project at NRL was to develop ways to improve the Navy's global communications network. As well as exploring the conventional types of communications, several 'odd-ball' projects were given official sanction, including one which would use the moon as a natural reflector to beam signals from one side of the earth to the other. The idea required a dish of exceptionally large size and, because of its obvious potential for radio astronomy, the communications and astronomy groups joined forces to develop the plan. An engineering study completed in 1956 claimed to prove the feasibility of a gigantic dish and, two years later, the Navy persuaded the US Congress to vote $79 million to fund the project.

To counter the effects of gravitational distortion, the design of the dish included a mechanism for self-compensation consisting of electronically controlled and hydraulically driven surface panels. The idea was the same in principle as the one originally proposed by Barnes Wallis for the Parkes dish, but later rejected as too complex and costly. From the outset the Navy's intention was to build a multipurpose instrument. Half the operational time would be devoted to radio astronomy, the other half to classified research in areas such as communications, radar and the tracking of remote spacecraft. With a collecting area more than eight times larger than Parkes, the giant dish would be capable of tracking spacecraft eight times further into space. In 1959 the sitework began at Sugar Grove in West Virginia, 50 km north–east of the Green Bank radio astronomy site, in preparation for the telescope which would rise as high as a seventy-storey building.

Perhaps enough pages have already been devoted here to the trials and tribulations of the pioneering entrepreneurs of large radio telescopes—Bernard Lovell in Manchester and Taffy Bowen at Parkes —but the events at Sugar Grove remain in a class of their own. A number of factors contributed to the debacle that followed, not least the sheer impossibility of the design. The weight of the telescope alone was calculated to be 36,000 tons, equivalent to a medium-sized battleship. For questionable reasons of security the project also became cloaked in secrecy, which excluded astronomers or other scientists from commenting on either the design or the desirability of the research program proposed for the telescope. Another factor was that advances in other areas of technology made several of the functions planned for the telescope either redundant or outmoded. One rumour was that the telescope would monitor Soviet radio transmissions by using the moon as a relay point, an idea that seemed positively hare-brained following the launch of the first US spy satellite in 1960.

The only part of the telescope that did rise was its estimated cost, increasing from the initial figure of $79 million in 1958 to over $200 million by 1961. Such a staggering escalation proved too much even for Congress, with its normally boundless indulgence in military research. Following the cancellation of the project in July 1962, Daniel Greenberg commented in the journal *Science*:

> 'At that point, expenditures totalled $96 million, for which the Navy had a lot of plans, a large clearing in the West Virginia mountains, a 17,000 cubic-yard concrete foundation, a 550 ton pintle bearing, and a few other monumental odds and ends that will surely baffle archeologists of some coming century unless a copious explanation is carefully left behind.' [3]

The Sugar Grove telescope was undoubtedly an over-reaction to the Soviet success with Sputnik, a dangerous conceit that huge budgets and American know-how would swiftly propel the US into the lead in the space race. The fiasco made NASA extremely wary on the subject of large telescopes and particularly eager to explore alternatives.

☆ ☆ ☆

Discussions with NASA on the way the Parkes telescope might be used in the space program began early in 1959, during a visit by Taffy Bowen to Washington. Collaboration with NASA held a number of attractions, not least the possible financial benefits as Bowen later reported:

> 'On the question of NASA and their possible use of the telescope, I agree that life would be quieter without them. However, the possible financial gains are too good to miss—always provided they go into the research activities of Radiophysics. As an example of what is happening, a Blaw–Knox 85 ft dish is being put up at Woomera at the present time, all at NASA's expense. The total cost is firmly stated as being £A2,000,000, of which £900,000 is a contribution to Woomera for their part of the construction program. With this kind of money going around, we cannot afford to be stand-offish.' [4]

Right Artist's impression of the 600 ft colossus planned by the US Navy. *below* The site for the telescope at Sugar Grove in West Virginia at the time of the project's cancellation in July 1962. By then $US96 million had been wasted on the project, enough to have paid for the Parkes telescope forty times over.

On another level it was also plain that considerable political mileage could be made from this cooperation with NASA. Indeed, as CSIRO minister Richard Casey carefully explained to fellow cabinet ministers in June 1959, the Parkes telescope was likely to strengthen Australia's hand in negotiations with the United States:

> 'In the US considerable sums of money are at present being expended on the launching of earth satellites and of space probes. The Americans now realise that if they are to accurately follow the projectiles that are launched into space for the purpose of making observations on the surface of the moon and for other similar purposes, they need large radio telescopes to enable them to receive the radio signals from these probes, transmitting information back to the earth's surface. [Our] contacts have so far been at the unofficial level but NASA has expressed the desire to have use of our telescope when it is built and has asked to be informed of progress made .... Although Australia is unlikely in its own right to enter this field of research ... nevertheless it is important nationally that we should take full advantage of [developments] here which would influence our position.' [5]

In February 1960 NASA formalised its interest in Parkes as part of the wider US–Australia agreement relating to the Woomera installations referred to above. The agreement focused on two points. First, NASA would enlist Parkes as a tracking station during crucial phases of missions when the dish would offer greater sensitivity and reliability than the Woomera antenna, at least until it could build a dish of similar size in Australia. NASA agreed to meet the operating expenses, provide the receiver equipment and give advance notice of when the telescope would be required. The second point involved using the Parkes dish as the prototype for the large dishes in the second-generation deep space network planned by NASA. Complete sets of technical specifications and drawings were turned over to NASA, with an open invitation extended to its personnel to visit the site during the telescope's construction. In return NASA agreed to fund a detailed study of the engineering characteristics of the telescope, starting with a grant of $US172,000.

In addition to NASA personnel, other visitors during the telescope's construction included scientists and engineers from the Jet Propulsion Laboratories (JPL) in Pasadena which carried out a large share of NASA's research and development. JPL had designed the Explorer I satellite, the 26 m Woomera dish and had then been contracted by NASA to develop the new deep space network. A number of other American organisations were also granted visiting rights at Parkes, such as Bell Labs and the Office of Naval Research. As Bowen commented shortly after the completion of the telescope: 'The Yanks are now with us putting a tape measure over everything, but as usual they are thinking of a gold-plated Cadillac rather than a utility model.' [6]

Another group from the Lincoln Lab at the Massachusetts Institute of Technology studied the structural features of the Parkes dish during the design of its 40 m telescope, completed in 1964 with $12 million in funds from the US Air Force. Known as the Haystack, this telescope was used mainly for radar studies of the moon and nearby planets. To protect the dish from the elements Haystack was enclosed in a large geodesic bubble or 'radome'. The same type of spherical structure later sprang up like mushrooms in the Australian outback at Pine Gap and other US installations.

NASA's interest in the engineering features of the Parkes dish led to the formation of a project team at Radiophysics known as the aerial group, led by Harry Minnett and with the task of studying the structural, electrical and radio properties of the telescope. One important program was an analysis of the surface accuracy of the dish. During the commissioning of the telescope in 1961, the dish had been simply parked in the zenith and the surface adjusted as closely as possible to a true parabolic shape using the standard tools of a surveyor—a theodolite and tape measure. No information was available on the amount of surface distortion caused by gravitational stresses as the dish actually tilts from the zenith. This led Minnett and Maurice Puttock to develop an automatic survey device installed at the centre of the dish which could photographically record these slight distortions. The results allowed the surface to be reset to minimise the effects of gravity and also indicated that even more accurate panels could be installed in the future. Eventually, this enabled the telescope to be operated at wavelengths much shorter than the original design specification.

These studies on the Parkes telescope provided the engineers at the Jet Propulsion Lab with a wealth of data and ideas to use in their design of the 64 m Goldstone dish, the first of the antennas in NASA's new deep space network. Goldstone adopted many features of the Parkes dish, such

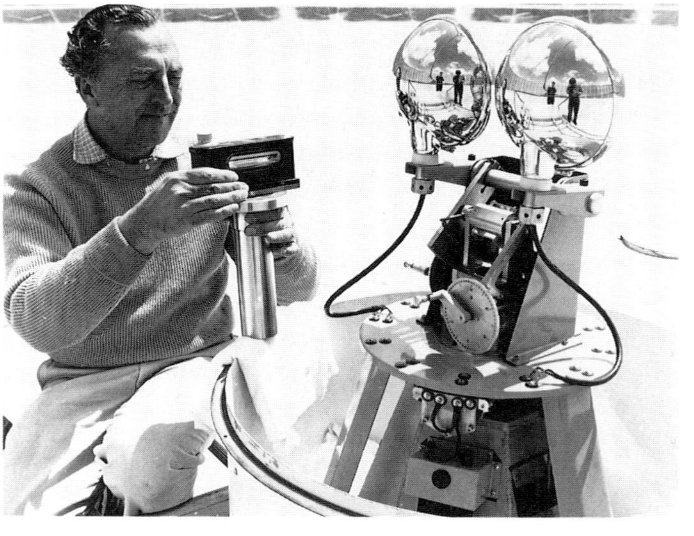

Engineering studies funded by NASA led to the invention of an automatic survey camera to measure the surface accuracy of the Parkes dish. Developed principally by Harry Minnett (pictured), the instrument could detect distortions over any part of the surface to within one millimetre, for any angle of tilt.

as the master equatorial unit and the servo-drive system, but by no means was it a direct replica. Most of the differences reflected the different requirements of a dish designed to both transmit and receive signals to and from nearby spacecraft, rather than one designed only to receive signals from distant astronomical objects. Goldstone was to be a more flexible and robust instrument, one that could tilt down to the horizon and operate in much stronger winds. The design originally allowed for operation in winds up to 100 km per hour, but when this sent the projected cost to well over $20 million, this specification was relaxed to 60 km per hour. Construction of the instrument began in 1963 at a site in the Mojave Desert in California and was completed in April 1966 at a cost of $16 million. The Goldstone and Parkes dishes had more in common than just their design features. Head of the Goldstone project Eb Rechtin reported that even the inauguration had a curious similarity with the one at Parkes: 'As is customary in such 64 m dedications, a slight wind came up. In our case, it managed to break the red ribbon that the Congressman was supposed to cut to dedicate the antenna!' [7]

The NASA-funded studies on the Parkes telescope generated a minor research industry. By 1966 over thirty research papers on the telescope's features and performance had been published in engineering and telecommunications journals, information made freely available to the designers of large dishes elsewhere. Parkes turned out to be not only the most advanced radio telescope of its type, but also the most extensively studied.

Exploring the Neighbourhood

Important though they were in perfecting the technology of large dishes, the NASA-funded studies at Parkes were hardly the stuff of newspaper headlines. If there was any real prestige to be gained for Australia's premier scientific instrument in the new space age, it would come from using the Parkes dish to track spacecraft. However, in sharp contrast to the engineering studies, by the mid 1960s the tracking program at Parkes had been a distinct anticlimax.

The agreement reached earlier in 1960 to allow NASA occasional access to the telescope had not been spelt out in any detail. NASA had also reached an agreement with Jodrell Bank, paying a rental of £400 each day the telescope spent tracking its spacecraft. The Jodrell Bank astronomers had scored a major triumph early in 1960 when they succeeded in transmitting signals which released the deep space probe Pioneer V from its final rocket stage. Outside its contract work a certain amount of freelancing was also carried out, with Jodrell Bank usually the first to announce the launch of a new Soviet or US spacecraft to a news-hungry world. Even though tracking only occupied a small fraction of the telescope's time, the intense public interest it created brought its own rewards. If the space program was a head-to-head contest between the Soviet Union and the United States, then at least with the Jodrell giant Britain could claim some share of the action, one item of space-age technology superior to anything possessed by the Russians or the Americans. National acclaim mounted rapidly, leading to a complete reversal in public and government support for the telescope. As recounted earlier, well after the completion of the telescope in 1957 the funding crisis had reached such proportions that Lovell had been threatened with prison. Two years later, largely in recognition of the telescope's role in the space program, a substantial grant from the Nuffield Foundation finally bailed the project out. In 1961 Lovell's personal efforts were at last recognised by a knighthood for his contributions to British science.

Although Bowen privately expressed reservations about the role played by Jodrell Bank in the tracking program, he too had been eager to become involved with NASA and for Radiophysics to share in some of the limelight. The first opportunity arose in mid 1962 when NASA gave notice that it would launch two Mariner probes to Venus and requested the use of Parkes for tracking during the encounter periods. Nothing substantial came of the request. In July Mariner I was purposely destroyed five minutes after launch because of a serious deviation in its flight path. The following month Mariner II struck the same problem, but mid-course corrections put it back on track and the probe successfully transmitted data back to earth as it flew by Venus. Mariner's signals were detected at Parkes for a further month after the encounter, though the exercise was basically a test observation and NASA made no use of the data. Next, NASA turned its attention to Mars. Mariner III was cancelled but then Mariner IV was successfully launched early in 1965. Detailed measurements were made by Doug Cole at Parkes of the probe's position, as well the strength and Doppler shift of its signals during the encounter period in July–August. Telemetry signals were also recorded, stored on magnetic tape and then added to NASA's own recordings which, in turn, were used to produce the first close-up photographs of the Martian surface. The results surprised astronomers for they revealed a heavily cratered planet with a cold thin atmosphere dominated by carbon dioxide, certainly not the hospitable planet suggested by telescope observations from earth.

By 1966 the actual amount of time spent tracking at Parkes had been minuscule, at least in comparison with the first few years at Jodrell Bank. Mariner IV was in fact the only occasion Parkes had been formally incorporated into NASA's tracking network. But even on this occasion the collaboration ran into difficulties. When NASA scientists published the findings of the Mariner IV mission, including the remarkably detailed photographs of the Martian surface, no acknowledgment was made of the role played by Parkes. Giving due credit is a sensitive issue in any area of science. In isolation this oversight would not have been an issue for the Parkes astronomers, but it is worth noting here that radio studies of the planets had developed into a small but particularly fruitful part of the research program.

As well as the work on Jupiter described earlier, some intriguing observations on the other planets had also been made during 1963–65, principally by Ken Kellermann visiting Parkes from Caltech:

> 'I was mostly working on extragalactic sources and knew very little about the planets. The policy was that daylight was reserved for telescope maintenance, except in special cases where the observations could not be made at night. I was living at Parkes and it bothered me to see the telescope idle, so I looked for an observing program that had to be done in the daytime. Mercury was an obvious choice. I anticipated that I would be able to detect Mercury only when the sunlight side was pointed towards the earth. When my observing time started the night-time side of Mercury was pointed towards the earth, but I began anyway in order to work out the best experimental procedure. I was very surprised to be able to detect a relatively strong signal. I followed the planet for months as it went around the sun and noticed that the day and night sides were at essentially the same temperature.
>
> It was clear that there were only two ways to explain how the night-time side could be as hot as the daytime side. Either the planet rotated or there really was an atmosphere which carried heat from the sunlight side to the dark side. All the textbooks said that the planet did not rotate, but the only published research I could find was about a hundred years old and was in French, which I did not read. On the other hand, someone had recently suggested there might be a thin argon atmosphere that could conduct heat from the sunlight to the night side, so I interpreted my results in terms of an atmosphere.' [8]

Radar observations made shortly afterwards at Arecibo in Puerto Rico showed that Kellermann had guessed wrongly. The planet was in fact found to be rotating very slowly, with the length of the Mercury day equal to 59 earth days.

The planetary program at Parkes brought other rewards. Kellermann was the first to detect radio emission from Uranus, the third most distant planet from the sun. Observations on Saturn, Venus and Mars also gave much improved values for the surface temperatures of these planets. The observations on Mars were of particular interest because they ruled out the possibility that the red planet possesses Van Allen belts similar to those surrounding the earth. The data telemetered from Mariner IV to Parkes confirmed some of the information found by Kellermann's direct measurements.[9]

In view of NASA's apparent indifference to Parkes as a tracking station, it came as a complete surprise late in 1966 when NASA put forward a proposal to incorporate the telescope permanently into its worldwide tracking network. Several factors lay behind this sudden change of mind. Until then the network of 26 m dishes had met most of NASA's requirements, but plans to send probes to more distant planets plus the manned Apollo missions to the moon demanded a network of larger dishes. For an organisation accustomed to exponential growth NASA had also just experienced its first budget cutback. One consequence was that NASA had been forced to postpone the construction of the second (Australia) and third (Spain) stages of its own 64 m network. To add to its troubles, its dish at Goldstone had encountered problems which had shut down operations for an extended period (see below). These developments began to make Parkes an increasingly attractive alternative. NASA proposed using the telescope for a minimum of one day a week over a period of three years and was prepared to pay a hefty rental for the privilege. The first tracking program would be to monitor the series of Pioneer probes, launched into orbit around the sun to form a network of solar weather stations.

Despite its attractions there were two drawbacks to the NASA proposal. On a scientific level, the experience with Mariner IV had already pointed out the dangers of becoming a junior partner in another organisation's research program. Added to that, not enough time on the telescope was available to meet the mounting number of observing requests from Australian astronomers. An increasing number would have to miss out. For both these reasons, the NASA proposal was turned down.

Crossing Live to the Moon

First words spoken on the moon:
'Houston, Tranquility Base here. The Eagle has landed.'
Neil Armstrong, Apollo 11, July 1969

Last words spoken on the moon:
'Let's get this mother out of here!'
Eugene Cernan, Apollo 17, December 1972

The decision early in 1967 that Parkes would not join the Deep Space Network seemed to spell the end of any further involvement with NASA. Two years later, however, NASA's Office of Manned Space Flight made another approach, one that would be difficult to refuse:

> 'The Apollo program has now progressed sufficiently that we are planning our first manned lunar landing mission in mid-69. This very complex mission will impose tracking and data acquisition requirements which exceed those of past space flight missions by a wide margin. One of these, the provision of live TV from the lunar surface, is feasible only with the addition of support by large antennas of 64 m or greater diameter. In view of this, I would like to request your consideration of the participation of the Parkes facility in support of the Apollo lunar landing mission.' [10]

Until then the question of whether to become involved in the Mariner or Pioneer programs could be argued on the grounds of whether the scientific rewards were great enough. The Apollo program immediately raised issues of much broader importance. John Bolton adds:

> 'In contrast to our earlier approaches from NASA, on this occasion we would be directly involved with the Jet Propulsion Lab at Pasadena. Our contacts with people like Lee DuBridge went all the way back to the early fifties and they'd helped us a lot. On several occasions during the commissioning of Parkes, components had broken down which couldn't be replaced by any Australian supplier. All we had to do was call JPL and they would send whatever we needed on the next military aircraft or via the diplomatic bag from Washington. We owed them a lot. Of course, this was only one factor. For a host of reasons it was obvious that we had to do the Apollo missions. We couldn't say no when the lives of people were at stake.' [11]

When Yuri Gagarin became the first man into space early in 1961, the Soviet Union again demonstrated its commanding lead in the space race. American pride took a second heavy blow, but the response was immediate. In May 1961, Alan Shepard made a successful suborbital flight in a Mercury capsule to become the first American into space. Although this was a much inferior technical achievement to the Gagarin flight, the world's media heralded the event as an outstanding success. Three weeks after Shepard's flight, the United States announced its ambitious plan to land a man on the moon by the end of the decade. The plan was formulated more for political reasons than military or scientific necessity. Although there was genuine concern that the Soviets might arrive on the moon first and establish a military base, the central purpose of the Apollo program was—as John F. Kennedy's words at the beginning of this chapter state—to provide startling proof to the world that the United States was in a position second to none.

Most of the American space program became focused on this single goal. First came the series of Mercury shots (1961–63) which tested the ability of astronauts to work in earth orbit, followed by the Gemini missions (1965–66) which used two-man spacecraft to refine the techniques of rendezvous, docking and spacewalking. During the same period the unmanned Ranger, Surveyor and Orbiter series amassed cartographical, geological and geophysical data about the moon. Overall, the size of the space program expanded enormously. Supported by an enthusiastic Congress, NASA's budget approximately doubled each year, peaking at $5.25 billion in 1965 when the number of people directly on its payroll or under contract reached half a million. NASA's program involved more than 20,000 companies, as well as 120 universities and research institutes. If the United States was to be the Great Society in the 1960s, then NASA would be its technological showpiece.

The rapid pace during the early part of the Apollo program kept NASA on schedule for a manned lunar landing by the end of the decade. But then, tragically, in January 1967 three astronauts died after becoming trapped inside their module which caught fire during a launch simulation. The enquiry into the disaster uncovered serious deficiencies in the design of the spacecraft, resulting in a delay that left the race to the moon wide open. The Soviet Union had also been making good progress with the series of one-man (Vostok) and two-man (Voskhud) missions which led to the development of the new Soyuz spacecraft. However, the first Soyuz mission in April 1967 also ended in tragedy when the spacecraft crashed after its parachute failed to open during re-entry. The accident caused a similar delay in the Soviet program. Manned flights by both countries resumed within days of each other in October 1968.

With the launch of Apollo 8 in December 1968, the first manned flight away from the grip of the earth's gravity, the United States demonstrated that it had opened up a substantial lead. Shortly afterwards the Apollo 9 module was test flown in earth orbit, and then Apollo 10 conducted a full dress rehearsal for a lunar landing. By mid 1969 the United States was on the verge of fulfilling the pledge made eight years earlier, while the Soviet Union slowly lost momentum. A telling factor was the development of the Saturn V launcher by the United States, by far the largest and most powerful rocket ever built. The Soviet attempt to build a comparable launch vehicle, crucial to the success of their lunar program, suffered a series of setbacks including three known occasions when the rocket exploded on the launchpad or shortly after lift-off. Instead, the Soviet Union decided on a radical change in direction, concentrating its effort on two main programs—unmanned exploration of the moon and the development of the earth-orbiting manned Salyut station. It was a bold and eventually successful strategy. By the end of the Apollo program in 1972, the Soviets had re-established their lead in space.

The flight of Apollo 11 and the first moonwalk was a dramatic interlude in the astronomical program at Parkes (for a step by step account see Tangent 10). The signals received at Parkes from the lunar surface were transmitted first to Sydney, relayed to a telecommunications station at Moree, beamed via satellite to California, and then sent to NASA's mission centre in Houston. Signals also came in from other parts of NASA's network, including a small dish at the Honeysuckle Creek tracking station near Canberra, but the signal from Parkes was by far the strongest and clearest, and the one selected by Houston for all but the first few minutes of the landing. From Houston the picture was beamed throughout the world, ultimately seen by an estimated 600 million television viewers in fifty countries. For the first time, the entire global telecommunications system was focused on an event of unique importance.

Shortly after Apollo 11 NASA made clear its desire to include Parkes in its plans for future missions. In September 1969 chief administrator Tom Paine sought help with the Apollo 12 mission:

> 'I am keenly aware of the role and importance of the Parkes antenna in support of the Australian radio astronomy program and regret the need for requesting that time be diverted from this program to support the Apollo program. However, you can understand my concern for Apollo since absence of this coverage, with associated loss of data, could abort the mission … . I would greatly appreciate your assistance in obtaining the necessary approvals for the Apollo 12 support.' [13]

Two months later George E. Mueller, associate administrator of NASA's manned space flight program, visited Australia and clarified the role Parkes would be expected to play in future Apollo missions, particularly during the critical stages of lunar landings. In return Mueller offered a sum of $US90,000 for the telescope's part in Apollo 11 and 12, with a

TANGENT 10

One Small Step

The preparations at Parkes for Apollo 11 began several months before the scheduled launch date of July 16, 1969. Only backup equipment was required because NASA had arranged for the crucial part of the mission, the first moonwalk, to occur at a time when the moon would only be in view from its main 64 m dish in California.

Round 1: Goldstone would take the first moonwalk.

Two months before the mission NASA altered the schedule by adding an eight-hour rest period to allow the astronauts time to sleep, in between landing on the moon and then attempting the first moonwalk. Goldstone would now be out of view and the status of Parkes had suddenly changed from backup to prime receiving station.

Round 2: Parkes would take the first moonwalk.

A month before the launch a team of NASA engineers led by Bob Taylor arrived at Parkes and began installing equipment, most of which was set out on makeshift racks and tables in the telescope's control room. Among the several tons of equipment was a prototype of the first home video recorder to capture the televised moonwalk. The astronomers at Parkes took responsibility for ensuring that the telescope's drive and control systems were in perfect working order. An analysis was made of the failure rates of every part of the telescope dating back to 1961, with identical backup equipment made ready should the need arise. Nothing was left to chance. If all else failed, teams of men had been organised to turn handles in the gear boxes and drive the telescope by their own muscle power!

The launch went according to schedule on July 16. Over the next four days the equipment at Parkes was thoroughly tested using signals sent from the command module Columbia during its outward journey. After ten orbits of the moon Neil Armstrong and Buzz Aldrin transferred to the lunar module Eagle, leaving Michael Collins in the command module. Then began the descent to their chosen point of touchdown, the Sea of Tranquility. Once the Eagle had landed, an excited Armstrong overruled the idea of a rest period and advanced the scheduled moonwalk by four hours. All the elaborate preparations now seemed in vain—Parkes would be on the wrong side of the earth.

Round 3: Goldstone would take the first moonwalk.

Then there was an unexpected problem. The two astronauts ran into difficulties fitting their spacesuits and the moonwalk was delayed by several hours. By the time Armstrong had descended the steel ladder on Eagle to utter his famous remark—'That's one small step for a man, one giant leap for mankind'—the moon had just come into view at Parkes.

With the moon almost directly overhead, the Goldstone dish was perfectly placed to receive the TV transmission of the moonwalk, as arranged. But events again took a strange twist. Twenty kilometres away from the dish there was a faulty connection in the cable carrying the TV signals from Goldstone to the command centre in Houston. Goldstone's picture disappeared into the ground.

Final round: Parkes took the first moonwalk.

Throughout the afternoon of Monday, 21 July the team at Parkes carefully tracked the moon to achieve the strongest possible signal. At 5.57 pm, after almost five hours of reception, the telescope was reluctantly shut down as transmission ceased from the telecast 400,000 km away.

To add to the drama, the operating conditions could not have been worse. For several days during Apollo's outward flight to the moon the weather had been perfect at Parkes. Incredibly, moments before the moonwalk began a violent squall could be seen approaching. With the dish down at full tilt it was in its most vulnerable position. In any other circumstances the dish would have been hurriedly driven to the zenith and stowed in its safety position.

The wind gust hit at 100 km per hour, tipping the dish back and badly damaging the first gear-tooth on the rack. Those in the control room felt the tower sway several inches. It was an extraordinary piece of bad timing. The gust was the strongest ever recorded during the telescope's eight years of operation.

The battering wind kept up throughout the afternoon, but despite the dangerous conditions the flow of signals from the moon to Houston via Parkes continued without interruption. A few days later the *New York Times* reported: 'The tracking crew at the Parkes antenna in Australia won praise today for its work in picking up signals of the Apollo 11 moonwalk. The flight controller, Cliff Charlesworth, said he wanted to give the crew a special bouquet for getting into operation quickly when the signals were suddenly shifted from the Goldstone tracking dish in California. 'It's obvious that the support they provided us was pretty outstanding', Mr Charlesworth said, 'because I think it will be years before anyone can beat that TV spectacular'.' [12]

The events best remembered during Apollo 11 were the first steps on the lunar surface and the ceremonial flag raising. At the time the scientific importance of the mission was largely overlooked. Armstrong and Aldrin spent most of the two hours during the first moonwalk establishing a small lunar laboratory. Geological samples were gathered from the surface, a foil spread to capture particles of the solar wind and a mirror device set up to enable precise laser measurements of the earth–moon distance. A seismograph installed to register moonquakes showed the moon to be a very inert object with little sign of seismic activity. Moonquakes release less than one-billionth the energy involved in a typical earthquake.

promise of further grants on a contract basis for later Apollo missions. These grants were used to fund a partial repanelling of the surface of the dish (see the next chapter).

The four months between Apollos 11 and 12 provided an opportunity to catch up on some repair work at Parkes. Two years earlier, the concrete grout surrounding the steel bolts which fix the azimuth roller track to the top of the telescope tower had begun to crack and crumble. The azimuth track had started to shift slightly under the load of the roller bogies which support the 1000 tonne structure above, raising serious doubts about telescope safety. A similar problem had already occurred with the Goldstone dish, shortly after its completion in 1966, all because the concrete had originally been poured with an incorrect mix of water. Repairs had been made at Goldstone with an epoxy resin, but this too had failed causing serious damage to the azimuth track and forcing a second major shutdown of NASA's prime station. At Parkes the replacement of the grout with a new type of expansion cement proved successful, but then a more serious problem was identified. The massive load moving round the roller track was causing a slight bulging of the tower. Cracks had appeared in the sloping roof of the tower, in a number of windows and at regular intervals around the internal wall of the control room. Strengthening the tower was the only solution. The outside of the tower was stripped back so that steel columns, running from the first floor up to the azimuth track, could be bolted to the tower wall. New pipes, cables, window frames, and airconditioning ducts were installed, the job being finished off with a new external cladding of concrete rendering. The work closed down operations for two months—the telescope's first major shutdown—and was completed just in time for the launch of Apollo 12 in November 1969.[14]

Apollo 12 had an unnerving start when shortly after lift-off the rocket was struck by lightning which blew the main fuses. The problem was quickly rectified and the remainder of the mission turned out to be a copybook flight. The astronauts were more adventurous than the Apollo 11 crew, roaming greater distances from the lunar module to gather rocks, set up scientific instruments and even recover pieces from the earlier unmanned Surveyor 3 spacecraft. At Parkes the mission proved uneventful. Only a few minutes of the telecast were received before one of the astronauts burnt out the camera by pointing it at the sun. The reception of television signals, however, formed only a small part of the task at hand. Monitoring the constant stream of information on the status of the spacecraft's numerous functions and on the biomedical condition of the astronauts was vital to the success of the mission, especially when the command or lunar modules entered rest periods and switched to transmission at very low power levels. While the NASA engineers operated the main receiving equipment in the control room, the Parkes staff took responsibility for the front-end equipment in the aerial cabin and also for ensuring the telescope kept the spacecraft in its sights. Data on the flight path supplied by Houston were often inaccurate, so that at the beginning of a tracking shift the spacecraft usually had to be located by square searching the most likely patch of sky.

The participation of Parkes had not been requested for Apollo 13 because the moon during this mission, scheduled for November 1970, was expected to lie too far north in the sky. The mission had however taken the devil's number. Midway into the outward flight the oxygen tanks on the command module exploded, damaging the spacecraft and severely cutting its power supply. With no possibility of turning back or of rescue by another spacecraft, the three astronauts transferred from the command to the lunar module and waited on help from earth. At Parkes, John Bolton and several of the staff had been listening in to the transmissions between Apollo 13 and Houston:

> 'As soon as we heard the distress signal we got into action. We didn't need to wait for an official invitation. NASA had a manual of operating and emergency procedures which I had pretty much learnt by heart. The plan would be to let the spacecraft continue round the moon and, hopefully, come right back to earth. At this stage of the mission the spacecraft would still be in our view so we started immediately to round up equipment. A

NASA engineers at Parkes during the Apollo 12 mission in November 1969. Most of the equipment was left permanently in the control room until Apollo 17 completed the program of manned lunar landings three years later.

charter plane brought some of it out from Sydney, and NASA engineers drove other bits up from Canberra. The telecommunications links to the United States also had to be opened up. We managed to get it all working by early evening that day, just as the spacecraft came into view. It had just enough power to transmit voice and telemetry signals. The signal was about a thousand times weaker than those we had received from Apollo 11 during the moonwalk. What little power they did have had to be conserved for crucial manoeuvres in the trajectory. By this stage I had done nothing for about six hours but stand in the control room shouting orders.' [15]

The NASA emergency manual allowed at best a minimum of 24 hours to prepare the telescope for spacecraft reception; the Parkes staff had managed it in less than six. For the next eight hours Parkes monitored the feeble signals from Apollo 13, a critical period in which NASA had to assess the various options and, most importantly, decide on the precise trajectory that would bring the crippled spacecraft home. Aside from the very real danger of crashing on the lunar surface, the angle and speed of approach had to be calculated exactly to avoid the possibility of the spacecraft being trapped in lunar orbit. Alternatively, the spacecraft had to avoid being propelled by gravity around the moon too fast and then being catapulted away into space in an uncontrolled direction. The tension rose at Parkes when radio communication stopped as the craft disappeared behind the moon—one occasion when the lunar occultation of a radio source was definitely unwelcome. Almost half an hour later Apollo 13 reappeared, roughly on course. Further corrections to the trajectory were made on the return journey. Eventually, the three astronauts splashed down safely in the Pacific, almost four days after the on-board explosion had occurred.

Parkes participated in the four remaining Apollo missions, but none matched the heroic levels of Apollos 11 or 13. Following routine support for Apollos 14 and 15, Parkes was again called in to provide emergency help with Apollo 16 when problems arose with the ignition system in the lunar module's rocket. Unlike Apollo 13, the emergency quickly passed. During Apollo 17, the last manned landing to date, the TV transmission again came through Parkes, just as it had during the first moonwalk three years earlier. On this occasion the transmission continued after the astronauts had rejoined the orbiting command module. NASA planned to observe the effects of the lunar module falling back and crashing on the lunar surface. The impact would provide information on the properties of the surface and also on the crust through the moonquake detectors left by the astronauts. Unfortunately the spectacular impact went unseen. The camera left behind on the mission's moonbuggy had been pointed in the wrong direction! Similar to the Apollo 11 mission, the operations at Parkes were carried out in a violent dust storm which lasted throughout the transmission.

The completion of the Apollo program in December 1972 marked the end of space tracking at Parkes on any regular basis. By then NASA was close to completing its own global network of large dishes and no longer depended on Parkes. As already mentioned, immediately after building the first deep space network of smaller dishes at Goldstone, Woomera and Johannesburg, NASA began planning the 64 m network with the first dish completed at Goldstone in 1966. Two other instruments came later at Tidbinbilla, near Canberra, and at Madrid in Spain (South Africa had been ruled out as too unstable politically). The Tidbinbilla dish was only one of the three main tracking stations established by NASA in the Canberra region. The first was at

Honeysuckle Creek about 60 km south–west of Canberra, consisting of a 26 m dish completed in 1967. Built primarily for the Apollo program, Honeysuckle Creek later supported the manned Skylab space station as well as a range of Pioneer, Viking and Mariner missions to the planets. Another station at Orroral Valley was established primarily for tracking earth-bound satellites, among them the Nimbus meteorological series, the Landsat program for studying earth resources, and the Explorer satellites for investigating the outermost regions of the earth's magnetosphere. Orroral Valley also provided major support for the space shuttle program in the 1980s. Outside the continental United States, Australia has played host to more American tracking stations than any other country in the world.

Initially NASA planned to build its 64 m dish at the Island Lagoon station near Woomera, but then decided on a site further east. For a time Parkes was considered until, in 1962, NASA chose Tidbinbilla, south–west of Canberra. Two years later a 26 m dish was installed, followed eventually in 1973 by the 64 m dish. Since then Tidbinbilla together with the large dishes at Goldstone and Madrid has carried out most of NASA's deep space tracking. In 1985 the 64 m dish was connected by a microwave link to the Parkes telescope, 275 km to the north, to form the Tidbinbilla–Parkes Interferometer, the first long baseline array of its type in Australia. The array has been used to study a variety of objects such as

The 26 m dish at left was the first instrument built by NASA at Tidbinbilla. NASA completed the 64 m dish in 1973, one of the three instruments forming its second-generation deep space network. In 1985 the large dish was connected by a microwave link to the Parkes telescope to form a powerful new long-baseline interferometer. The Tidbinbilla–Parkes link has been used for radio astronomy and for tracking distant spacecraft such as Voyager 2 during its encounters with Uranus and Neptune.

the cores of unusual galaxies, galactic radio stars, pulsars and interstellar molecular clouds. The array was also used to test many of the ideas and techniques which were later incorporated into the design of the Australia Telescope.

Voyages and Visitations

> Damn the solar system. Bad light; planets too distant; pestered with comets; feeble contrivance; could make a better one myself.
>
> Lord Jeffery, circa 1800

In a break with tradition the Parkes telescope was contracted in 1986 to carry out tracking for the European Space Agency, the European counterpart to NASA. The mission on this occasion was to cover the encounter by the Giotto space probe with the comet named after Edmund Halley. Of the five probes planned to encounter the comet, Giotto was to be the most ambitious and, as it turned out, by far the most successful.

Giotto was equipped with ten miniaturised instruments designed to measure a range of features such as the molecular composition of the comet and the strength and shape of its magnetic field, information that might finally answer age-old questions on the origin and nature of comets. Most of the instruments were to lie dormant during the long outward journey and then be activated a few hours before the encounter. The main instrument was an elaborate colour television camera designed to transmit images back to earth as Giotto approached the comet.

Giotto was launched from French Guiana in July 1985. Eight months later it closed in on its target with a velocity approaching 100 km per second. At Parkes the countdown began early on the morning of Friday, 14 March 1986. Three hours before the encounter the probe registered the first effects of the comet, a decrease in the strength of the solar wind and fluctuations in the magnetic field. One hour later, Giotto began passing through clouds of ionised gas containing an unexpectedly high proportion of heavy elements such as iron. After another hour the first dust particles were detected at a distance of 200,000 km from the comet's nucleus. With ten minutes remaining Giotto entered uncharted territory as it penetrated the coma, the luminous cloud of gas and dust that surrounds the nucleus. The composition of the coma rapidly changed from an ionised gas to neutral molecules. The television picture received at Parkes began to show a clear image of the comet's nucleus. With two minutes before closest encounter a tiny dust particle penetrated Giotto's protective shield but the impact was absorbed. The magnetic field then vanished as the probe entered the comet's magnetopause. The density of the comet's atmosphere began to increase rapidly. The best television

Halley's Comet photographed by the Schmidt telescope at Siding Spring during its 1986 visitation. The data received at Parkes from the Giotto probe provided a wealth of new information on the nature of comets.

pictures of the nucleus were now being received. At a distance of just 600 km from the nucleus Giotto hit a small chunk of matter, sending the probe into an uncontrollable tumble.

Thirty minutes later, Parkes reestablished contact with Giotto. The badly damaged probe was no longer capable of useful work but the mission had already succeeded beyond expectation. The colour pictures of the nucleus revealed a potato-shaped object about 15 km long and 10 km wide. The surface appeared rough with hills about 100 metres in height. The most surprising feature was the almost jet-black colour of the surface. Astronomers had predicted the whitish-grey of a dirty snowball, but instead found an object better described as a snowy dirtball. The tarry substance coating the surface apparently contains organic molecules. The discovery heightened speculation that primitive lifeforms might exist in the nucleus of comets such as Halley's.

Throughout the encounter Giotto transmitted a busy stream of information back to earth 145 million km away. After collection and amplification at Parkes these signals were relayed via a satellite tracking station at Carnarvon in Western Australia to the European Space Office Centre at Darmstadt in West Germany. From there the telecast of the encounter with Halley's Comet was relayed to TV networks throughout the world. Seventeen years after Apollo 11, Parkes had again played a central role in a scientific and technological event of great importance.

☆ ☆ ☆

NASA's principal reason for installing the microwave link between Parkes and Tidbinbilla in 1985 was not for radio astronomy, but for tracking its spacecraft. Earlier in 1977 NASA launched Voyagers 1 and 2 which had carried out successful flybys of Jupiter and Saturn during 1979–81. After its encounter with Saturn, Voyager 1 was lost from service when the planet's gravitation swung the probe out of the plane of the solar system, leaving Voyager 2 to continue on alone to the next planet, Uranus. Tidbinbilla combined with Parkes was chosen to be the prime receiving system for the flyby in January 1986, the first occasion an array had been used for tracking spacecraft. The increased sensitivity enabled Voyager 2 to transmit a much higher volume of data during its brief encounter with Uranus. The reconstructed photographs of the planet's surface and its moons were of exceptionally high quality. The Parkes telescope had been the first to detect natural radio emission from Uranus in 1964, and the first to receive man-made signals from the planet 22 years later.

Voyager 2 continued on its journey to rendezvous with Neptune in 1989. Normally, Neptune is the second most distant planet from the sun, but because of Pluto's present position in its highly elliptical orbit, Neptune is now the most distant until the year 2007. Again, Parkes became part of NASA's tracking network, with the dishes at Tidbinbilla, Goldstone and Madrid enlarged from 64 to 70 metres especially for the encounter. In March 1989 the Tidbinbilla–Parkes array began gathering data sent from Voyager which revealed new information on Neptune's complex cloud systems and violent storms, together with a strong magnetic field surrounding the planet. Whereas only two moons were previously known, Voyager discovered another four as well as a partial ring system around the planet. Voyager made its closest approach to Neptune in August 1989, and now journeys on beyond our solar system.

CHAPTER ELEVEN

Gazing at North Goobang

And he sees the vision splendid of the sunlit plains extended,
And at night the wondrous glory of everlasting stars.

Banjo Paterson

In astronomy circles it is often remarked—mostly by envious northerners—that God, in creating the universe, perversely located all the most interesting regions of our galaxy in the southern hemisphere, but all the astronomers in the north.

Neville Fletcher

The time devoted to tracking spacecraft has only made up a small fraction of the operations at Parkes, although one that has usually brought the telescope to prominence in the public eye. In between the Apollo, Giotto and Voyager missions the telescope's agenda returned to full-time astronomy. More highlights of this research are presented in Chapter 12, but first there are a number of other features of the astronomy program worth examining.

We begin by looking at how the astronomy program is organised at Parkes. Questions such as who would operate the telescope, how to decide which research problems should be tackled, who would be allowed access to the telescope, were major issues to be settled during the early years. Certain aspects of the astronomy program have changed significantly over the years. One example has been the increasing use of the telescope by international visitors. Astronomy itself became increasingly international, as well as increasingly competitive. In 1961 Parkes followed the Mark I at Jodrell Bank as the second large steerable dish. By 1971 no fewer than eight other large dishes had been constructed around the world, all operated by groups competing for a significant share of astronomical discoveries. For the Parkes astronomers to remain competitive, the telescope's performance had to be continually improved and upgraded. Resurfacing the dish, the introduction of new receivers and the installation of computers are some of the steps that have enabled Parkes to maintain its position at the forefront of astronomy in the 1990s. Thirty years is a remarkable lifespan for any scientific instrument.

Who Uses the Telescope?

One of the first issues to be settled after its construction was who would take responsibility for controlling the telescope. For contractual reasons sole authority for driving the dish during the commissioning phase had rested with John Bolton. Once the observational program began in March 1962 Bolton proposed that a small group of senior staff should be schooled in control of the telescope who would, in turn, train and 'license' other astronomers as the opportunity arose. Taffy Bowen disagreed with this procedure, favouring instead the routine adopted at Jodrell Bank where a former naval commander had been recruited to take charge of a team of five specialist controllers. Bowen drew a parallel with the research carried out by Radiophysics in cloud physics. For obvious reasons a qualified pilot was entrusted to fly the aircraft, leaving the scientists to concentrate on the operation of the experimental equipment.

Initially, responsibility for controlling the telescope was shared between a small number of senior astronomers and some of the technical staff, all of whom had carefully studied the operating procedures. Wind was the most critical factor. A windspeed meter prominently displayed on the control desk enabled the controller to keep track of sudden changes in the weather. Chart recorders also registered the electrical current in the telescope's drive motors which indicated the wind loads on the dish. After two years of operation, the data from these recorders had revealed a number of occasions when astronomers, probably in single-minded pursuit of some celestial object, had pushed the telescope beyond its safety limits. With his overriding concern for telescope safety, Bolton decided to entrust the driving solely to the technical staff. From then on, a clear division of labour was established between the astronomers and the controllers. Over the years these controllers accumulated a detailed knowledge of the inner workings of the telescope, a valuable asset in view of the increasingly transient population of astronomers.

The completion of a new and advanced facility in any area of science holds out the promise of major discoveries. Inevitably, there is a healthy degree of competition among the possible users of the new facility. At Parkes it was clear from the start that the demand for observing time would always exceed the time available. The preparation of reasonable guidelines for the allocation of observing time therefore became an important issue. The task would not always be straightforward. Bernard Lovell has recorded: 'It seems extraordinary that normally unemotional colleagues have shown passion and anger about their allocation of time on the Mark I. Groups who may already have been using the telescope day and night for two or three weeks have frequently descended on me in force to argue and plead for another 24 hours.' [1]

The allocation of time at Parkes was initially placed in the hands of a program planning committee. First, the committee divided the observing year into four three-month periods. About six weeks before the beginning of each quarter the committee issued a bulletin on the operational condition of the telescope and the types of observing programs which would be feasible. The astronomers then submitted requests for observing time to the committee. These requests usually stated the astronomical significance of the observations, the type of equipment to be used, as well as an estimate of the number of nights and the number of hours each night required. Finally, if some unforeseen problem should arise, the request usually suggested a possible backup program of observations. The committee's task was to shortlist the best requests and then allocate the appropriate observing times. If an astronomer lost a substantial amount of time because of equipment failure or bad weather, the request would be deferred and given priority at the start of the following quarter. Once these general guidelines had been established, the committee membership was rotated at regular intervals among senior staff and astronomers visiting Radiophysics on short-term appointments. Dick McGee volunteered to become permanent secretary of the committee. In theory, the various competing research interests would receive a fair representation.

The scientific merit of a request before the planning committee was of paramount importance, but a number of other factors also came into play. One can be called the timeliness of the proposal. Earlier in Chapter 9 we saw how proposals which could immediately exploit the superiority of the new telescope and start the observing program 'with a splash' were to receive top priority. The early studies on the polarisation of radio sources belonged to this category. Besides the great astronomical importance of these observations, there was also the technical innovation which enabled the feed to be rotated in the telescope's aerial cabin. This was a feature simply waiting for someone to press the right button on the control desk. The observations on 3C 273 were also timely. Not only were quasi-stellar objects one of the most intriguing problems in astronomy, but a favourable occultation of this source by the moon would not occur again for twenty years.

Another factor in allocating time revolved around the availability of suitable equipment, particularly receivers. Usually this meant deciding whether the equipment actually suited the program proposed and whether or not the observations should be deferred until better equipment became available. More often than not, the number of observing requests could be trimmed back because of a question mark over equipment, rather than because of doubts about the scientific merit of the requests. One further factor, potentially the most controversial, was the desire to share time as democratically as possible among the Radiophysics staff, as well as an increasing number of astronomers from other institutions. Over thirty astronomers at Radiophysics depended on the Parkes telescope to provide their scientific livelihoods.

While these guidelines are applied to most observing proposals, a small fraction cannot be judged by the usual rules. Some proposals are basically 'wildcat' ideas where the likelihood of success is small, or difficult to assess, but which might lead to an outstanding discovery. An extreme example is the search for extraterrestrial intelligence (SETI)—the chances of success might seem hopelessly small, but the consequences of immeasurable importance. Other proposals are sometimes those that lie outside the mainstream of radio astronomy. Although they may lack the glamour of conventional topics, some can eventually lead to significant discoveries. One example has been the long-term program at Parkes on flare stars, unusual stars which occasionally erupt with outbursts of radio energy up to a million times more violent than flares observed on the sun. Flare stars were the first true 'radio stars' discovered, words once used incorrectly to describe the first radio sources detected in the 1940s. To catch one of these flare events as it occurs requires patience, luck and, most importantly, plenty of observing time. Bruce Slee writes:

> 'These flare-star programs were not popular with my colleagues who believed that I was wasting valuable observing time. Such an attitude has since proved shortsighted by the fact that some 20% of the time on the Very Large Array in the United States is now allocated to radio star research and it will be an important area of research for the Australia Telescope. The detection of radio flares from these stars is important in showing that other stars undergo star-spot cycles similar to our sun, except that the flares can be several orders of magnitude more energetic.' [2]

In one respect radio astronomers have a head start over optical astronomers. With operations restricted to night-time, optical telescopes are hampered by rain, cloud and fog, even a bright moon can spoil the observations. Paradoxically, poor atmospheric conditions can sometimes be an advantage. The observatory high on Mt Wilson often had its best nights when a thick layer of smog blanketed the city lights below in Pasadena and Los Angeles. Overall, the amount of observing time for optical telescopes is usually limited to about 200 nights a year. In contrast, radio astronomy is full-time astronomy. The Parkes telescope can operate around the clock, seven days a week, every week of the year, with shutdowns only when heavy rain or high winds force the dish to be stowed in its safety position. In practice, however, most observing is done at night, partly because the sun is a strong source of radio interference and partly because a number of other essential activities are best carried out during daylight hours. These activities include the installation and testing of equipment such as receivers, plus a range of other housekeeping tasks. Similar to any large item of precision machinery, regular maintenance and servicing are an essential part of the daytime routine. Bolton relied on his wartime experience in the navy:

> 'We started with what was basically a ship-at-sea routine. All electric motors, generators and moving parts were stripped down every three months. Parts

were machined, new brushes put in the motors and so on. This routine got us by in the navy and I felt that if we did this at Parkes we would be on the safe side. After a few overhauls the interval was relaxed to six months, and after several years overhaul of the azimuth drive motors was extended to one year. Our caution paid off in the long run. We had no serious malfunction in the main drive system during the first twenty years of operation.' [3]

Before the beginning of each observing quarter Bolton would estimate the maintenance time needed and, similarly, the receiver group would specify the time required for the installation and testing of new equipment. This time was automatically granted by decree and the remaining time then handed over to the planning committee for allocation to the astronomers. On average the telescope has been available for astronomy 18 hours a day during the week, and up to 24 hours a day at weekends. Very little of this time has been lost. In fact, during the first ten years of operations, less than 5% of observing time was lost because of bad weather or because of malfunctions in the telescope or its equipment.

Despite this high level of reliability, there would always be times when Murphy's law intervened. Marc Price recalls one occasion:

> 'At Parkes we 'discovered' the moon. Many radio astronomers have done this! We were very excited. We were doing the 400 MHz survey and all of a sudden this uncharted source came booming through. We very carefully measured its position and then started trying to map it, and it turned out that indeed it appeared to be extended in right ascension. As time went on, we found the source was a ridge of emission, it wasn't simply a single source. There was so much excitement that the astronomers had even taken over the driving of the telescope. About this time we measured the position again. No, it wasn't simply extended, it was moving! The telescope driver, who had gone outside to avoid all this noise, came back in and said, 'Gee, it's a beautiful night out there. The moon is full. It's the prettiest thing I ever saw, there's the shadow of the aerial cabin right in the middle of the dish!'.' [4]

To commemorate such occasions some of the astronomers decided to award a medal for acts of 'gross stupidity'. One astronomer received the medal for also 'discovering' the moon, another for almost driving the rim of the dish into the ground.

Prime time for observing at Parkes occurs in the winter months when the nights are longest and the Milky Way is directly overhead. Competition for observing time is then strongest but, over a whole year, most astronomers could expect to receive about five observing nights. In practice, astronomers usually collaborate in small groups, so this means several observing sessions each year, totalling perhaps twenty nights. The weeks or months in between are spent back at the Radiophysics Lab analysing the data and preparing for the next round of observations. Radio telescopes such as Parkes gather data at such a prodigious rate that usually a few nights of observing requires a few months of analysis before the results are ready for publication.

During an observing period the astronomers spend the night-time hours in the telescope's control room working through the schedule, checking that the equipment is operating smoothly and perhaps taking a preliminary look at the data collected. In the daytime—aside from the number one priority of sleep—time is often set aside for browsing through the astronomical literature or for writing a research paper describing the results gathered during an earlier visit. In the early days, meetings were occasionally held in the observers' quarters where one of the astronomers would give a brief report on recent work. These talks were informal—serious scientific discourse could wait until the return to Sydney or until the next major conference. In this particular shire of New South Wales, a more light-hearted approach was in order. Again, Marc Price:

> 'Perhaps the spirit of the times could be expressed best by saying that we used to have a luncheon group at Parkes that we called the North Goobang Philosophical Society. Every lunchtime we'd all get together and discuss the weighty matters of the day: which feed didn't work, which new supernova remnant needed to be looked at, and which were the latest results from all over the world that we felt we should follow up on. Of course being in a Commonwealth country, we felt that we had to have a royal charter, and so we had to seek patrons. The two people we sought were a senior scientist in the Laboratory and the Astronomer Royal. That is how the Australians got the reputation for having an organisation that was indeed Wild and Woolley! We did establish the Society and even sought the patronage of these gentlemen. However, they caught on to our trick and squashed it in the early days! So it went unchartered.' [5]

Doug Milne adds:

> 'We formed the North Goobang Philosophical Society as a hit at conservatism and the establishment. 'North Goobang' as the site of the telescope, rather than 'Parkes', was part of the attack. Dick McGee even referred to the North Goobang Shire Telescope project in public lectures, whilst I believed that it should have been the Alectown Radio Telescope. Marc Price got a friend to tool a coat-of-arms in bull skin, which said something about our creed. This coat-of-arms still hangs in the dining room at Parkes.' [6]

During the first few years Radiophysics astronomers had almost exclusive use of the telescope. After waiting six years for its completion they expected, with some justification, to have first pick of scientific fruit promised by the new dish. Towards the end of the 1960s the number of outside astronomers gradually increased, with about two-thirds of the visitors coming from overseas institutions and one-third from Australian universities such as the ANU, Sydney and Tasmania. The astronomers from the University of Tasmania were accorded a type of favoured status, partly because they had no major instrument of their own, and partly because of their close links with Radiophysics extending back to the

The North Goobang Philosophical Society was an informal group formed by the astronomers at Parkes. Its shield depicts the source Centaurus A with its familiar dark dustband.

1940s. Until 1960 Hobart had been the only other centre active in radio astronomy in Australia. Although this group had specialised in very low-frequency observations, there had always been a strong overlap between its research interests and those of Radiophysics.[7]

In most cases these outside astronomers carried out observing programs in collaboration with Parkes staff, rather than on their own. With the exception of the Hobart group which built some of its own equipment, observing requests from outsiders usually involved the use of receivers built at Radiophysics. Long hours had been invested in the design, construction and testing of these complex devices, so it was only fair that Radiophysics staff should share in the rewards of this research. While outside astronomers were often the ones to put forward an exciting new proposal, Radiophysics staff were generally the ones required to put the idea into practice. Ray Norris makes the comment:

> 'I think it is interesting that time allocation at Parkes was still far more open than at Jodrell, where time was allocated by the director. I remember as a young post-doc knocking on Sir Bernard's door to ask for more observing time. No proposal was needed—I think we were evaluated by his personal opinion of our worth and what use we had made of previous time allocated—but Sir Bernard would just consult his diary and say something like 'Ah yes, Norris, I think we could give you a week in March.' Very little time on Jodrell telescopes was allocated to non-Jodrell users, far less than at Parkes.' [8]

Towards the end of the 1960s pressure mounted to increase the number of outside observers granted access to the telescope. Several astronomers began to view the Parkes telescope as a closed shop —difficult for outsiders to gain time, but easy for insiders. Criticism focused on the program planning committee which consisted exclusively of Radiophysics appointments. Some critics believed that the committee should have a broader membership representing the national astronomical community as a whole, a procedure adopted at several facilities elsewhere. The National Radio Astronomy Observatory in the United

States was one such model favoured. Astronomers would submit a research proposal to the director, who would send the proposal out to independent and anonymous referees. The referees then returned a written report on the feasibility of the proposal and its possible importance to astronomy. Of course proposals still got knocked back, but an astronomer could at least take comfort that the proposal had been given a fair hearing by his or her peers. Similar procedures were followed when submitting a paper to one of the astronomical journals and, in most cases, when applying for a grant from a funding body. Because most outside astronomers applying for time at Parkes were known personally to the program planning committee, difficulties were bound to arise in the long term.

Part of the problem arose from the fact that Parkes had no clearly defined national role. The American NRAO with a permanent core staff of astronomers had been established with the express purpose of providing a facility for institutions and universities not in a position to construct large telescopes of their own. In contrast the Parkes telescope, located at a site called the Australian National Radio Astronomy Observatory, was a national observatory in name only. The telescope and site remained first and foremost the property of one institution. It had been a long road to raise funds, design, construct and then commission the telescope. When it came to allocating observing time, the Radiophysics astronomers believed with some justification that first priority was theirs. For outsiders, observing was a privilege not a right.

Times had changed by the mid 1970s. Astronomers came in increasing numbers from Australian universities, from the well-established groups in England, Holland and the United States, and from some of the new powers in astronomy such as West Germany. The increase in the number of outside observers came partly from a change in policy on the appointment of Paul Wild as chief of Radiophysics, following Taffy Bowen's retirement in 1971. One of Wild's first acts was to rewrite the time allocation guidelines to encourage more outside observers. The participation by Australian universities had been small from the beginning and had actually dropped by the end of the 1960s. For astronomers from other countries, the participation had been much higher. During the first few years of operation, one in four of the research papers from Parkes had an overseas astronomer listed among its authors. By 1970 this had grown to one in two papers. These numbers can, however, be deceptive. Most of these overseas astronomers were on extended visits as postdoctoral fellows and were really short-term members of the Radiophysics staff. The actual number of observing requests which came directly from overseas institutions never exceeded more than about 10% of the total.

The opening up of the Parkes telescope to outsiders during the 1970s reflected a major transformation in the way astronomers practised their profession. Rather than long-term visits, it became common for astronomers to visit Australia specifically to carry out an observing program at Parkes and then to return directly to their home institutions. The Radiophysics astronomers also began to make frequent short-term visits to overseas facilities. Others have commented on this new style of 'suitcase' astronomy:

> 'Later growth has allowed radio astronomers to participate in a mobility that would earlier have been both impossible and inconceivable—working successively with a variety of groups, of both radio and optical workers. In the late 1960s, in this situation of proliferating groups, increasingly sophisticated techniques, and ever more complex problems, some radio astronomers have 'gone international', travelling from site to site to carry out observations and coauthoring papers with members of a number of scattered groups.' [9]

The increasing internationalism of the 1970s marked a new and third phase in the development of radio astronomy. At Radiophysics, the first phase from 1946 to 1960 had been characterised by small and fairly stable groups which designed, built and operated their own telescopes. The second phase began with the introduction of large precision-built telescopes such as the Parkes dish, where larger groups could feed off the one instrument. The principal users usually belonged to the institutions that funded and operated the instrument. The astronomers who had 'gone international' in the 1970s broke away from this practice of a single large group tied to one instrument. The same process was already well advanced in other areas of science like particle physics, where physicists had become increasingly reliant on visits to the giant accelerators at locations such as CERN in Geneva. In one respect, this third phase was similar to the first in that small and fairly informal collaborations again became the accepted practice. However, in striking contrast to the period before 1960, astronomers no longer had to design and build their own instruments, or even know too much about how they worked. Instead, the contemporary astronomer seeks out observing time on a telescope, wherever in the world that might be, the one best suited to pursue his or her research interests.

When Big was Beautiful

> The prudent thing to do is to keep below 250 ft as we have done. Anybody who tries to go above this—300, 400, 500 or 600 ft—does so at his peril. He'd better have an enormous supply of money or else indulge in some very special research in engineering design and materials.
>
> Taffy Bowen, 1961

As mentioned in the previous chapter the telescope at Parkes became something of a mecca for the designers of large dishes elsewhere in the world. American groups borrowed on the Parkes design and some of its features were incorporated into the three large tracking instruments built by NASA at Goldstone, Tidbinbilla and Madrid. The Parkes telescope also became the favourite prototype for designers planning instruments solely for radio astronomy. The first of these was a 45 m dish constructed by the National Research Council of Canada, a research organisation similar to the Australian CSIRO. Freeman Fox carried out the design study and two firms subcontracted for the construction were firms that had worked at Parkes.[10] The Algonquin telescope came into operation at Lake Traverse in Ontario in 1966. Basically it is a scaled down version of the Parkes telescope, the main difference being that it was designed to operate at wavelengths as short as 2 cm.

In 1965 the Astronomical Institute at the University of Bonn in West Germany announced plans to build the largest steerable dish yet attempted. The steelmakers MAN and Krupp carried out the design, and construction of the 100 m telescope was completed in 1971 at Effelsberg in the mountains south–west of Bonn. The telescope has some similarities with the Parkes dish, but in many respects it is a superior instrument. Even though 50% larger in diameter, the Effelsberg dish was engineered more accurately with the central region of 60 m diameter capable of operation down to a wavelength of 3 cm, and over the whole surface to 5 cm. This improvement in accuracy was achieved by using computers to calculate the complex set of deflections in the dish before finalising its optimum design. The same computational firepower had not been available to the designers at Freeman Fox in the 1950s. However, with the completion of the Parkes telescope, precise measurements of the deflections over the dish could then be made and this information used in the computer calculations by the German designers. The data from Parkes played an important part in the design of the larger and more accurate Effelsberg telescope.

Two other large telescopes came into operation in the 1960s, though neither belongs in the category of fully steerable dishes. The first was the 91 m transit telescope built by the NRAO at Green Bank in 1962. This dish could only be tilted in the north–south direction and relied on the earth's east–west rotation to cover the radio sky. The second telescope, which dwarfs all others in its physical dimensions, is the 305 m diameter fixed dish at Arecibo in Puerto Rico. The dish consists of a vast bowl, spherical in shape rather than parabolic, suspended in a large natural hollow. Built with funds from the US Department of Defense in 1963 and operated by Cornell University, the instrument was originally intended for radar studies of the earth's ionosphere but then was adapted to radio astronomy. Arecibo simulates the operation of other large telescopes, not by moving the dish itself, but by tilting the position of the feed up to twenty degrees from the zenith. Although the sky coverage is limited it does include the plane of the Milky Way. Arecibo has played a leading part in the discovery of pulsars (see next chapter).

Two large telescopes influenced by the Parkes design: *left* The 45 m telescope constructed at Algonquin, Ontario in 1966. The dish could be tilted down to the horizon and stowed during heavy snowfalls. *below* The 100 m telescope completed at Effelsberg, West Germany in 1971. In contrast to Parkes, the entire structure turns in azimuth on a large circular track, similar to the Mark I at Jodrell Bank.

The world's largest reflector, the 305 m telescope at Arecibo in Puerto Rico. The wire mesh forming the surface was later replaced with aluminium panels to allow observations at much shorter wavelengths.

During the 1960s several other groups planned telescopes at least as large as the 100 m steerable dish then underway at Effelsberg. In California a cooperative venture by Caltech, Berkeley and Stanford worked out a preliminary design for a 122 m dish. Not to be outdone, a collaboration on the east coast involving Harvard, MIT and the Smithsonian Astrophysical Observatory planned a 135 m telescope to be built near Boston. Across the Atlantic, the Swedish Natural Science Research Council proposed a 100 m dish near Göteborg. In each case these organisations were freely provided with information on the Parkes telescope.

The Jodrell Bank group had not been idle since the completion of the Mark I in 1957. Two years later, in a policy paper prepared for the Department of Scientific and Industrial Research, Bernard Lovell mounted a case for a much larger telescope:

> 'As a result of this paper, the DSIR agreed to make a grant of £10,000 for a 'design study to be made of a steerable radio telescope of significantly greater size than the 76 m instrument now in operation at Jodrell Bank'. By 'significantly' I had in mind a telescope vastly bigger than the Mark I —extending for 300 metres or more. Husband was bothered about the height problem particularly in relation to differential wind effects with

height. We therefore conceived the idea of a telescope elliptical in shape with the height limited to manageable dimensions of 90 to 180 m but with the horizontal dimensions extended to 300 m or more.' [11]

After his earlier experiences Lovell had no illusions about the likely success of such an ambitious project. Lovell did at least succeed in obtaining funds to build a smaller prototype of the elliptically shaped telescope. With dish dimensions of 38 m (horizontal) and 25 m (vertical) the Mark II was erected in a field adjacent to the Mark I in 1964. This more accurate instrument enabled observations down to 6 cm. Its success led to the construction two years later of a similar instrument, known as the Mark III. Though of the same size and same elliptical shape as the Mark II, the Mark III was located at a site 25 km from Jodrell Bank and used to partner the Mark I in long-baseline interferometry.

In contrast to the tortured history of the Mark I, the construction of these next two telescopes at Jodrell Bank was virtually trouble-free and within budget. By 1967 Lovell felt encouraged to revive the idea of a very large telescope and design studies were carried out for a dish 120 m in diameter. Three years and nearly £1 million were spent developing the design, but the proposal went no further. The estimated cost of £22 million was considered far too expensive and, finally, in 1974 the project was abandoned.

The Mark II telescope at Jodrell Bank with its unusual elliptically shaped dish. The Mark I is in the background. The Mark II was intended to be the prototype for a gigantic telescope with dimensions almost ten times larger.

When big was beautiful: the era of large dishes in radio astronomy

Telescope	Institution	Year of completion	Diameter	Type	Sky coverage	Minimum wavelength[A]	Cost[B]	Comments
Mark I Jodrell Bank	Manchester University	1957	76 m	Parabolic	Complete	40 cm	$1.8 mill.	Surface upgraded in 1971 for 6 cm wavelengths. Renamed the Lovell Telescope
Parkes NSW	CSIRO	1961	64 m	Parabolic	60° from zenith	10 cm	$1.8 mill.	Progressive upgrade of surface to millimetre range. Integral part of Australia Telescope
Green Bank West Virginia	NRAO	1962	91 m	Parabolic	Transit	17 cm	$0.9 mill.	Upgraded early 1970s for 6 cm wavelengths. Collapsed and totally destroyed in 1989
Arecibo Puerto Rico	Cornell University	1963	305 m	Spherical	20° from zenith	50 cm	$9 mill.	Upgraded in 1974 for 6 cm wavelengths
Mark II Jodrell Bank	Manchester University	1964	38x25 m	Elliptical paraboloid	Complete	6 cm	$0.7 mill.	Used both as stand alone telescope and as interferometer with the Mark I
Green Bank West Virginia	NRAO	1965	42 m	Parabolic	Complete	1.5 cm	$13 mill.	Only large dish with polar mount. Others have alt–az form
Mark III Cheshire	Manchester University	1966	38x25 m	Elliptical paraboloid	Complete	30 cm	$0.4 mill.	Movable on track as partner to Mark I. Now part of MERLIN
Algonquin Canada	NRC	1966	45 m	Parabolic	Complete	2 cm	$4 mill.	Now mothballed
Effelsberg West Germany	Bonn University	1971	100 m	Parabolic	Complete	4 cm	$8 mill.	Remains the highest performance fully steerable reflector

Notes: Not included are other reflectors such as the cylindrical paraboloids at Illinois, Ohio and Nançay, or other large antennas not used specifically for radio astronomy, such as the NASA 64 m deep space instruments.

[A] Minimum wavelength at time of construction. The value given is where distortions in the reflecting surface reduce the collecting efficiency to about one-half the optimum (long-wavelength) value.

[B] Approximate cost in $US at time of telescope's completion (may include equipment and associated building costs).

An important factor leading to the project's cancellation was a growing realisation that such large sums of money could be better spent in other ways. Advances in technology were beginning to overtake the usefulness of very large 'stand-alone' dishes. The Jodrell Bank astronomers decided instead to develop the techniques which they had already played a large part in pioneering—long-baseline interferometry. Three 25 m dishes were bought for the bargain price of £1 million each from the United States, where these dishes were being mass produced for the Very Large Array (VLA) in New Mexico. The three dishes were positioned at widely separated sites and then linked to the existing Jodrell Bank telescopes to form a single large array known as MERLIN, an acronym for multi-element radio-linked interferometer network. MERLIN in Britain and the VLA in the United States were unquestionably the telescopes of the future. In the mid 1970s the proposal to build a similar large multi-element array in the southern hemisphere led eventually to the Australia Telescope.

Like the proposed 120 m telescope at Jodrell Bank, the plans to build mega-dishes in the United States, Sweden and elsewhere in the 1960s progressed no further than a few grand designs. The 100 m Effelsberg telescope remains the single largest steerable dish ever built. The table on the page opposite summarises various details of the major dishes. Beginning with the Mark I at Jodrell Bank in 1957 and ending with the Effelsberg telescope in 1971, this was the period when single large reflecting telescopes made their greatest impact on astronomy. The Parkes telescope was the second of these large instruments constructed and in several respects it provided the benchmark for later telescopes. Although smaller in size than the Mark I, the Parkes telescope allowed observations in the wavelength region of greatest interest, below 21 cm. The 42 m Green Bank telescope completed four years after Parkes also targeted this region of short wavelengths. However, because of inherent difficulties associated with its polar mount and the sheer expense of the instrument, the Green Bank telescope was ruled out as the model instrument for radio astronomy groups elsewhere. In short, the Parkes telescope demonstrated the feasibility of combining the two major requirements—sheer size and a dish of high surface accuracy—all at modest cost.

Time for Renewal

> The Parkes telescope is probably one of the most successful research instruments ever built.
>
> John Bolton, 1973

Bolton's statement which appeared in a *Nature* article surveying the future of Australian astronomy was no idle boast by one who had been the telescope's director. After twelve years of virtually trouble free operation, the size, accuracy and versatility of the telescope had enabled the Parkes group to make major contributions to most areas of radio astronomy. The total of almost 500 research papers from Parkes provided solid justification for the original decision to build the telescope.

By 1973 that might well have been the end of the story. Some of the newer telescopes built in the northern hemisphere had begun to overtake the Parkes dish in technical supremacy. Although Parkes maintained an advantage by its access to the southern skies, it seemed only a matter of time before Parkes would slip from the major league of radio astronomy centres. This was not to be the case. Beginning in the late 1960s a program of improvements helped Parkes to keep pace with the advances in radio telescopes elsewhere. These improvements were made in three main areas: resurfacing parts of the dish to improve its accuracy, the installation of a new generation of highly sensitive receivers, and the introduction of computers to control the telescope and to process the data collected.

The single most important feature of any large dish is its resolving power, given by the ratio of the observing wavelength to the diameter of the dish. The value of this ratio determines the telescope's power to resolve individual radio sources or, alternatively, the structure of large extended sources. Put more simply, the resolving power determines how sharply the picture of the radio sky can be focused. The resolving power can be enhanced by increasing the dish diameter but, as noted already, by 1970 several proposals to go larger than 100 metres were abandoned for engineering and financial reasons. In any case, other advances had enabled smaller dishes spaced at large distances to be linked to form arrays with resolving powers that could never be matched by single large dishes.

The only other way to improve the resolving power was to build dishes with surfaces of increasing accuracy, to operate at shorter and shorter wavelengths. As shown in the table, telescopes such as those at Green Bank, Algonquin and Effelsberg were designed specifically to improve on the minimum operating wavelength (10 cm) of the Parkes dish. Besides the desire to increase the resolving power of these telescopes, astronomers also began to realise that many interesting astronomical objects could only be studied at shorter wavelengths. As Taffy Bowen noted in 1970:

> 'There have been two recent developments in radio astronomy of the utmost significance. The first of these is the discovery of new and exciting line radiation in the 1 to 5 cm wavelength range from materials like water vapour, formaldehyde and the recombination lines of hydrogen. The second is the fact that certain of the more distant quasi-stellar objects appear to emit more and more energy at this end of the radio spectrum. There is no doubt, therefore, that radio astronomy is extending into the centimetre and

millimetre wavelength region and the most exciting developments are likely to occur in this wavelength range in the next ten or twenty years.'[12]

In broad terms, from the decametre waves of the 1930s, the metre waves of the 1940s, the decimetre and centimetre waves of the 1950s and 1960s, radio astronomy had begun in the 1970s to move into the region of millimetre waves.

In line with this trend a program began in 1970 to upgrade the surface accuracy of the Parkes dish. Originally the dish could only be operated down to 10 cm, but then fine tuning of the surface in the early 1960s had allowed observations as low as 6 cm. Wavelengths even lower had been shown to be feasible through studies of the surface carried out with the automatic survey device (see Chapter 10). New perforated aluminium panels were then developed with much higher surface accuracy, and with much less leakage of the short wavelength radiation through the surface than the existing coarse mesh panels. After wind tunnel tests had given the all clear on their aerodynamic suitability, the new panels were installed in two stages in 1970 and 1972. The resurfaced section extended from the existing central steel plate out as far as the tripod legs at a diameter of 37 m. The new section of the dish could operate down to 1.3 cm, a wavelength eight times shorter than the original minimum.

After this first successful 'reskinning' of the surface, plans were then drawn up to replace the central section of steel plate. Under the supervision of Don Yabsley, over 300 aluminium alloy panels were carefully laid, each coated with highly reflecting white paint to reduce the effects of thermal expansion by sunlight. The new inner section extending to a diameter of 17 metres enabled operations to wavelengths as short as 4 millimetres, with performance characteristics equal to those of any of the large northern dishes. Below 4 mm, however, other new and smaller precision telescopes which had been specifically designed for this new field of millimetre astronomy came into their own, such as the 14 m instrument operated by the University of Massachusetts and the 11 m telescope at Kitt Peak in Arizona.

☆ ☆ ☆

The progressive upgrading of the dish surface brought with it a demand for new receivers to exploit this short wavelength region. During the 1960s the Radiophysics group constructed an impressive collection of receivers, ranging from relatively simple transistor devices for long wavelengths to advanced parametric amplifiers for shorter wavelengths. Up to five of these receivers could be installed on the telescope at any one time, so that observations could switch from one receiver to another with a minimum amount of downtime. If required, a particular radio source could be observed at several wavelengths during the course of a single night. The most successful of these receivers was one known as the

The first reskinning of the Parkes dish took place in two stages in 1970 and 1972. The original wire mesh was replaced by perforated aluminium panels out to a diameter of 37 m. NASA funded the new surface in return for use of the telescope during the Apollo missions. In 1982 the aluminium panels were extended out to a diameter of 44 m, on this occasion by courtesy of the European Space Agency.

11 cm paramp, which achieved high sensitivity and stability even though it operated without any special protection from the widely varying air temperatures at Parkes. After its installation in 1967 the paramp was used to compile the 11 cm (2700 MHz) Parkes catalogue, the second major sky survey and most important program of the 1970s. The Radiophysics group established a reputation for producing some of the best receivers for radio astronomy in the world.

The push to shorter wavelengths in the 1970s raised several formidable problems. As a rule of thumb, observing at wavelengths ten times shorter means it is ten times harder to construct a receiver with the required sensitivity and stability. Temporary help came with the loan of a 6 cm receiver from the NRAO group at Green Bank, a highly sensitive paramp cryogenically cooled to operate at 20 K (twenty degrees above absolute zero). The Radiophysics group then embarked on the development of a similar device to operate at 3 cm. Despite a major effort, the performance of the new receiver fell far short of a similar device developed in the United States and available commercially. Technological advances had begun to overtake the types of receivers pioneered specifically for radio astronomy. The rapid expansion of the space and communications industries encouraged several large corporations to make massive investments in these devices and cryogenic equipment for cooling them. A new closed-circuit refrigerator ideal for a remote field station such as Parkes was now being produced in commercial quantities. By using compressed helium gas, the refrigerator could cool to 20 K, suitable for the operation of low-noise paramps and field-effect transistor (FET) amplifiers. A later version produced 4 K, the temperature required for the operation of maser receivers.

The research and development of devices on this scale was beyond the resources of Radiophysics and most other radio astronomy groups elsewhere. It now proved easier to buy receivers, rather than attempt to design and build them in-house. Head of the receiver group Brian Cooper comments on the transition:

> 'By the early 1970s liberal funds were being provided to us by NASA for the purchase of receivers and we were able to avail ourselves of skills developed in the United States in the course of the space communications program. A 6 cm cooled paramp was bought off-the-shelf in 1970, and a 2 cm cooled paramp in 1977. Observing in the 1 cm region began in 1971. In the early 1980s the first maser amplifier was used at Parkes, and recently cooled FET amplifiers have also arrived on the scene.' [13]

The availability of commercial receivers marked a further departure from the early years when virtually every component of a radio telescope had to be designed and built by radio astronomers themselves. Even so, much work remained to be done adapting the new receivers to radio astronomy, as well as improving other components of the telescope involved in the collection, amplification and recording of radio signals.

One important breakthrough was the development of a new type of corrugated feed horn by Harry Minnett and Bruce Thomas. The feed horn must collect the radio signals focused by the dish as efficiently as possible and funnel them to the receiver without responding to unwanted sources of noise, such as natural radiation from the ground. The corrugated feed horn developed for Parkes was later adopted for satellite tracking stations and for the Australia Telescope.

Another major innovation was the development of a device known as the 1024 channel digital correlator. Unlike conventional receivers, which are tuned to observe at a single sharp wavelength, the digital correlator provides a way of observing how the intensity of a radio source varies over a certain wavelength range or frequency bandwidth. Previously, observations of this type had been made with two simpler 48 and 64 channel instruments. The 1024 channel digital correlator, designed principally by Brian Cooper and Jon Ables, was installed at Parkes in 1972. This correlator along with similar instruments built in the northern hemisphere led to the surprising discovery of a wide range of molecules in interstellar space (see Chapter 12).

The continual improvement in the precision of the telescope and its instrumentation received a major boost with the installation of a computer in 1968. Computers had already begun to make a significant impact on many areas of science and technology in the 1960s, radio astronomy included. The Mark II telescope constructed at Jodrell Bank

Alan Wright and Lyn Newton at the main control desk at Parkes.

in 1964 had been the first to have its drive system controlled by computer. Similarly, the 100 m telescope at Effelsberg owed its high precision to detailed computer calculations carried out during the design stage. More important than the design and control of telescopes, the great potential of computers lay in their capacity to record, process and analyse the vast amounts of data collected by the new telescopes. Pencils, paper and pen-recorder charts—the stock in trade of the early radio astronomers—gave way to a new era of digital recording and high-speed electronic computing.

As mentioned in an earlier chapter, computing research had made a flying start at Radiophysics with the construction of the CSIRAC in 1951, one of the world's first computers. CSIRAC, however, made no impact on radio astronomy. Paradoxically, computing research had been wound down and its resources diverted to the rapidly growing radio astronomy program. The first attempt to provide computing support for the radio astronomers was not made until shortly after the completion of the Parkes telescope. A system was developed for recording information on the position and strength of radio sources on paper tape, which could then be analysed later by a computer back in Sydney. This recording system had only limited success, basically because of its inability to detect and correct systematic errors in the data. The best solution appeared to be to install a computer at the telescope itself. John Shimmins took up the challenge:

> 'Digital computers had already been brought into use at Cambridge and Jodrell Bank. The Americans had also gone in for data reduction by computer much earlier than us, partly because computers were more readily available to them. In 1965 very few people in Australia knew anything about computers. After returning from a trip to Canada I looked into the matter and we decided to buy a PDP-9, one of the very first models. We could record and process in real time all sorts of information gathered during observations, basically all those things that otherwise had to be read directly from the dials on the control desk. Until then it took hours of calculation to extract useful information from the mass of data. I would estimate that for every hour's observing time, another ten hours of tedious data reduction would be required back in Sydney. The computer meant less of our time was spent in this drudgery and—at least I hope—more time spent thinking!' [14]

This fundamental change in the way astronomy is done was not confined to Parkes. Astronomer Ben Gascoigne has described a similar transformation at other observatories, radio and optical alike:

> 'This change in style goes deep. While observers now work in warm, well-lit control rooms, watching their observations unfold on video screens, they are distanced from the telescope and instruments—spectrographs or whatever—by computers and resident technicians, and not only the old-time observers' skills but the observers themselves are steadily becoming redundant. They are already so at the UK Schmidt and the Molonglo

telescope. More than ever the bulk of the astronomer's work is done at his home institution, much of it at a computer with the help of one of the massive, many-faceted programs that are so characteristic of contemporary astronomy. These will have taken thousands of hours to write, and be culled from the work of many people in many countries; while an astronomy department can do without a telescope, it certainly cannot do without a computer.' [15]

The introduction of computers, new receivers and the progressive resurfacing of the dish have been the three major developments which have upgraded the performance of the Parkes telescope. Numerous other innovations have also contributed to maintaining Parkes at the forefront of astronomy. One has been the continuing program to improve the pointing accuracy of the dish. Alan Wright reports:

'Parkes is now the most accurate fully steerable dish as regards pointing of *any* in the world. This was clearly demonstrated during the Voyager flyby of Neptune in August 1989 when the telescope tracked the spacecraft over its full hour angle range (nine hours) using essentially 'blind' positioning. At no time was the telescope more than 10 arcsec from the desired position, a precision better than any of NASA's own dishes. We all felt this was a very satisfactory culmination of the years of hard work that had gone into the computer modelling of the dish deformations.' [16]

Another development was the installation of an 18 metre telescope to form an interferometer with the main dish, though in this case the innovation only brought mixed success. The small telescope is in fact older than its larger partner. As early as 1955 the Radiophysics group had decided that, because of the long wait expected for the 64 m dish, a smaller instrument would be useful as an interim measure. The 18 m dish was chosen, one of a number being mass produced at this time for the US Army by the D. S. Kennedy company in Boston. At first there were plans to install the dish at the Parkes site for use as a type of practice instrument, but instead the telescope was erected at the Fleurs field station in 1959 and used in conjunction with the Chris Cross for both solar and hydrogen-line studies.

Early in 1963 the telescope was dismantled and taken by truck to Parkes. A system of railway tracks was laid out for the telescope, consisting of two arms running north and east of the main dish, with the end of each arm connected by a further circular segment of track. The position of the dish along the tracks, together with its direction of pointing, were controlled via a system of cables running from the main dish. After the installation had been completed late in 1965, the interferometer was used by Ron Ekers, Doug Cole and Doug Milne to map the structure of extended sources which had already been located by the 64 m dish and entered in the first Parkes catalogue.

Although used successfully to study several hundred sources, the interferometer had one major drawback. Fine structural details of a

The 18 m telescope was installed at Parkes late in 1963. The dish partnered the main telescope in studies on the internal structure of radio sources, but has now fallen into disrepair.

source could be determined, but the interferometer could not accurately pinpoint the actual position of a source. The problem lay in the cable which trailed across the ground connecting the two telescopes. Changes in the air temperature during the course of observations were enough to interfere with the signals relayed through the cable, a type of 'gremlin' known as a phase instability. Without phase stability, information on the position of sources, but not their structure, is lost from the signal. By the time the interferometer was brought into operation a great deal of information on the structure of the most interesting sources had already been scooped by the long-baseline interferometer group at Jodrell Bank. As described in Chapter 9, a more pressing task at Parkes was to determine accurate positions for sources so that they could be identified with optical objects. The shortcomings of the interferometer meant that the optical identification program had to rely on source positions measured with the 64 m dish alone. The large dish could achieve position accuracies as good as any similar dish in operation, but an effective interferometer free from the gremlin of phase instability could have yielded positions with perhaps twice the accuracy. Optical identifications would have been that much easier.

Despite this drawback the interferometer was used in a number of important studies. One was an extensive investigation of the astrophysical

TANGENT 11

Public Interest, Public Image

No other science generates public interest in the same way as astronomy. The awe and wonderment people feel while gazing upwards to the heavens, the realisation that astronomy grapples with the most fundamental questions of existence, combine to set astronomy apart from the other sciences. This special fascination is reflected in the abundance of books and magazines on astronomy, which far outnumber those devoted to any other field of science. The potential of astronomy to enlighten and entertain people, whether fiction or fact, extends to the other media as well. The program *The Sky at Night in Britain*, hosted by astronomy savant Patrick Moore, began in 1959 and became the longest running television program of any type in the world.

Observatories are part of this fascination in astronomy. No matter how remote, large telescopes attract a constant stream of visitors. There is a similarity with religious pilgrims who travel long distances to experience the aura surrounding a mountain-top monastery or shrine. Soon after its completion in 1948 the Hale telescope at Palomar became a similar type of mecca, attracting over 300,000 visitors a year. Closer to home, the cluster of telescopes on Mt Stromlo attracted over 50,000 visitors a year by the early 1960s. Traffic congestion became so bad that a request had to be issued to Canberra radio stations to stop promoting the observatory as a tourist attraction.

Large radio telescopes make more spectacular viewing than optical telescopes enclosed within their dome buildings. At Parkes, the sight of the first scaffolding rising from the the Goobang Valley in 1960 was enough to bring visitors from all points of the compass. The new landmark led to bundles of letters requesting guided tours of the telescope, some from the most unlikely groups of people. One came from a Cambodian tennis team visiting Australia, another from the organisers of a large group attending the Annual Country Diggers Swimming Carnival!

In response to these requests the telescope was at first opened to the public for two hours every Sunday afternoon. Visitors could inspect the tower and see the dish put through a driving demonstration. Station manager George Day spent these Sundays on duty: 'The astronomers at the observers' quarters would cooperate with the local staff in shepherding the long queues through the telescope and answering their questions. The usual route was through the tower, up the internal spiral staircase, emerging at the turret. The number of people who suddenly found they couldn't stand heights was always a concern—there was no turning back!' [18]

These regular Sunday afternoons became increasingly tedious. Equipment had to be locked away from prying fingers and there were the inevitable grumbles from astronomers forced to interrupt their observing schedule. The Sunday sessions were then replaced by a series of open weekends, held once every three months. However, the problems continued: 'Often we'd get up to 6000 visitors on those weekends. Then there was one disastrous time when it rained heavily.

Mud was tramped through the tower, which of course soon turned to dust. With so much sensitive equipment about—there were over a thousand gold-plated relay contacts alone in the control desk—it tock us weeks to get the telescope working reliably again. At this stage there were plans to install a computer and the tower had to be absolutely dust free. We had no option really.' [19]

The cancellation of the open weekends created an instant public relations problem for the staff at Parkes. Complaints had already been received about the long queues waiting to inspect the telescope and from tourists passing through the area at the wrong times. The Parkes chamber of commerce and local tourist groups took up the issue, eager to develop the potential of what had already become the area's number one attraction.

The obvious solution was to provide a visitors' centre which could house a permanent exhibition on astronomy and allow the public to view the telescope without disrupting its operation. Plans were already underway for a similar centre at Jodrell Bank, known as the Concourse Building. Tourist and business groups were in favour of such a centre but then, predictably perhaps, a disagreement arose over where the funds were to be found. One side argued that it was local business which stood to gain the most from the increased tourist trade. The argument carried some weight. The county authority near San Diego, which built the public road to Palomar, estimated that the tourist dollars generated by the Hale telescope had paid for the road ten times over.

The tourist and business groups on the other side argued that a visitors' centre would be part of the telescope's facilities and should be paid for by government funds. An approach by these groups to Canberra late in 1964 failed to make much headway. The science minister replied that no funds were available for the centre, but the government could probably see its way to build a public toilet block!

Three years later the government relented. Building plans were drawn up and the construction completed in August 1968. The following February, the visitors' centre was inaugurated with almost as much fanfare as the telescope itself in 1961.[20]

Television crews have been frequent visitors to the space-age set at Parkes. The occasion here was filming for the ABC documentary, *The Astronomer*, made in 1966.

properties of interstellar space by Radhakrishnan and Miller Goss, a study now regarded as a minor classic. Another was the discovery of the Magellanic Stream by Don Mathewson and others (details are in the next chapter). In the late 1970s the interferometer was revamped with the installation of a new control system and a major extension of the east–west track. The new instrument, known as the two element synthesis telescope (TEST), has been used to map the structure of spiral galaxies.

Finally, there has been another addition to Parkes familiar to many people, one quite separate from the telescope and its instrumentation. From the time of its completion, the telescope has been a major attraction for visitors. Aside from the professionals who come to carry out observations, many more astronomers have come simply to inspect one of the world's great instruments. A guest book kept in the observers' quarters records some of these visits—its pages read like a who's who in astronomy. Many other leading scientists have visited the telescope including, during the first ten years alone, seven Nobel Laureates.[17] Apart from scientists, the telescope has always been a special attraction to the general public. Interest proved so strong during the early years that the telescope created a major public relations problem of its own. This led to the decision to build a separate visitors' centre at the site in 1968 (see Tangent 11).

Over the years the visitors' centre has staged a series of displays on astronomy, highlighting the research carried out at Parkes. The most recent features a cleverly crafted audio-visual called *The Invisible Universe.* Each year about 70,000 visitors are informed and entertained at the centre, the peak periods occurring in school holidays. This number is less than some northern observatories; Jodrell Bank for example has registered well over 100,000 visitors every year since 1957. Yet, considering the location of Parkes in a sparsely populated part of rural Australia, this number is impressively large. Overall, the total number of visitors to Parkes since 1961 is estimated to be more than three million.

CHAPTER TWELVE

Laboratory of the Extremes

For them the Milky Way system is a giant laboratory in which the Good Lord conducts experiments (free of charge!) under conditions so extreme that we cannot hope to find—or duplicate—them here on earth.

Bart and Priscilla Bok

Astronomy provides us with the largest physics and chemistry laboratory imaginable, where matter exists in extremes of density and temperature not found on earth. In neutron stars, we study matter a million, million, million times denser than lead, while in interstellar space matter is a million, million, million, million times less dense than air.

Ray Norris and Jim Roberts

In this final chapter we return once more to the astronomy carried out at Parkes. In Chapter 9 several areas were reviewed, beginning with the studies on the polarisation of radio sources. These studies provided a compass to explore the magnetic field in the Milky Way and in other galaxies and, closer to home, the unusual properties of the planet Jupiter. Next we continued with quasars, the most distant and mysterious objects in the universe, and then concluded with the extensive sky surveys at Parkes which have catalogued the basic data on thousands of radio sources. These research areas formed a substantial part of the observing program at Parkes in the 1960s and 1970s. In the present chapter we survey a number of other areas which have formed a large part of the research program from about 1970 onwards.

The various types of radio sources discussed in Chapter 9 have one feature in common. Whether local sources such as Jupiter, galactic objects such as supernova remnants, or the extragalactic powerhouses such as quasars, all are detected by their continuous emission of synchrotron radiation. Here we concentrate on other types of radio sources which emit radiation of a different kind. Pulsars are one example. Their emission is beamed, like a lighthouse, and they are observed by their rapid series of pulses. This type of pulsed or discontinuous emission calls for different types of receiver equipment and different observational techniques. There are also other types of radio sources which are observed by their line radiation. Chapter 4 described how the discovery of the 21 cm hydrogen line in 1951 opened up a major new area in astronomy, one which led to the mapping of the spiral structure of the Milky Way. For many years the hydrogen line was the only example of line radiation known but then, in the mid 1960s, the floodgates opened. Numerous lines were detected, revealing the presence of an astonishing variety of molecules spread through interstellar space. Many of these molecules are the 'prebiotic' material essential for life, which has led astronomers to wonder about the part these molecules might have played in the origins of life, here or elsewhere in the universe.

Beams, Blips and Beacons

> Could the signals, we speculated, originate from a civilisation on another planet right outside our solar system? This was the beginning of the story of the 'Little Green Men' which we put out almost as a joke. We referred to the signals as LGM and the possibility that they were signals beamed to us from another civilisation was not quite so ridiculous then as it seems now.
>
> Martin Ryle

The discovery of pulsars is one of the most extraordinary episodes in radio astronomy. Similar to a number of other landmarks in the new science, the discovery was made by serendipity, not by design. The detection of pulses which arrived with such incredible regularity was inexplicable in terms of any known astronomical object. Could the pulses instead be a signal from an extraterrestrial civilisation? In 1968 one of the discoverers of pulsars, Antony Hewish, wrote in *Scientific American:*

> 'Almost exactly a year ago a small group of workers operating a new radio telescope at the University of Cambridge were surprised to find that weak and spasmodic radio signals coming from a point among the stars were, on closer inspection, a succession of pulses as regularly spaced as a broadcast time service. With scepticism bordering on incredulity, the Cambridge group began systematic observations intended to reveal the nature of these strange signals even as they undertook to explain them away in terms of man-made radio interference. After all, seasoned radio astronomers do not make the mistake of supposing that every queer signal on their records is truly celestial; in 99 cases out of 100 peculiar 'variable radio sources' turn out to be some kind of electrical interference—from a badly suppressed automobile ignition circuit, for example, or a faulty connection in a nearby refrigerator.' [1]

The new instrument referred to by Hewish looked more like an elaborate array for training vines at a horticultural station than a radio telescope. The array had been built specifically to study a phenomenon known as interplanetary scintillations. Radio signals from sources with small angular size undergo irregular fluctuations in intensity when passing through the solar wind in our region of interplanetary space. This causes the radio source to 'scintillate', an effect much the same as the twinkling of starlight, though in that case the cause is turbulence in the earth's atmosphere. The Cambridge group planned to study these scintillations to obtain information on the size of the newly discovered quasars, as well as other types of compact sources. By sheer luck the new array, designed to operate at a wavelength of 3.7 metres, proved an ideal pulsar detector. Pulsars generally have their strongest signal at a wavelength of several metres, beyond the normal operating range of large dishes such as Parkes.

After its completion in July 1967, the task of analysing the chart recordings produced by the array fell to a young graduate student, Jocelyn Bell. A month later Bell noticed a new radio source with fluctuations in its strength apparently unrelated to interplanetary scintillations or to any type of local electrical interference. Further observations by Bell and Hewish established that the source lay in the plane of the Milky Way with signals consisting of a succession of regular pulses. The signal varied erratically in the intensity of each pulse and, on some occasions, faded out completely. But the astonishing feature was the extraordinarily precise timing between individual pulses—exactly 1.3372795 seconds.

Within weeks three other Little Green Men had been discovered, each with a different pulsation rate but each with the same incredible regularity. The timing was measured to an accuracy of one part in ten million, a precision only matched by the most advanced types of crystal clock. The observations continued in secret for several months as the Cambridge group worked to establish a natural, physical origin for the signals. One of the Parkes group, Max Komesaroff, recalled:

> 'In late 1967 there were rumours that the Cambridge group had made a spectacular new discovery. The nature of the discovery was evidently a dark and closely guarded secret. Privately and unofficially I was told that Cambridge had been receiving 'signals from outer space'. Neither cajolery nor threats could extract anything further from my informant; evidently that was all he knew. Naturally I was startled, but then, thinking it over, decided that my initial reaction had been excessive, and that a conservative interpretation of my friend's remark was possible—after all surely all astronomy involves observing 'signals from outer space'. When Cambridge finally went public on February 24, 1968 it turned out that they had not in fact been receiving morse-coded messages from Alpha Centauri or from the Andromeda Nebula, but their story was only slightly less fantastic.' [2]

Once the possibility of super civilisations had been ruled out, astronomers began to speculate on the type of object that could give rise to pulses of radiation. Until then continuity in emission had been one of the keystones of astronomy. Sporadic and intense outbursts from

Above Antony Hewish and the Cambridge array used to discover the first pulsars. One of the three dishes forming the one-mile telescope is in the background. *right* The Australian $50 note illustrates a train of pulses from the first known pulsar, PSR 1919+21, recorded at Parkes shortly after its discovery at Cambridge. The strength of the pulses varies erratically but the timing between them is accurate to less than a billionth of a second.

objects such as supernovas and variable stars were well known, but these were still emissions of continuous radiation. The discovery that this new type of object produced pulses of radiation presented a fundamental challenge. Compact stars such as white dwarfs were the first candidates proposed, but then the interest of theoretical astronomers focused on neutron stars, a hypothetical object first discussed by Walter Baade and Fritz Zwicky in 1934. According to the theory, a star about the size of the sun near the end of its life undergoes gravitational collapse to form a new state of matter consisting solely of neutrons. The collapsed object is perhaps only one-millionth of its former size and its material is incredibly dense. Astronomers had considered these hypothetical objects to be an interesting prediction of quantum physics but, until 1968, one that really belonged in the realm of science fiction.

Neutron stars now seemed to offer an explanation for the mysterious pulsed radiation. One suggestion was that the radio pulses arise from continuous vibrations in the surface of the star. Another theory envisaged a type of binary system where a neutron star rapidly orbits a normal star or, perhaps, two neutron stars are in rapid rotation about each other. By the end of 1968, most astronomers favoured a theory put forward by Tom Gold which pictures the pulsar as a neutron star spinning extremely fast on its axis. A combination of intense magnetic fields and relativistic effects in the high energy plasma surrounding the neutron star leads to a beam of radiation that spins around with the star. If the pulsar is suitably aligned in space, a pulse of radiation will be observed every time the beam sweeps across the earth. The effect is the same as the flashing signal a navigator sees from a distant lighthouse.

While theorists worked towards an acceptable physical model for pulsars, astronomers all over the world set off on a hectic race to discover more of these objects. In 1968 alone more than one hundred papers reporting observations of new pulsars or their interpretation were published. The large dishes at Jodrell Bank and Arecibo were particularly successful in terms of the number discovered, but several other observatories also claimed one or two pulsars including such less familiar places as Pushchino (USSR) and Ootacamund (India). Initially, the most successful instrument in the pulsar hunt was the SuperCross at Molonglo, operated by the University of Sydney. The Cross proved ideal because of its large collecting area and its relatively long operating wavelength. By the end of 1972, no less than 34 of the 84 known pulsars had been discovered at Molonglo. The Molonglo pulsars provided a sample large enough to establish beyond doubt that these objects are our relatively close neighbours in the Milky Way. The most important for astronomy turned out to be the third, detected in October 1968. Bernard Mills remembers the surprise:

> 'Two had already been discovered (the first ones to be found in the southern hemisphere) and my colleague Michael Large was just beginning a systematic search for more. One morning he rang me from the Observatory in great excitement—a pulsar had shown up with a repetition rate of 11 pulses per second, very much faster than any known at that time. On checking his position for the pulsar I found an even more interesting result. The pulsar was located close to the centre of a more or less circular region of complex radio emission some five degrees across. This extensive radio source was known to have the properties of a supernova remnant, resulting from a star which must have exploded over 10,000 years ago.' [3]

This supernova remnant was well known to astronomers as the source Vela X. Detailed studies of the remnant at Parkes had already established its distance to be about 1500 light-years, the same distance Mills and his group derived for the new pulsar. The identification of the pulsar with the small massive remnant of an exploded star provided the first step in understanding how pulsars are formed. The link between pulsars and supernova remnants was strengthened shortly afterwards by American astronomers with the discovery of a pulsar at the centre of the Crab Nebula, the spectacular supernova of 1054 AD. The pulsation rate measured for the Crab pulsar was an extraordinary 30 times per second, making it the fastest pulsar then known.

☆ ☆ ☆

In the search for pulsars the Parkes telescope was easily overshadowed by the Molonglo Cross. However, because it is a fixed instrument, the Cross was best suited for detecting pulsars, but not for studying them in depth. Detailed observations of a particular pulsar require an instrument which can track the object for long periods of time. Here the Parkes telescope came into its own in terms of actually learning what makes a pulsar tick.

Pulsar observations began at Parkes in March 1968, a few weeks after the announcement of the discovery at Cambridge. A pulsation rate of 1.3373011 seconds was measured for the first Cambridge pulsar, a value which corrected a small bookkeeping error in the rate already published. The Cambridge group had counted one pulse too many per day and calculated a rate too short, by a minuscule 0.00002 seconds! Next the Parkes group began a study of the Vela pulsar, prompted by the discovery at Arecibo that the pulsation rate of the Crab pulsar had slowed down by a small amount over a period of a few months. In December 1968 a similar slowing down for the Vela pulsar was measured at Parkes, the time between pulses increasing each day by 10 nanoseconds, that is, one-hundredth of a microsecond. This is an extremely small change over a single day, but on a cosmic time scale a change large enough to eventually decelerate the pulsar to a slow spin.

The Vela and Crab observations provided strong support for the rotating neutron star model put forward by Tom Gold. The slowdown in the pulsation rate was predicted by this theory, but not by competing theories such as the vibration model or the binary model of two stars in rapid orbit, both of which predicted that the pulsation rate would gradually speed up. The Indian astrophysicist Radhakrishnan writes:

> 'In the first fifty or so papers on pulsars, white dwarfs vibrating in various ways were still the favourite candidate, although Gold had already proposed rotating neutron stars. The discoveries of the Vela and Crab pulsars with their very short periods (by the standards of the day), and the measurement soon after of the lengthening of these periods, as predicted by Gold, put the rotational neutron star hypothesis way ahead of the rest of the field. But the reason for the beaming of the radiation as from a lighthouse was as obscure as ever. The spectrum of the radiation had been shown quite early to be highly non-thermal, but the question of an emission mechanism was

generally avoided. The theoretical papers of the time that attempted to address this issue now read like science fiction, all attesting to the strangeness of the objects whose nature we were struggling to fathom.' [4]

A crucial step was the discovery at Parkes that Vela's pulses of radiation are strongly polarised. Further, over the brief duration of each pulse, the plane of polarisation was found to sweep through a large angle. The next pulse repeats the same behaviour. The effect could be understood by assuming that the rapidly spinning pulsar has its magnetic axis inclined at a large angle to its rotation axis. A narrow beam of radiation would then be emitted out along the magnetic axis, with the beam rotating as the neutron star spins. This improved version of Gold's theory, known as the polar cap model, was developed by Radhakrishnan and Max Komesaroff using the Parkes observations. The polar cap model forms the basis of the present-day theory of pulsars.

Another peculiar property of pulsars was also revealed by the Parkes observations on Vela. In February 1969 Radhakrishnan and Dick Manchester carried out timing observations and confirmed the slowdown in pulsation rate of 10 nanoseconds per day discovered the previous December. Two weeks later the observations were repeated but, unaccountably, Vela's pulsation rate in the meantime had actually speeded up by almost 200 nanoseconds. Simultaneous observations using the NASA tracking dish at Goldstone confirmed this sudden change in the pulsation rate. The American astronomers coined the word 'glitch' to describe the abrupt change. Three or more of these glitches in Vela's rate were subsequently detected at Parkes between 1971 and 1978, while American astronomers reported similar events for the Crab pulsar in 1972 and 1975. The fifth and most revealing of the Vela glitches occurred in 1981, observed by a group from the University of Tasmania. The glitch was shown to occur virtually instantaneously, after which the pulsar returned to its normal behaviour of gradual slowdown.[5]

The cause of these glitches is still not clear. One early idea was that a type of 'starquake' causes the neutron star to shrink. By conservation of angular momentum the spin rate, and thus the pulsation rate, suddenly increases. To produce a glitch of the size observed, the radius of the neutron star would only need to shrink about one centimetre. This theory had to be abandoned, however, because it predicted the occurrence of glitches about once every several hundred years, rather than once every few years as observed for the Vela and Crab pulsars. The theory currently accepted proposes a sudden transfer of angular momentum from the interior of the neutron star to its outer crust. The interior material is thought to consist of a type of quantum superfluid which rotates more rapidly than the solid neutron crust. A sudden increase in the coupling between the crust and the interior superfluid will speed up the rotation rate of the crust. This model seems to explain both the magnitude and the frequency of the Vela glitches.

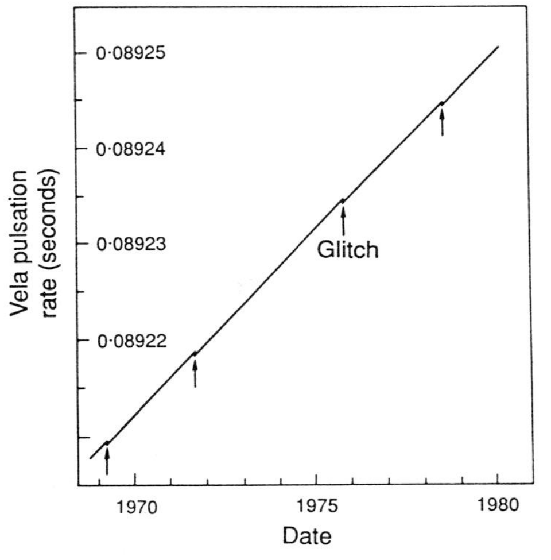

Pulsation rate of the Vela pulsar measured from 1969 to 1980 at Parkes. The rate steadily increases as the pulsar gradually slows down, apart from four glitches when the rotation of the pulsar suddenly speeds up. After each glitch, the pulsar enters another long phase of gradual slowdown until the next glitch.

An intriguing feature of some pulsars is that they act as true galactic lighthouses, beaming radiation at visible wavelengths. Early in 1969 astronomers at the Steward Observatory in Arizona detected light pulses from the Crab pulsar. The light pulses are emitted at the same rapid rate as the radio pulses, so rapid in fact that astronomers had always assumed the flickering source to be a star with continuous emission of light. Later, X-ray pulses were also detected from the Crab pulsar by using detectors launched by rocket. The X-ray power emitted is about 100 times stronger than that at optical wavelengths, which in turn is about 100 times the radio power emitted by the pulsar. Early in 1977 Vela became the second known optical pulsar, detected by a collaboration of Australian and British astronomers using the Anglo-Australian telescope at Siding Spring. In contrast to the Crab, the Vela optical pulses are extremely weak. Over eight telescope hours were required to register a single image. This made Vela the faintest optical object ever detected by any telescope. Curiously, the optical signal actually consists of two pulses, neither of which coincides with the radio pulse. The radio pulse appears first, followed 20 milliseconds later by the first optical pulse, then a further 22 ms later by the second and stronger of the two pulses. Every 89 ms this sequence of one radio pulse then two optical pulses is repeated.[6]

So far optical emission has been observed from only three pulsars, although several hundred pulsars are now known. Each of these pulsars has a very fast pulsation rate compared with most pulsars, so it appears that optical pulses are only emitted by either very young or rapidly rotating pulsars. The reasons for this are not yet fully understood.

While the Vela and Crab pulsars have provided a goldmine of information for both southern and northern observers, a number of other pulsar properties have been revealed by studying the statistical features of the population as a whole. To compile a sample as large as possible, a group led by Dick Manchester began a major finding survey of the southern sky using the Molonglo Cross. The Parkes telescope then followed up by making measurements on the pulsation rate, the pulse shape, the slowdown rate and the possible occurrence of glitches for each new pulsar. Estimates of the distance to each pulsar were also made, as well as a plot of their distribution throughout the Milky Way. Upon its completion in 1978, the Molonglo–Parkes survey had catalogued a total of 155 new pulsars, more than the worldwide total of 149 known before the survey began.[7]

As mentioned earlier, the discovery of pulsars associated with the Crab and Vela supernova remnants provided a vital clue to how pulsars are formed. A supernova explosion can occur when a star several times more massive than the sun has evolved to a point where it has exhausted all of its nuclear fuel. Without this powerful source of radiant energy inside the star, the outward radiation pressure can no longer support the star's outer layers. Consequently, the star undergoes gravitational collapse with pressures building up to the stage where a catastrophic explosion blows the outer layers, comprising perhaps as much as half the star's original mass, out into space in the form of vast clouds of fiercely incandescent gas. The temperature of the outwardly moving layers is so high that light elements manufactured earlier in the star are fused into heavier elements, releasing even more energy.

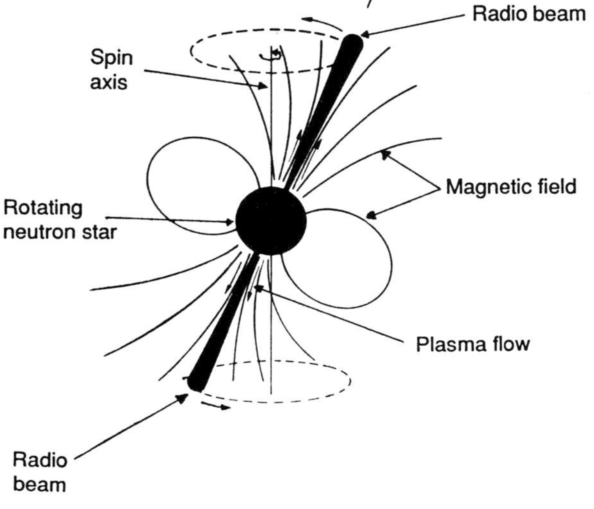

A pulsar has its magnetic axis inclined to its spin axis. The magnetic field of a pulsar is about a trillion times stronger than the earth's, producing enormous forces that strip plasma material from the surface of the pulsar. The plasma is accelerated to nearly the speed of light causing it to emit intense radiation. Most of the radiation is beamed out along the north and south poles where the magnetic field is strongest. Every rotation a pulse is observed as the radio beam sweeps across the direction of the earth.

In the opposite direction, the colossal force of the explosion creates an implosion crushing the remaining material in the star's core into a pure form of nuclear matter, leaving a remnant perhaps no larger than 20 kilometres in diameter. One cubic centimetre of the material is estimated to weigh about one billion tons, a density almost impossible to imagine. This is the density which would be reached if the entire Australian continent, above sea level, were to be compressed into the space occupied by a cricket ball! Throughout the catastrophe, two physical characteristics of the star are preserved—its angular momentum and its magnetic energy. What began as a large, slowly spinning star has been reduced to a highly condensed, rapidly spinning dynamo.

Although this scenario is generally accepted, the evidence supporting the supernova origin of pulsars is hardly overwhelming. Several hundred pulsars and well over one hundred supernova remnants are currently known, but only a handful of pulsars have been definitely identified with a supernova remnant. The great majority of pulsars show no sign of an accompanying supernova shell; similarly, it seems that some supernovas can lead to the formation of other objects just as bizarre as pulsars. One example is a new class known as X-ray stars, sources which give off intense, continuous emission of both X-rays and radio waves. These sources are binary systems where material is drawn off from a supergiant star and consumed by a neutron star, or possibly even a black hole, in orbit around its more massive companion. One of the best known of these X-rays stars is Circinus X-1, a source that has been watched closely at Parkes since its discovery in 1975. Evidence suggests that the source formed about 10,000 years ago when the smaller star in the binary system underwent a supernova explosion, leaving a non-pulsing radio source orbiting a supergiant star. Circinus X-1 provided the first example of how a compact object other than a pulsar could result from a supernova.[8]

One of a number of Aboriginal rock carvings near Broken Hill in New South Wales. Astronomers wonder whether the carving might record the spectacular appearance of the Vela supernova some 10,000 years ago.

In view of the likely association between supernovas and pulsars, there has been intense interest in a supernova which burst into view in February 1987 in the Large Magellanic Cloud. This was the first time astronomers had the opportunity to observe a supernova as it actually occurred since Johann Kepler recorded a similar event in our own galaxy in 1604. A concerted effort has been mounted at many observatories, including Parkes, to witness the birth of a pulsar from this event. Although several observatories recorded the initial radio burst from the supernova, known as SN1987a, attempts to detect the telltale pulsed radiation have so far been without success. The detection of subatomic neutrinos from this supernova suggests that a neutron star has been formed in the explosion. However, whether or not the remnant is a pulsar remains to be seen. Early in 1989 a group of Berkeley astronomers reported optical pulses from the remnant but, tantalisingly, the signals did not return. It now seems that the signals were no more than interference caused by a nearby piece of laboratory equipment. Possibly conditions have not been favourable for the formation of a pulsar on this occasion or, possibly, some other exotic remnant such as a black hole has resulted from the explosion. Alternatively, a pulsar might have formed already but be hidden behind the expanding shell of stellar debris, or be aligned in such a way that its radio beam does not sweep across our line of sight.

The study of pulsars has formed one of the major research programs at Parkes since the discovery of these amazing objects over twenty years ago. Two other pulsar 'firsts' made at Parkes are well worth mentioning. One was the first pulsar outside our galaxy, PSR 0529–66, discovered in 1983 during a survey of the Large Magellanic Cloud. More recently, a

Tom Gold (left) at Cornell University was the first to suggest that pulsars are rapidly spinning neutron stars. Venkataraman Radhakrishnan (right) extended this idea by working out how the pulsar's radiation is beamed like a lighthouse. Radhakrishnan made his first visit to Australia sailing his trimaran *Cygnus A*.

new type of binary pulsar located by the Parkes group in the globular cluster known as 47 Tucanae has revealed properties quite unlike any other pulsar. Although binary pulsars had been known for some years, the unusual feature of PSR 0021–72A is its exceptionally short period of 0.0045 seconds, equivalent to over 200 revolutions per second. The pulsar is in orbit around an invisible companion with an orbital period of 32 minutes. The movement of these two massive bodies in such a tight, fast orbit, combined with the extremely rapid spin of the pulsar around its own axis, creates unusual effects in the gravitational force between these two bodies. These effects will enable a number of demanding experimental tests to be made of Einstein's theory of general relativity.[9]

For all that has been learnt about pulsars since their discovery in 1967, it remains a wide open area of astronomy. The detection of pulsars in binary systems, the existence of pulsars with periods of only a few milliseconds, and the possibility that many pulsars may form from a collapsing white dwarf rather than a supernova have forced theorists to formulate pulsar models of increasing complexity. In 1988 a new survey commenced at Parkes in collaboration with Jodrell Bank astronomers aimed at examining the region along the galactic plane and also concentrating on a large sample of likely objects, such as supernova remnants and globular clusters, which may contain undetected pulsars. The survey yielded over fifty new pulsars, including an unusual 'eclipsing binary pulsar', only the second of its type known. The binary system lies in a rate of 86 times per second. The white dwarf is aligned so that it blocks out the pulsar signal as it orbits the pulsar once every 109 minutes—and so its name, eclipsing binary pulsar.[10]

Studies of systems such as these might one day reveal a new and the unusual feature of PSR 0021–72A is its exceptionally short period of standing their extraordinary physical properties. In the meantime, the pulsar pioneer Radhakrishnan has summarised the current situation: 'Some major questions like the nature of the object and the reason for the periodicity of the radio emission were answered within a year or so of the discovery of pulsars. But a total understanding of the emission mechanisms and a satisfactory explanation of the great diversity in their behaviour continues to elude us even after twenty years.'[11]

Hunting for Hydrogen

Twinkle, twinkle little star,
I don't wonder what you are
For by spectroscopic ken
I know that you are hydrogen.

Anon.

From pulsars we turn now to another of the major research programs at Parkes, studies of interstellar hydrogen. The contrast between the two states of matter could not be more extreme. In pulsars, matter is 10^{14} times more dense than lead; in interstellar space, matter is 10^{24} less dense than air. There is also a basic difference in the type of radio emission from these two states of matter. Pulsars emit continuous radiation which we observe as a series of pulses because of the lighthouse effect. Interstellar hydrogen, on the other hand, emits discrete radiation at its characteristic wavelength of 21 cm.

The discovery of the hydrogen line at Harvard in 1951 presented astronomers with a powerful technique for investigating the large scale structure of the Milky Way. By 1958 a collaboration between Leiden and Sydney radio astronomers had completed mapping the Milky Way, providing clear evidence for the spiral structure of our galaxy. Immediately after the completion of the Parkes telescope these studies were continued by Frank Kerr, the chief 21 cm specialist at Radiophysics in the 1950s and who later transferred to the University of Maryland. Kerr's new survey revealed a great deal of fine detail in the spiral structure. The information helped to clear up several problem areas in the earlier Leiden–Sydney map, which had spliced together two separate surveys carried out with different telescopes in different hemispheres. A survey of the centre of the galaxy also showed that the structure and motion of vast clouds of hydrogen are far more complicated than previously thought, indicating that there are violent streaming processes in and around the galactic core.

The Large and Small Magellanic Clouds have been a frequent target of the 21 cm studies at Parkes. Known as the 'crown jewels' of the southern sky, these small satellite galaxies of the Milky Way are of special interest because they provide a laboratory closeby for observing how stars and gas clouds interact within a galaxy, and how this might lead to the formation of particular galactic shapes. Observations on the small cloud have shown that its interstellar hydrogen is distributed fairly evenly throughout a slightly flattened system, rotating in a plane edge-on to our line of sight. The most surprising discovery came from measurements of the Doppler shifts in the 21 cm line. The small cloud was shown to consist of two independent masses of stars and gas rapidly moving apart at a velocity of about 50 km per second. The Small Magellanic Cloud is in fact a double object. The southern sky actually has three crown jewels rather than two. The smallest of the three is now known as the Mini Magellanic Cloud.

Observations on the small clouds by Jim Hindman and others were carried out in parallel with studies of the large cloud by Dick McGee and Janice Milton.[12] Unlike its smaller double companion, the large cloud contains interstellar hydrogen clumped into a number of extensive complexes. Each complex forms a separate stellar nursery with young stars imbedded in the hydrogen cloud, which itself is slowly coalescing and giving rise to the birth of new stars. The Parkes observations also provided the first evidence of a simple spiral structure. Previously, both the large cloud and its smaller companion had been classified as 'irregular', one of the small fraction of amorphous galaxies with no clearly defined shape. The great majority of galaxies have either the familiar spiral shape (80%) or are ellipticals (17%). The large cloud appears to have two large spiral arms lying in one plane, while in another plane inclined to it by about twenty degrees, there are two other shorter spiral arms. Although lacking the majestic catherine-wheel symmetry of our own galaxy, the Large Magellanic Cloud does display a primitive but discernible spiral structure.

More recent explorations of the Magellanic Clouds at Parkes by Don Mathewson have led to an even bigger surprise—the discovery of the Magellanic Stream. This is a vast filament of hydrogen trailing behind the clouds and extending in a large arc across the southern sky. One suggestion is that the stream is a kind of turbulent wake, left behind by the clouds as they glide through a thin halo of hydrogen gas surrounding our own galaxy. Another theory by Mathewson and his group at Mt Stromlo suggests that the Magellanic Stream arose from a collision between the two Magellanic Clouds about 400 million years ago. The collision ripped apart the small cloud into the two parts we now observe and left trailing behind the long gaseous filament. It now appears that the Magellanic Stream formed from a combination of effects described by both the collision and turbulent-wake theories.[13]

These studies at Parkes on the structure and dynamics of both the Magellanic Clouds and the Milky Way were a major advance on the original 21 cm observations by groups in the 1950s. There was, however, an even bigger bonus. The greater sensitivity and resolving power of the Parkes telescope also made it possible to study hydrogen emission from far more remote galaxies.

A number of large and relatively close southern galaxies came under scrutiny first. The observations provided information on the internal dynamics of these galaxies, the distribution of hydrogen within each galaxy and the relative amount of hydrogen remaining for the formation of new stars. Next, a group led by Brian Robinson began a study of the distant galaxies which comprise the Virgo cluster, a large family of galaxies at a distance of 35 million light-years from the local group which contains the Milky Way. These galaxies are too distant to resolve individual structural features, but measurements can still be made of the total amount of interstellar hydrogen remaining within each galaxy. The amount gives an excellent measure of the age of a particular galaxy. For one of the large ellipticals M87 (also known as the radio source Virgo A), only a slight trace (0.1%) of hydrogen could be detected showing that the galaxy has converted virtually all of its material into star building. In some of the spirals, stellar evolution is less advanced with between 2% and 10% of the primordial hydrogen still available for star formation. In comparison, interstellar hydrogen accounts for 2% of the total mass of the Milky Way, indicating that our galaxy is mature age. The corre-

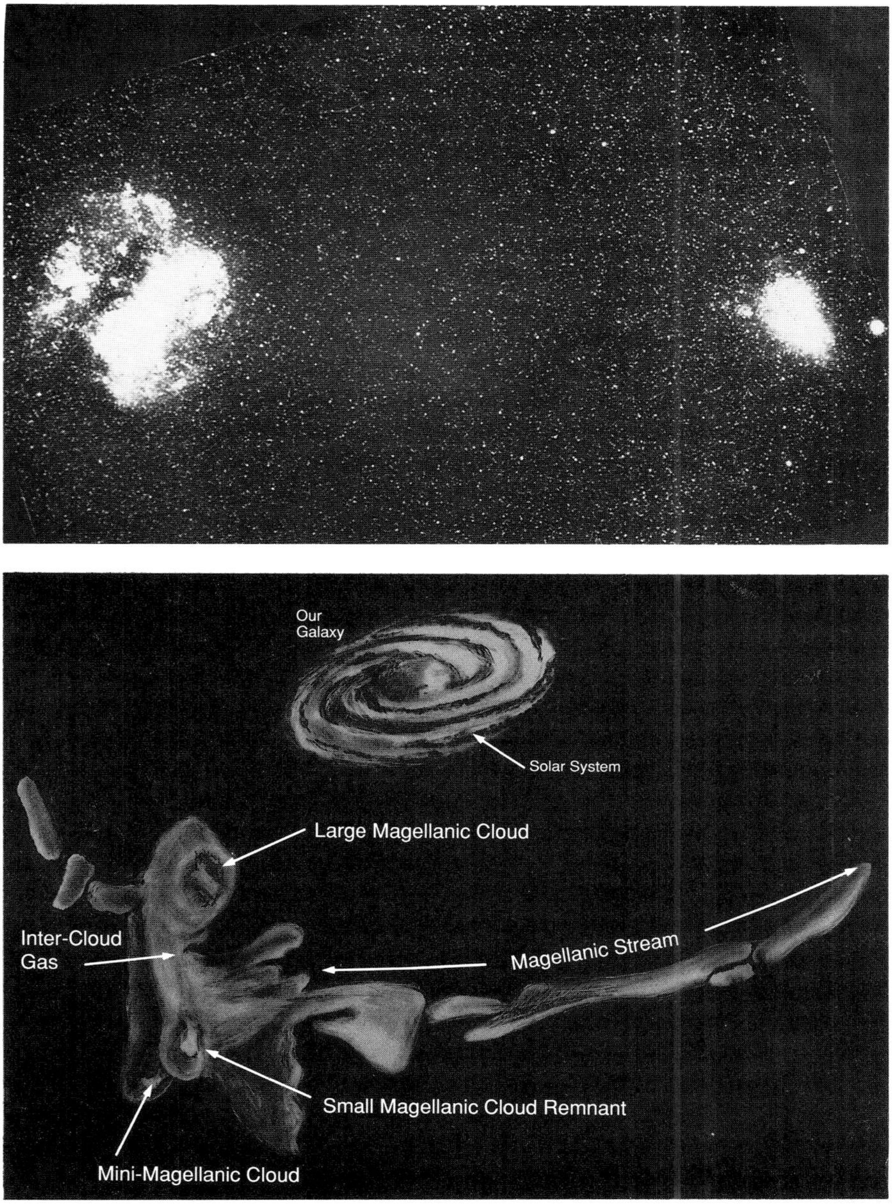

The Magellanic Clouds viewed from earth (above) and from an imaginary vantage point outside the Milky Way (below). Our galaxy is a majestic spiral with a nucleus and pinwheel arms. The clouds are smaller, less symmetrical and contain a higher proportion of interstellar hydrogen. The Mini-Magellanic Cloud, discovered at Parkes in 1983, is moving away from the Small Magellanic Cloud but remains connected by a hydrogen bridge. From the earth, the mini cloud lies hidden behind the small cloud. The Magellanic Stream stretches out in a vast arc behind as the cloud system moves in slow orbit around the Milky Way.

sponding figures are 10% and 30% for the Large and Small Magellanic Clouds, so that star formation is still a going concern in these adolescent galaxies. Recent surveys at Parkes, made in collaboration with the UK Schmidt telescope at Siding Spring, have extended these studies on the hydrogen content to several thousand galaxies.

Another observation on the Virgo cluster yielded the first information on a fundamental problem in astronomy. Two questions can be posed. Aside from the *interstellar* hydrogen available within galaxies for further star information, can hydrogen be detected in the vast reaches of space between galaxies? If so, is there enough of this *intergalactic* hydrogen for new galaxies to form at this present stage of the universe? The observations at Parkes answered yes to the first question, no to the second. Intergalactic hydrogen could be detected in the Virgo cluster but its distribution was found to be incredibly thin. By way of illustration, there are about 10^{19} molecules per cubic centimetre in the earth's atmosphere and only about one hydrogen atom per cubic cm in the interstellar space of the Milky Way. In the Virgo cluster, the intergalactic density is a million times thinner again; that is, on average only about one hydrogen atom per cubic metre. Even though this gas is extremely rarefied, the size of the Virgo cluster is so vast that the total amount of hydrogen is enough to make ten large galaxies. And yet, because this hydrogen is so rarefied, it seems highly unlikely that this gas will ever coalesce to form new galaxies. Most probably, the galaxies we now see in the Virgo cluster and elsewhere formed early in the universe when the gas was far more concentrated, before the cluster, and the universe itself, expanded to their present size.[14]

As well as ruling out the likelihood of further galaxy formation, these observations on the density of intergalactic hydrogen had an important bearing on a central question in astronomy: What will be the eventual fate of the universe? Since its origin in the big bang some 15 billion years ago, the universe has been expanding with the most distant quasars and radio galaxies receding at the greatest velocities. Whether or not gravitational attraction will eventually halt this expansion, so that what follows is an epoch of steadily quickening contraction, depends basically on a single parameter—the total mass of the universe. Astronomers were already aware that by adding up the mass of all the billions of stars in a single galaxy, and then adding up the mass of all the billions of visible galaxies there would not be anywhere near enough mass in grand total to halt the expansion. The observations at Parkes on the Virgo cluster indicated that there would still not be enough of this previously unseen hydrogen to make up the mass required.

Recent estimates based on both optical and radio studies give a figure of only 20% of the mass needed to halt the expansion. It now seems highly likely that a large amount of 'dark' matter remains undetected in the universe, perhaps in the form of exotic subnuclear particles postulated by some physicists, or perhaps in the form of massive 'brown' dwarf stars favoured by some astronomers. Otherwise, it seems that without this missing mass the universe will go on expanding forever. There will be no epoch of contraction and no opposite of the big bang; no 'big squeeze' as George Gamow once called it. Instead, the universe will slowly cool as it expands, the stars fading out one by one. When the universe is perhaps ten times older than its present age, there may be only a cold, dark, and desolate void.

Murmurs from the Stellar Nursery

> It is an amazing thing that the human brain, which is a tiny subset of the material universe, composed of hydrogen atoms created in the big bang, oxygen atoms synthesised inside long-gone stars, and some impurities, can observe the rest of the universe. How can the part comprehend the whole, let alone begin to consciously manipulate it, as we are doing?
>
> Ron Bracewell

Studies of the 21 cm hydrogen line have added a new dimension to astronomy. Hydrogen's telltale emission has made it possible to explore how this basic building block of the universe is distributed throughout our own galaxy and in distant clusters of galaxies as well. In view of the rich source of information provided by the hydrogen line, the question arose whether there might be other lines in the radio spectrum. The search involved a number of difficulties, as Brian Robinson wrote in *Scientific American*:

> 'Soon after Ewen and Purcell of Harvard University discovered the hydrogen emission at the 21 cm wavelength in 1951, radio astronomers began to search for radio emissions from other atoms in interstellar space. They realised that detection of any such signals would be difficult, because radio emissions from atoms are rare and extremely weak when they do occur. To begin with, there is the rarity of matter in space: the amount of hydrogen averages less than one atom per cubic cm, and all other atoms are very much sparser. Secondly, atomic processes that could give rise to detectable radio emissions are considerably less common than those that produce emissions of light. The energy of a quantum of radiation depends on its frequency, and a quantum at radio frequencies is a million times weaker than a quantum at the much higher frequencies of visible light. Consequently radio emissions can arise only from atomic processes that involve comparatively slight changes in the atoms' energy state. Furthermore, because the probability of emission drops sharply with decrease in frequency (or increase in wavelength) of the radiation, any radio signal from a collection of atoms is very weak.' [15]

Despite the presumed scarcity of other interstellar atoms, searches began in the 1950s for signs of their presence. The best candidate appeared to be deuterium, the naturally occurring heavy isotope of hydrogen with a nucleus consisting of one proton and one neutron. Laboratory measurements had already demonstrated that deuterium emits line radiation from the same type of spin-flip transition by its electron, but in this case at a wavelength of 92 cm. On earth there is only about one deuterium atom for every 600 hydrogen atoms, but astronomers reasoned that because the hydrogen-line signals are relatively strong it should be possible to detect the deuterium line. A number of painstaking searches were made, including one by the Radiophysics group at Dover Heights in 1954. Another by the NRAO at Green Bank in 1962, which ran to almost 2000 hours of observing time, also failed to detect interstellar deuterium.[16]

These discouraging results for deuterium and the fact that very few radio lines from other atoms were known from laboratory measurements persuaded most astronomers to follow more profitable areas of research. There was, however, an exception: the oxygen–hydrogen molecule OH, known as the hydroxyl radical. Unlike hydrogen with its single line at 21 cm (frequency 1420 MHz), physicists had shown that OH can radiate at four different wavelengths, corresponding to four possible ways the molecule can vibrate in its lowest energy state. The wavelength of each of these four lines was calculated to be close to 18 cm (1666 MHz). Then, in 1959, a group at Columbia University in New York succeeded in measuring the first precise laboratory values for two of these lines, at the frequencies of 1665 and 1667 MHZ. Oddly, four years were to go by before astronomers followed up this work by the physicists. The first search made in 1963 by Sandy Weinreb and a group at the Massachusetts Institute of Technology brought immediate success. The MIT group was able to detect interstellar OH not by its *emission*, but by two *absorption* dips in the spectrum of the strong source Cassiopeia A. The absorption dips proved that there must be thin clouds of OH lying between this source and the earth, filtering out radiation at the characteristic line frequencies of 1665 and 1667 MHz. The relative strengths of the two lines were in the ratio of 5 to 9, exactly as the calculations had predicted. Twelve years after the discovery of the 21 cm hydrogen line, two new lines had at last been added to the radio spectrum.

The MIT group cabled news of the discovery to other radio astronomy centres around the world. At Parkes a suitable receiver was rapidly assembled and observations began on Sagittarius A, the strong radio source known to be close to the centre of the galaxy and lying behind several spiral arms rich in gas clouds. Within a month the Parkes group had confirmed the discovery of OH absorption at both 1665 and 1667 MHz, only a matter of days before similar announcements were made by groups at Berkeley in California and Harvard in Massachusetts.[17]

The Parkes observations of the region near the galactic centre showed several unusual features. The OH appeared to be concentrated in small self-contained clumps around the centre, often in regions where very little hydrogen could be detected. Doppler shifts in the two OH lines away from their exact laboratory values also revealed that most of these OH clouds are streaming in towards the galactic centre, whereas most hydrogen clouds were known to be either streaming outwards or swirling around the centre. Both OH lines proved much stronger than those found for Cassiopeia A by the American groups, and this encouraged the Parkes group to hunt for the third and fourth OH lines. These two minor or 'satellite' lines had been too weak to detect in the earlier laboratory studies at Columbia. In April 1964 the Parkes group found both, again by searching in the direction of the galactic centre, one at 1612 MHz, the other at 1720 MHz. This was the first occasion astronomers had measured the emission lines of an atom or molecule before physicists in the laboratory. The number of lines in the radio spectrum had now increased to five.[18]

There was one puzzling feature of the OH signals the Parkes group could not explain. The strengths of the two main lines at 1665 and 1667 MHz were not in the expected ratio of 5 to 9. For several OH clouds near the galactic centre, the 1665 MHz line was definitely stronger. The explanation was soon provided by Harold Weaver and the Berkeley group after an examination of several HII regions. As well as the usual absorption lines from these clouds of ionised gas, strong emission from OH could also be detected. In fact, the emission at 1665 MHz was so intense that the Berkeley group hesitated at first in attributing it to OH. On the slim chance that they had stumbled upon some totally new form of interstellar matter, the Berkeley astronomers gave it the name mysterium.

Further work at Harvard, MIT and Parkes quickly confirmed the discovery and proved conclusively that mysterium was none other than OH. Only one emission mechanism could explain the exceptionally strong intensities and very narrow bandwidths of the OH lines. These particular OH clouds are acting as a giant maser. Earlier, we described how the first laboratory masers developed in the mid 1950s had made use of a quantum process known as stimulated emission of radiation to amplify weak radio signals. The maser soon led to the laser. Yet, great skill was required to build and operate these devices in a laboratory. The realisation that the same type of delicate tuning occurred naturally in chaotic gas clouds was completely unexpected. Once again radio studies had sprung one of the great surprises in astronomy.

The Parkes group had come within an ace of making the discovery, as Brian Robinson recalls:

> 'We were bitterly disappointed to have missed it. Our old records of the Sagittarius B2 source showed clear evidence of the OH emission, but we had

failed to recognise it for what it was. We had been looking for the absorption lines of OH. At one point the absorption dip for one of the lines dropped almost to zero, but we thought such a pronounced effect was probably caused by some problem with our receiver. When we re-examined this dip later with an improved receiver, the absorption at this frequency shot right though to a negative value. Of course, negative absorption just means emission. We could easily confirm the discovery by the Berkeley group, simply by using the data we had obtained a year earlier than them.' [19]

Despite missing the discovery, observations since then have brought adequate compensation to the Parkes group. In 1973 John Whiteoak and Frank Gardner reported the first extragalactic OH signals from the southern galaxy NGC 4945. Surveys too of the Milky Way carried out by Jim Caswell and Ray Haynes have located over half of the approximately one hundred OH masers presently known. Similar to the first OH maser discovered, the majority of these sources are found in HII regions, though a small number appear to be associated with supernova remnants or unusual late-type stars. One source discovered at Parkes, OH351.78–0.54, is currently the strongest OH maser known. Its extremely intense emission at 1665 MHz indicates that it is amplifying radiation at this frequency by almost a factor of 100 billion—a natural amplifier of extraordinary efficiency.

Most OH clouds are relatively small in size, possibly no larger than our solar system. The clouds are thought to be a brief, intermediate stage in the contraction of a hydrogen cloud to form a 'protostar'. Earlier in its history, the cloud is more diffuse and probably can only be detected by its hydrogen-line emission. As the cloud contracts it becomes warmer, producing the infrared radiation required to pump the masing molecules to higher energy levels. The OH molecules then start to act in concert, switching on their intense maser radiation and signalling the birth of the protostar. Later, further contraction leads to steadily increasing temperatures, reaching the point where nuclear fusion begins

Charles Townes and his group at Berkeley detected the telltale signals from interstellar ammonia and water, the first of a growing list of molecules in space.

between the hydrogen atoms. Visible light is then switched on from the newly formed star.

These OH observations continue to provide surprises. A recent discovery is that not just protostars in our galaxy, but whole galaxies can act as exceptionally powerful sources of maser radiation. Over thirty of these sources—known as megamasers—have been discovered, the most powerful being detected in 1988 by a collaboration of astronomers at Parkes and Siding Spring. Whether or not these new megamasers are also murmurs from some vast galactic nursery is yet to be revealed.[20]

Little Green Molecules

> Measure for measure, we are to an atom
> what a star is to us.
>
> Paul Davies

> I, a universe of atoms,
> an atom in the universe.
>
> Richard Feynman

The hydroxyl radical OH with its four radio lines clustered around a wavelength of 18 cm opened up a way of exploring previously inaccessible regions of star formation. No less important, the discovery of OH spurred astronomers on to hunt for other molecules in interstellar space. Again the challenge was to search for the characteristic radio lines associated with slight changes in the rotational and vibrational states of these molecules. Advances in two closely related fields were crucial to the success of this program. First, precise laboratory measurements of the radio lines were essential. Searching at random through the radio spectrum would otherwise be as productive as looking for the proverbial needle in a haystack. Second, most of these radio lines were predicted to be at short wavelengths, perhaps a few centimetres or less. This meant new types of radio telescopes with greater surface accuracies than those available in the 1960s, new types of receivers for these short wavelengths, as well as new data handling techniques capable of identifying the extremely faint signals expected from interstellar molecules.

Charles Townes and the astronomers at Berkeley were the first to master these new techniques. In December 1968, using a small dish at the Hat Creek Observatory especially designed for short wavelengths, this group detected weak emission from ammonia (NH_3) in a dense gas cloud near the galactic centre. Two weeks later, the same group announced the discovery of interstellar water at its characteristic wavelength of 1.35 cm, on this occasion from the Orion Nebula.

The discovery of these thin mists of water and ammonia proved that molecules containing three and four atoms could exist in space. Radio astronomy groups elsewhere immediately joined in the search, setting off what one astronomer described as 'an unbelievable rush to observe new molecules, a bandwagon that is still rolling.'[21] There was an avalanche of discoveries in the years that followed. The next molecule found was formaldehyde (HCHO), once commonly known as embalming fluid, followed in 1970 by carbon monoxide (CO), cyanogen (CN) and hydrogen cyanide (HCN). By the end of 1972 the list of interstellar molecules had grown to thirty, with the groups at Green Bank and Kitt Peak claiming most of the discoveries.

The Parkes group opened its account in 1971 with the detection of thioformaldehyde (HCHS), the first interstellar molecule discovered by a group outside the United States. A year later methanimine (CH_2NH) was found in Sagittarius B2, a vast cloud near the galactic centre which has since been shown to contain the most varied collection of interstellar molecules known in the galaxy. Two further discoveries at Parkes shortly afterwards were methyl formate ($HCOOCH_3$) and vinyl cyanide (CH_2CHCN). Another was methylamine (CH_3NH_2), its signals detected from both the Sagittarius B2 and Orion sources. Methylamine exists on earth as a colourless gas and is used in dyestuffs and insecticides, even as an additive in rocket propellant. The discovery of methylamine and the other four molecules at Parkes was made in collaboration with two chemists at Monash University in Melbourne. Ron Brown and Peter Godfrey made laboratory measurements of the characteristic radio lines of these molecules, before the search began at Parkes.[22]

Although the Parkes–Monash collaboration discovered a reasonable share of these new molecules, against fierce competition from the American groups, this was achieved only by pushing the Parkes telescope to its operational limits. Each of the five molecules has its characteristic wavelength in the range from 1 to 10 cm, and so could be detected only by using the inner steel plate section, the most accurately configured part of the dish. Most of the new molecules detected elsewhere, however, have radio lines shorter than 0.5 cm (5 millimetres), well below the operational capability of the Parkes dish.

To keep pace with this new field of millimetre radio astronomy, plans took shape at Radiophysics to construct a new telescope with a far greater surface accuracy. A proposal put forward early in 1973 called for the construction of a 25 m dish which could operate at wavelengths as short as 2 millimetres. The site for the telescope would be either at Parkes or at Culgoora, alongside the Radioheliograph. Although smaller than the Parkes dish, the telescope's exceptionally high precision presented a formidable engineering challenge. The projected cost of $3 million was close to the actual cost of the Parkes telescope ten years earlier. Predictably, the question of who would fund the project became the central issue. One proposal to build and operate the telescope as a joint Australian–Canadian facility looked particularly promising, but then in mid 1974 the National Research Council in Ottawa announced its withdrawal from the project. The proposal also faced stiff competition from two other expensive projects at this time—Sydney University's

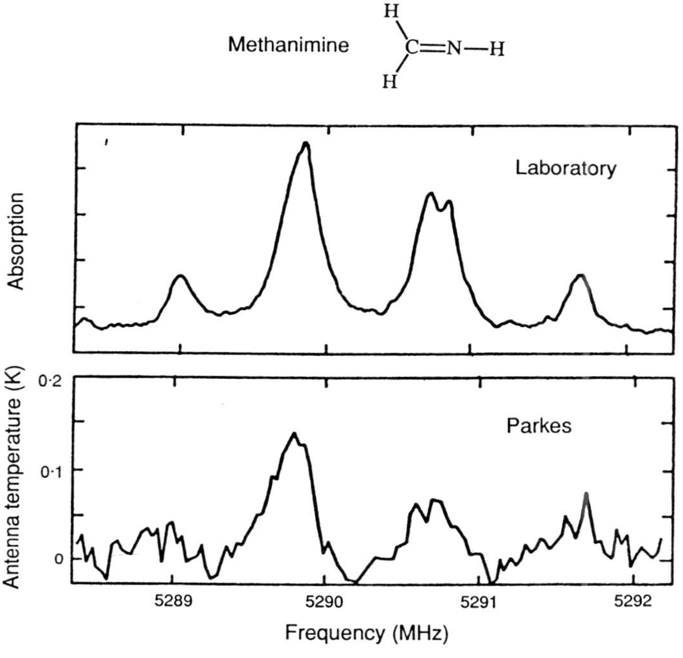

The pair of records that led to the discovery of interstellar methanimine in May 1972. The laboratory spectrum of the molecule measured at Monash University (above) matches closely with the spectrum of the source Sagittarius B2 recorded shortly after at Parkes.

stellar interferometer at Narrabri and the Australian National University's 1.5 m optical telescope planned for Siding Spring. Finally, a significant number of Australian astronomers believed that the next big project should be a large synthesis telescope, an idea that grew eventually into the Australia Telescope.

The proposal was shelved which meant simpler alternatives had to be pursued. One was the decision to continue reskinning the surface of the Parkes dish in successive stages (see Chapter 11). Upon completion of the third stage in 1975, the central section could operate down to wavelengths as short as 5 millimetres. Another plan led to a small dish designed especially for millimetre astronomy. Built by Krupp Industries of West Germany, the 4 metre dish came into operation in 1976. This telescope has been used in independent investigations at wavelengths as short as 2 millimetres, but in most cases these studies have been carried out in conjunction with the research program at Parkes.

Although the pace of new discoveries slowed significantly after the early part of 1970s, over seventy interstellar molecules are now known. A large proportion of them have been studied at Parkes, ranging from the simple diatomic and triatomic molecules through to the larger multi-atom molecules. Some such as the cyanoacetylene series are mysterious

The 4 metre telescope used for millimetre radio astronomy. The telescope is located in the northern Sydney suburb of Marsfield, the site of the Radiophysics Lab since its shift from the grounds of the University of Sydney in 1969. The operation of millimetre telescopes is limited more by local weather conditions than by man-made sources of radio interference, so there is not the same need to locate them in remote areas.

molecules which do not occur naturally on earth. Part of the research program has been to identify the family of radio lines associated with each molecule. Similar to the OH radical which has four lines, most molecules have more than one characteristic line. The seventy or so molecules currently known in fact give rise to almost one thousand radio lines, a sizable number of which were first detected at Parkes. Some lines have wavelengths as short as 1 millimetre, others as long as 40 centimetres. When we recall that until 1963 only one radio line was known—the famous 21 cm hydrogen line—the study of line radiation from interstellar matter has undergone an explosive growth.[23]

A number of these radio lines arise because of the presence of different atomic isotopes within a molecule. As one example, a research program at Parkes has been to study the family of lines from formaldehyde (HCHO). The wavelengths of its lines depend on whether the molecule contains the normal carbon-12 or the rarer carbon-13 isotope, the normal oxygen-16 or the rarer oxygen-17 or oxygen-18. The relative abundance of these rarer isotopes provides an insight into the types of nuclear processes occurring in a particular region of the galaxy. One interesting finding is that the rarer isotopes are more abundant near the centre of our galaxy than in the neighbourhood of the solar system. The normal isotopes such as carbon-12 and oxygen-16 were made long ago. Then, when interstellar matter containing the normal isotopes was processed through stars, the rarer isotopes were manufactured from the normal ones. The increased abundance of the rarer isotopes near the centre of the galaxy indicates that the interstellar matter in the inner parts has been recycled through more stars than matter in the outer parts.

Among such a long list it is no real surprise that several molecules give rise to the same strong maser radiation found for the OH radical. One is silicon oxide (SiO), another is the newly discovered methanol (CH_3OH) maser. For astronomers, the most important is water which emits intense radiation at 1.3 cm. Surveys of the galaxy at Parkes by Jim Caswell, Bob Batchelor, Rick Forster and Kel Wellington have located over half the H_2O masers presently known.[24] Most H_2O masers occur in association with an OH maser, though they are usually displaced slightly in their position. Similar to the OH masers, it seems highly likely that they pinpoint the position of embryonic stars in the same stellar nursery, but at a slightly different stage of development. Both types of masers are also found in the gaseous envelopes of old stars, and so they provide information on stellar evolution ranging from the youngest to the oldest stars.

Despite their extremely weak signals several simple molecules have been detected in other galaxies. The first organic molecule was formaldehyde, discovered by Frank Gardner and John Whiteoak at Parkes in the two galaxies NGC 253 and NGC 4945. Other molecules such as carbon monoxide, hydrogen cyanide, ammonia and the CH and HCO radicals have since been detected in members of our local group of galaxies. More recently, observations at Parkes have led to the detection of a H_2O maser in the Large Magellanic Cloud, one of the first extragalactic masers. So far very few complex molecules have been found, but there is every indication that other galaxies contain the same rich variety of interstellar material found in the Milky Way.[25]

The discovery of this wide variety of interstellar molecules has opened up a new science known as astrochemistry, a field that studies the chemical reactions occurring in gases of extremely low density, pressure and temperature, conditions which cannot be achieved in any terrestrial laboratory. The formation of the larger molecules takes long periods of time, the first step perhaps being triggered randomly by the ionisation of an atom by an energetic cosmic ray. Unlike simple molecules such as OH and H_2O, the stepwise building of the complex molecules can only occur within dark gas clouds. Here these molecules are shielded from the ultraviolet radiation of interstellar space which would rapidly disrupt the newly formed chemical bonds. These clouds are usually very large, with some containing material equivalent to one million solar masses. To the surprise of many astronomers, it appears that these clouds consist almost exclusively of hydrogen in its molecular form H_2. As distinct from atomic hydrogen with its characteristic 21 cm line, molecular hydrogen has no radio (or visual) lines and thus is invisible to astronomers. Recent estimates suggest that a substantial part of the mass of the Milky Way might exist in the form of invisible clouds of molecular hydrogen. These clouds might also account for some of the 'missing mass' of the universe referred to earlier. By studying the line radiation from a variety of complex molecules harbouring within a cloud, much can be learnt about its size and properties.

The most common molecule embedded within these giant clouds is usually carbon monoxide. Surveys carried out with the 4 m telescope at Marsfield have used the CO line as a tracer to detect the presence of these invisible clouds. Once located, the clouds can then be examined in detail for the presence of more complex molecules. The temperatures deep within the clouds are believed to be only about 50 degrees above absolute zero. The gas is also extremely rarefied within the clouds, with densities far below those achieved in the best laboratory vacuums. And yet, the vast size of these clouds means that even the rarest complex molecules are usually present in extraordinarily large amounts. To illustrate the point, on a not-so-sober note, mention can be made of the molecule ethyl alcohol, first detected by a group at the NRAO. The American group introduced their paper in the *Astrophysical Journal* with the observation:

> 'Ethyl alcohol has been of interest to mankind since the dawn of the earliest civilisations. During early October of 1974 we detected a truly astronomical source of ethyl alcohol located in the general direction of the centre of our galaxy. Preliminary estimates indicate that the alcohol content of this cloud (Sagittarius B2), if purged of all impurities and condensed, would yield approximately 10^{28} bottles at 200 proof. This exceeds the total amount of all man's fermentation efforts since the beginning of recorded history.' [26]

It is unlikely that any future astro-corporation will plan to tap into such a bountiful supply. The same cloud contains far greater quantities of water. The alcohol is in fact 'bottled' in the cloud at a mere 0.002 proof, an extremely feeble brew!

☆ ☆ ☆

The study of interstellar molecules through their characteristic line radiation has become a major field in contemporary astronomy. As we have seen, surveys of atomic hydrogen and simple molecules such as OH, H_2O and CO have shed new light on the *astrophysics* of the Milky Way and other galaxies. Similarly, studies of the larger interstellar molecules have given birth to the exciting new field of *astrochemistry*. More recently, the discovery of molecules of increasing complexity has given rise to the highly speculative field of *astrobiology*. A high proportion of interstellar molecules, some consisting of chains of eight or nine atoms, are carbon-based. In short, they are organic molecules. This has raised the intriguing question whether even larger organic molecules can form within these vast molecular clouds, leading perhaps to the existence of primitive organisms in interstellar space. In announcing the discovery of methanimine in 1972, the Parkes group wrote: 'Methanimine itself might be classified as a prebiological molecule, since it seems plausible that it could react with formic acid, HCOOH, also a constituent of interstellar clouds, to produce the simplest amino acid, glycine NH_2CH_2COOH.' [27]

If so, the next steps may not be so farfetched. Amino acids are the elementary building blocks of proteins which, in turn, are the key elements of living organisms. Several attempts to detect interstellar amino acids have been made at Parkes and elsewhere, so far without success. Another discouraging result has been the failure to find any of the benzene-type ring molecules. Nevertheless, the possible existence of interstellar biological material raises an issue of profound importance. Astrochemist Ron Brown has written:

> 'The question arises then, did the early chemical steps in the development of life on earth await lightning discharges in a primitive atmosphere or did the original nebula contain significant amounts of pre-biological material, perhaps more complex molecules than we have discovered so far? And if so, did they survive the rather torrid process of planetary condensation and accretion and finally dominate the production of biological materials?' [28]

The possible abundance of interstellar biological material has led some astronomers to go one step further and argue that life itself originated in space. One idea is that the earth was seeded several billion years ago after passing through a molecular cloud rich in simple organisms and other organic material. These tiny organisms would have gently floated down through the atmosphere, touched down in any number of places conducive to life, and begun the long chain of evolutionary history leading to the remarkable diversity and richness of life forms currently populating this planet.

The endpoint of this speculation goes far beyond even a question as fundamental as how life originated on earth. If in fact this prebiotic broth is so widespread, has intelligent life evolved elsewhere in the cosmos? Have these little green molecules given rise to little green people? A number of astronomers have attempted to find out directly, as we see Tangent 12.

The evidence is already overwhelming that the elementary ingredients of life are commonplace throughout our galaxy and, most likely, the universe beyond. Whether or not the second and third steps in the following sequence are also commonplace remains to be seen:

interstellar atoms	→	complex organic molecules	→	primitive organisms	→	advanced civilisations

Some time in the future it may be that a telescope such as the one at Parkes might, either by design or serendipity, detect a strange new signal which will answer the question. Perhaps, this is the most important of all challenges facing astronomy.

TANGENT 12

Seti, not Yeti

Speculation about life on other worlds is as old as astronomy itself. One famous false alarm was the belief that the canals visible on the surface of Mars were a vast irrigation project engineered by the planet's inhabitants. Radio astronomers have also had occasion to wonder whether they have stumbled upon signals from some extraterrestrial civilisation. The Cambridge astronomers were only half-joking when they labelled the first pulsars Little Green Men. It took several weeks of observations to rule out the possibility that these signals came from an unseen planet revolving around a distant star. A few years earlier, the discovery of OH masers led to similar speculation. Could these intense signals with their extremely narrow bandwidth arise in any natural way, or were they messages from distant civilisations?

Many astronomers have been in favour of a direct search for extraterrestrial intelligence, known as seti for short. Seti is not unlike the search for the half-man, half-animal supposed to inhabit the Himalayas. Everyone would like to believe in the yeti, but no one can produce the evidence.

The first seti proposal was made in 1959 by Giuseppe Cocconi and Philip Morrison. These two Cornell physicists estimated that, within a range of fifty light-years from earth, there might be as many as a hundred stars which possess habitable planets such as our own. The best chance of success seemed to be to tune in on the 21 cm hydrogen line, the fundamental radio line known presumably to any extraterrestrial astronomer.

Cocconi and Morrison wrote in *Nature*: 'The reader may seek to consign these speculations wholly to the domain of science fiction. We submit rather that the foregoing type of argument demonstrates that the presence of interstellar signals is entirely consistent with all we now know and that if signals are present the means of detecting them is now at hand. Few will deny the profound importance, practical and philosophical, which the detection of interstellar communication would have. We therefore feel that a discriminating search for signals deserves a considerable effort. The probability of success is difficult to estimate; but if we never search the chance of success is zero.' [29]

One astronomer to take the idea seriously was Frank Drake at Green Bank. In 1960 Drake made a series of observations at 21 cm using a 26 metre dish directed at a number of nearby stars. The project was named Ozma from the story 'The Wizard of Oz' about a faraway land inhabited by strange beings. Ozma of course failed to reveal messages from extraterrestrial civilisations, but the attempt marked the beginning of astrobiology.

The first seti attempt was made at Parkes in 1979. Trevor Cole and Ron Ekers suggested that instead of searching at sharp frequencies such as the hydrogen line, it might be better to search for pulsed signals. Much of our own radar and radio communication is by pulsed rather than continuous radiation. Over 50 hours of observing time were devoted to several globular clusters, X-ray sources, regions in the Magellanic Clouds, and the galactic centre. There was only one event which could not be readily explained as either a natural signal or some local source of man-made interference. In the direction of 82 Eri, a star 20 light-years distant, Cole and Ekers recorded a signal which they wrote is 'not attributable to lightning and is the only non-explicable event seen in three days of observing'. Later, a special four-hour session was devoted to 82 Eri but failed to register any further signals and the search at Parkes was abandoned.[30]

In 1989 a team including Ray Norris, Kel Wellington and Alan Wright started a new program, one which monitors a number of stars at several frequencies. Other searches are under way elsewhere, and undoubtedly even more technically advanced searches will be made in the future. Whatever the chances of success, and all agree they must be slim, there can be no doubt about the ultimate importance of the quest.

Over four centuries ago, Copernicus overthrew established dogma with the idea that mankind and the earth were not at the centre of the solar system. Astronomers from Galileo onwards then showed that the solar system is not at the centre of the Milky Way, dethroning the sun from the central position assigned to it by Copernicus. Early this century we learnt that the Milky Way is not at the centre of the universe, but only one of billions of galaxies. Astronomy continues to teach us humility. If seti is ever successful, it will be the next fundamental teaching of astronomy.

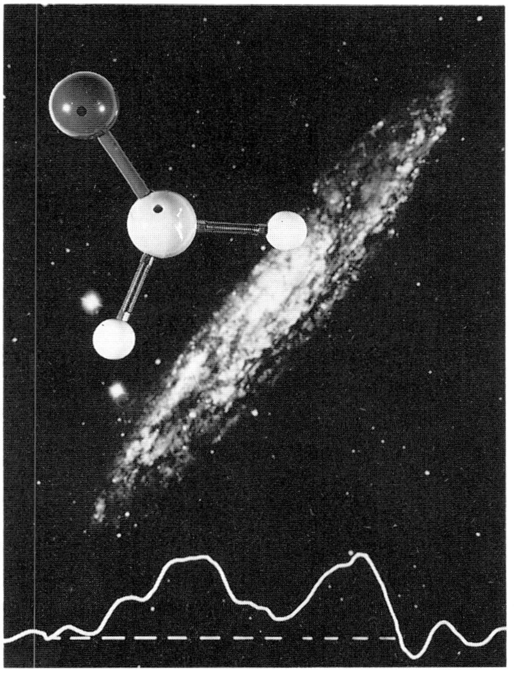

In 1974 Parkes astronomers detected signals from formaldehyde in NGC 253—the first organic molecule discovered in another galaxy.

Grand Old Lady

At the time of its inauguration in October 1961, the Parkes telescope was a spectacular advance on the aerials used in the early days of radio astronomy. Inventive arrangements of metal rods and wires had been enough to give Australia a flying start in the new astronomy. A small group of wartime radar specialists soon found itself a world leader in recording and deciphering the eerie radio noise from the heavens. Very little now remains from these early days. The rough and ready aerials were dismantled long ago, the field stations dotted around Sydney closed, and the pioneers have long since retired. Parkes was once the telescope of the future, but it is now the closest link we have with these early days, the golden era of Australian radio astronomy.

The place of these pioneers has been taken by a new generation of astronomers. Some like to refer to the Parkes telescope as the 'Grand Old Lady'. It is a name used with a hint of mischief, much affection, and is one not too far wide of the mark. Parkes is now the oldest radio telescope in operation in Australia and, after the Lovell telescope at Jodrell Bank, the next oldest large dish in the world. This ripe old age has been reached only because of a long line of improvements. The surface of the dish has been reskinned in several stages, facelifts which have enabled Parkes to keep pace with the younger telescopes of the world. Almost all of the electronic instrumentation and control system has been replaced in a series of surgical operations designed to improve the telescope's performance.

Although Parkes has enjoyed a full and active life, growing old gracefully, it was never expected that all this renewal and regeneration would keep the telescope at the frontier of astronomy indefinitely. The technology of radio telescopes was moving much too rapidly elsewhere. One of the most important innovations in the 1970s was the linking together of telescopes separated at large distances to give a dramatic increase in resolving power. The MERLIN array constructed by the Jodrell Bank group was the best example of this technique, known as long baseline interferometry. Another major development was the introduction of aperture synthesis, a technique where a number of small aerials are used to mimic a dish of very large diameter. The aerials are connected in interferometer pairs and moved, step by step, to all the spacings required to simulate a large aperture. The process is time consuming but, once all the signals are collected and processed by computer, radio maps are produced of exceptionally high resolution. The Cambridge group was the first to build these new synthesis telescopes, while the NRAO at Green Bank planned the largest one of all. The Very Large Array at Socorro in New Mexico can synthesise radio maps with a resolution equivalent to a single dish 27 kilometres in diameter!

In 1975 the Australians responded to these striking developments in the northern hemisphere with a proposal for a new instrument, known as the Australian Synthesis Telescope. The AST planned to use Parkes as its prime instrument, together with a number of medium-sized dishes mounted on carriages and movable along rail tracks. Despite the cost advantage of using the existing Parkes dish, the AST proposal faced an all too familiar problem. The government was asked to fund the whole project, the first occasion that this had happened. Until then each of the major radio telescopes built in Australia—Parkes, the SuperCross at Molonglo and the Radioheliograph at Culgoora—had been funded either wholly or partly by American foundations. Although the projected cost of the AST was much lower than any of the new synthesis instruments elsewhere, it was still too much for the government of the day to accept. A request for funding in 1981 was unsuccessful.

Australia's community of professional astronomers numbers about 200. It is a small group compared with the leading powers in astronomy in the northern hemisphere, small too compared with most other scientific and professional groups within Australia. Numbers aside, Australian astronomers have an impressive record when it comes to obtaining support for large and expensive telescopes. Although it has had its share of setbacks, internal divisions and disagreements, the astronomical community has presented a united front on the occasions when it counts. It has been possibly the most effective lobby group in Australian science.

The Australian Synthesis Telescope was a proposal to link the Parkes dish with a number of medium-sized dishes which could be moved along rail tracks. This sketch shows an early version of the proposal, an idea that eventually evolved into plans for the Australia Telescope.

When the Australian Synthesis Telescope failed to win government support in 1981, one might have expected the proposal to have been quietly put aside until a more opportune time, or perhaps to have been pruned back to a simpler, less costly alternative. Instead, Australian astronomers formulated a new and far bolder plan. Rather than building new dishes to accompany the main telescope at Parkes, a completely new array was planned for the site further north at Culgoora. The array would consist of five 22 metre dishes set out along a rail track with a sixth dish positioned six kilometres away. While these dishes would form a powerful synthesis telescope in their own right, this array could also be linked to a seventh dish at Siding Spring, as well as to the Parkes telescope 350 km to the south. Other dishes elsewhere in Australia could eventually be introduced to the array to form baselines almost the length and breadth of the island continent. The complete array, to be known as the Australia Telescope, would in most respects out-perform any of the northern arrays.

The boldness of the idea was the key to its success. The proposal won immediate support and was adopted by the government as one of several projects planned for the Australian Bicentenary. Now is not the time to recount the years of planning and construction—that might make an interesting book for someone to write in the future. The Australia Telescope was completed in time to celebrate the bicentennial year 1988, though the commissioning required to make the entire array fully operational continued for several years after.

The new array has consolidated Australia's position at the cutting edge of radio astronomy, a position which had been steadily eroded after the 1960s in the face of increasing competition from groups elsewhere and the rapid development of new techniques. With the completion of the Australia Telescope, now is a good opportunity to return to a question posed earlier in this book—should the Parkes telescope have been built? Was Taffy Bowen right?

The decision to proceed with Parkes in the 1950s shaped the course of Australian radio astronomy for the next three decades. In the showdown between two opposing philosophies, Parkes had been a victory for Bowen and those who favoured the versatility of a giant dish, at the expense of Joe Pawsey and those who favoured the development of new arrays with higher resolution. Few would now question the outstanding success of Parkes, but it came at the cost of winding down a line of development which Australia had so brilliantly pioneered. Since the 1960s, most of the revolutionary innovations have been the achievements of northern astronomers.

The construction of the Australia Telescope helps us to look at the decision to build Parkes in a new light. The new array uses the power of modern computers to merge the technology of large precision dishes with the techniques of aperture synthesis and long baseline interferometry. The two lines of development, at one time in conflict, turned out to be complementary. In the end, both Taffy Bowen's dream and Joe

Pawsey's dream were necessary to build the ultimate telescope of the 1990s.

As this book goes to press, the integration of Parkes into the Australia Telescope as its eighth and largest dish is still proceeding. Although its new role is bound to provide a new lease of life, much remains for Parkes to achieve on its own. As one example, astronomers at Parkes recently announced the discovery of a swarm of millisecond pulsars in the globular cluster 47 Tucanae, doubling the known number of these extraordinary objects. Another program has been a sensitive all-sky survey in collaboration with American astronomers. The survey at a wavelength of 6 cm has discovered thousands of new radio sources, vastly extending the number catalogued during the telescope's younger days. Parkes has now discovered and studied more radio sources than any other telescope.

The affectionate 'Grand Old Lady' may be premature. It seems Parkes is living life to the full, enjoying a busy and creative middle age. The Parkes Telescope stands as an enduring reminder of the dreamers who made it a reality and of the astronomers who continue to make it a success.

Notes to Part Three

CHAPTER 9

Epigraphs: Sciama, *Modern Cosmology*, p. vii; Gingerich, *The Early Years of Radio Astronomy* (Ed. W. T. Sullivan), p. 404; Silk, *The Big Bang*, p. 199; Hubble, see H. P. Robertson, *Scient. Am.* **195**(3) (1956), p. 81; Kraus, *Big Ear*, p. 166.

1 No recordings were kept of the first signals from Fornax A. It was however a cause for celebration, as Bolton recalls: 'At 3 am Brian Cooper and I drove back to town to Tatts Hotel and as we entered the town Brian said 'I hope there's someone still around to give us a beer.' As we turned the corner our prayers were answered—the lights were still on in the bar. Mine host and three large gentlemen were present. The publican legally served us our beer, being residents, and then said 'Let me introduce you to these gentlemen—Superintendent X of the Licensing Branch, Inspector Y of Betting and Gaming and Sergeant Z of the Vice Squad.' Needless to say, all four were breaking the law!' (letter to author 18/6/90).

2 RPA file A1/1/1, memo by Pawsey to radio astronomy group 29/6/59.

3 RPA file A1/3/11/17, memo by Westerhout 3/1/61.

4 F. Hoyle, *Galaxies, Nuclei and Quasars*, p. 48. Aside from the discouraging result with Cygnus, most polarisation searches failed because they were made at long wavelengths where the amount of polarisation falls away. Many of the stronger radio sources examined were later shown to be only weakly polarised.

5 Note 3, 'Research program for 210 ft radio telescope', Sept. 1959.

6 See e.g. R. N. Bracewell, B. F. C. Cooper and T. E. Cousins, 'Polarisation in the central component of Centaurus A' *Nature* **195** (1962) 1289; F. F. Gardner and J. B. Whiteoak, 'Polarisation of 20 cm wavelength radiation from radio sources' *Phys. Rev.* **9** (1962) 197.

7 Cooper to author 28/3/84. The discovery was reported in B. F. C. Cooper and R. M. Price, 'Faraday rotation effects associated with the radio source Centaurus A' *Nature* **195** (1962) 1084–5.

8 See e.g. F. F. Gardner and J. B. Whiteoak, 'Polarisation of radio sources and Faraday rotation effects in the Galaxy' *Nature* **197** (1963) 1162–4.

9 Interview with author, April 1984. An important part of these studies consisted of polarisation observations of the extended radio emission from the spiral arms of our galaxy. This led to the determination of the direction of the magnetic field in our local spiral arm—see D. S. Mathewson and D. K. Milne, 'A pattern in the large-scale distribution of galactic polarised radio emission' *Nature* **203** (1964) 1273–4. Similar observations were made by the Leiden group, providing independent evidence that continuum radiation from the galaxy is polarised and thus due to synchrotron emission.

10 B. F. C. Cooper, R. M. Price and D. J. Cole, 'A study of the decimetric emission and polarisation of Centaurus A' *Aust. J. Phys.* **18** (1965) 589–625.

11 See e.g. F. F. Gardner and J. B. Whiteoak, 'The polarisation of extended radio sources at 6 cm wavelength; I, II' *Aust. J. Phys.* **24** (1971) 899–911; 913–24.

12 For the first in the Parkes series on Jupiter see J. A. Roberts and M. M. Komesaroff, 'Evidence for asymmetry of Jupiter's Van Allen belt' *Nature* **203** (1964) 827–30.

13 T. A. Matthews, J. G. Bolton, J. L. Greenstein, A. R. Sandage and G. Münch, Report to the 107th American Astronomical Society, December 1960. See *Sky & Telescope* **21** (1961) 148.

14 The lunar occultation technique had first been applied in radio astronomy by Arthur E. Covington in Canada. During a solar eclipse in 1946 Covington showed

that strong radio outbursts emanate from the region of sunspots. For radio sources other than the sun, the technique requires a giant steerable dish which can track the moon across the sky for several hours at a time. As early as August 1955, in a report on possible research programs for the proposed Parkes telescope, Frank Kerr had suggested the study of 'occultations of radio sources by the moon'. Several years later the suggestion contributed to the discovery of quasars. See RPA file A1/3/11/2, Part B, 12/8/55.

15 Interview with author, March 1984.

16 Lovell, *Out of the Zenith*, p. 38.

17 For the four papers cited see *Nature* **197** (1963) 1037–41.

18 Interview with author, April 1984. See also J. G. Bolton, 'Radiophysics in exile: The 40th anniversary of extragalactic radio astronomy' *Proc. Astron. Soc. Aust.* **8** (1990) 381–3. Bolton makes the point that 3C 273 had not been examined at Caltech in 1960 because it lay too close to the strong source M87 which was used as a daily calibrator for the Owens Valley interferometer. If it had been observed, then because of its much lower redshift than 3C 48, the great distances of quasars might have been established in 1960, rather than 1963.

19 Hazard to author 28/3/91 and 27/6/91. See also Hazard in *Active Galactic Nuclei* (Ed. J. E. Dyson), pp. 1–19.

20 C. Hazard, M. B. Mackey and W. Nicholson, 'Additional identifications of radio sources with star-like objects' *Nature* **202** (1964) 227–8. The lunar occultation technique was also used with brilliant success in 1964 by Herbert Friedman to measure the angular size of the newly discovered X-ray source in the Crab Nebula. The detection of quasars by lunar occultations continued at Parkes for several years, principally by Rupert Clarke, a recent graduate from Cambridge. Eventually the technique fell from favour because of its inherent disadvantages. The technique is limited to those sources located in the narrow strip of sky traversed by the moon and, even then, years might pass before the occurrence of a suitable occultation. Further, a strong radio source such as 3C 273 is required otherwise the signal will be swamped by thermal radio emission from the moon itself. The exceptional pointing accuracy of the Parkes dish (approaching 6 arcsec) enabled most of the quasars subsequently discovered at Parkes to be detected by conventional means (see later in this chapter).

21 J. Kraus, *Big Ear*, p. 181.

22 A. Eddington, *The Expanding Universe*, p. 10.

23 F. Hoyle, *Nature of the Universe*, p. 112.

24 See note 18, 'Radiophysics in exile', p. 382.

25 H. C. Arp, J. G. Bolton and T. D. Kinman, 'A quasi-stellar object with a redshift of 2.22 and an unusual absorption spectrum' *Astrophys. J.* **147** (1967) 840–5.

26 B. A. Peterson, A. Savage, D. L. Jauncey and A. E. Wright, 'PKS 2000–330: A quasi-stellar source with a redshift of 3.78' *Astrophys. J.* **260** (1982) L27–9.

27 J. L. Greenstein, 'Quasi-stellar radio sources' *Scient. Am.* **209**(6) (1963) 54–62.

28 F. Graham-Smith, *Radio Astronomy*, p. 145.

29 F. J. Kerr, '210-ft radio telescope's first results' *Sky & Telescope* **24** (1962) 254–60.

30 J. G. Bolton, F. F. Gardner and M. B. Mackey, 'The Parkes catalogue of radio sources declination zone –20° to –60°' *Aust. J. Phys.* **17** (1964) 340–72.

31 RPA file BOL/3, Bolton to Bowen 2/1/65.

32 See e.g. J. G. Bolton, A. E. Wright and A. Savage, 'The Parkes 2700 MHz survey (14th part): Catalogue for declinations –4° to –15°, right ascensions 10h to 15h' *Aust. J. Phys. Astrophys. Suppl.* **32** (1979) no. 46.

33 See e.g. R. F. Haynes, J. L. Caswell and L. W. J. Simons, 'A catalogue of 5 GHz galactic plane sources' *Aust. J. Phys. Astrophys. Suppl.* **32** (1979) no. 48.

34 During the 1960s the identification work between Mt Stromlo and Parkes had been confined to a small number of plates taken on the 2 m telescope at Bolton's request,

as well as a number of photoelectric observations made with a 1 m telescope operated by the ANU at Siding Spring. The major area of collaboration between Mt Stromlo and Parkes during the 1960s focused on hydrogen-line studies of the galaxy, tying in with Bart Bok's principal area of interest, the structure and dynamics of the Milky Way.

35 S. C. B. Gascoigne in *Australian Science in the Making* (Ed. R. W. Home), p. 362. Gascoigne adds (p. 364): '... the AAT owes much to the Parkes telescope and perhaps even more to Taffy Bowen. Bowen was the first chairman of the AAT Board. He set up the AAT project office (on much the same lines as that at Parkes), guided the project through its first years, and had much to do with resolving the political problems of the early 1970s.' It was a fine swan-song to Bowen's career. The first project manager for the AAT was Mike Jeffery, the outstanding engineer who had taken on the same task at Parkes. Following Jeffery's premature death in 1969, Harry Minnett stepped into the position.

36 Note 35, p. 364. For a detailed history of the AAT, see Gascoigne *et al.*, *The Creation of the Anglo-Australian Observatory.*

37 Whiteoak to author 9/7/84.

CHAPTER 10

Epigraphs: Kennedy, see E. M. Walters, 'The partnership philosophy', p. 140, Ph.D. Thesis, Univ. Georgia, 1970; Armstrong, see *The Kennedy Space Center Story*, p. 133 (NASA, Florida, 1970); Jeffery, see A. L. Mackay, *Harvest of a Quiet Eye*, 81.4 (Inst. of Physics, Bristol, 1977).

1 Lovell, *The Story of Jodrell Bank*, p. 196.

2 See e.g. I. Southall, *Woomera*; F. I. Ordway (Ed.), *Advances in Space Science and Technology*, Vol. 7.

3 D. S. Greenberg, 'Big dish: How haste and secrecy helped Navy waste $63 million in race to build huge telescope' *Science* **144** (1964) 1111–12. Greenberg quoted a figure of $63 million but the actual expenditure on the project was $96 million; see e.g. *Sky & Telescope* **24** (1962) 132.

4 RPA file Z1/7, Bowen to White 29/7/60.

5 CSA files TZ797/31/3, Casey to Treasurer 16/6/59; TZ 797/31/2, Casey to Minister of Trade 29/7/59.

6 RPA file A1/3/11/9C, Bowen to D. W. R. McKinley 7/12/61.

7 RPA file A1/3/11/9, Rechtin to Bowen 10/5/66.

8 Kellermann to author 21/11/90.

9 See e.g. K. I. Kellermann, 'The thermal radio emission from Mercury, Venus, Mars, Saturn and Uranus' *Icarus* **5** (1966) 478–90.

10 Note 7, G. Truszynski to Bowen 27/11/68.

11 Interview with author, April 1984.

12 *New York Times* 25/7/69.

13 Note 7, Paine to Kenneth Anderson (Minister of Supply) 29/9/69. The Ministry of Supply had responsibility for providing support for NASA installations in Australia. Paine's request required the approval of Malcolm Fraser, at that time CSIRO minister.

14 Mike Jeffery from Freeman Fox supervised the roller track repairs. The remedial work on the tower two years later was also planned by Freeman Fox and carried out by Concrete Constructions, the Sydney firm which had built the tower. The work was supervised by Macdonald, Wagner and Priddle who later (minus Priddle) played a major role in the design of the Australia Telescope.

15 Note 11.

CHAPTER 11

Epigraphs: Paterson, *Clancy of the Overflow*; Fletcher, 'The history of radio astronomy—an Australian perspective', unpublished paper delivered to 1988 ANZAAS Congress; Bowen, 'Birth of a radio telescope' *Hemisphere* **5**(6) (1961), p. 10; Bolton, 'Prospects of astronomy in Australia' *Nature* **246** (1973), p. 283.

1 D. O. Edge and M. J. Mulkay, *Astronomy Transformed*, p. 235.
2 Slee to author 22/3/84. The first conclusive proof for strong radio bursts from flare stars was obtained at Parkes in November 1962 with observations on V371 Orionis (Wachmann's star). The flare was also observed optically with the Baker–Nunn camera at Woomera, and by two amateur astronomers in New South Wales. See O. B. Slee, L. H. Solomon and G. E. Patson, 'Radio emission from flare star V371 Orionis' *Nature* **199** (1963) 991–3.
3 Interview with author, April 1984.
4 R. M. Price in *Serendipitous Discoveries in Radio Astronomy* (Eds K. Kellermann and B. Sheets), p. 303.
5 Note 4, p. 302.
6 Milne to author 3/12/90. Alectown is the town closest to the telescope, not Parkes.
7 The collaboration between the Radiophysics and Hobart groups began with the mathematicians J. C. Jaeger and K. C. Westfold on the theoretical side and with the physicist G. R. A. Ellis on the observational side, and then continued with astronomers either based or trained in Hobart such as A. G. and K. B. Fenton, P. A. Hamilton, R. F. Haynes and P. M. McCulloch. As mentioned earlier, Grote Reber initiated much of the low frequency work in Tasmania after his arrival in 1955. Reber was however content to pursue his research in isolation and never applied for observing time at Parkes.
8 Norris to author 12/10/90.
9 Note 1, p. 66. Australian astronomers have difficulty participating in this internationalism. Although the 'tyranny of distance' is not as great as it was in the early days (see Tangent 3 in Chapter 4), travel involves much more time and expense. Northern astronomers routinely have their travel expenses included in their research grants, but often this is not the case in Australia. It is not uncommon for Australian astronomers to pay for overseas travel out of their own pockets.
10 The two firms were Associated Electrical Industries and Askania-Werke. Once again Mike Jeffery supervised the construction, while John Shimmins from Parkes spent six months at Algonquin as a trouble shooter during the commissioning of the telescope.
11 B. Lovell, *Out of the Zenith*, p. 69. The DSIR paper referred to on 'The future of radio astronomy in Great Britain' was coauthored in 1959 with Martin Ryle at Cambridge.
12 RPA file A1/3/11/9T, Bowen to Huruhata (Tokyo Astronomical Observatory) 23/4/70. For more on the discovery of the new line radiation from interstellar molecules, see Chapter 12.
13 Cooper to author 28/3/84.
14 Interview with author, March 1984.
15 S. C. B. Gascoigne in *Australian Science in the Making* (Ed. R. W. Home), p. 367.
16 Wright to author 24/10/90.
17 Six of the Nobel Laureates to visit Parkes up to 1970 were the physicists Luis Alvarez (Berkeley), Patrick Blackett (London), John Cockcroft (Cambridge), William Fowler (Caltech), Edward Purcell (Harvard) and John Van Vleck (Harvard). The seventh was Australian pathologist Howard Florey.
18 Day to author 9/4/84.
19 Bolton interview with author, April 1984. The visitors' cars were another problem—the ignition systems of Volkswagens and some trucks were particularly bad sources

of interference. On one occasion Bolton complained: '... nothing annoys me more at Parkes than having to wait until the last car leaves lovers' lane before I know whether I have equipment trouble or not' (see RPA file F1/4/BOL, letter to Bowen 12/12/65).

20 The opening of the $50,000 visitors' centre on 14 February 1969 was performed by the newly appointed science minister, Malcolm Fraser, and attended by the local federal member for Calare, J. A. England, a party of other parliamentarians, local officials, and prominent Australian scientists.

CHAPTER 12

Epigraphs: Bok, *The Milky Way*, p. 1; Norris & Roberts, Radiophysics Information Sheet 1984/44; Ryle *et al.*, *Search and Research*, p. 39; Bracewell, *Proc. Astron. Soc. Aust.* **6** (1986) 523; Davies, *Superforce*, p. 18 (Counterpoint, London, 1986); Feynman, *What do You care?*, p. 243 (Unwin Hyman, London, 1990).

1 A. Hewish, 'Pulsars' *Scient. Am.* **219**(4) (1968) 25–35.

2 Letter to author 7/5/84. For the first Cambridge paper on pulsars see A. Hewish, S. J. Bell, J. D. H. Pilkington, P. F. Scott and R. A. Collins, *Nature* **217** (1968) 709. The Cambridge group's decision to delay the announcement for several months caused almost as much uproar among astronomers as the discovery itself. A number of American groups were particularly outspoken in their criticism. Even after the announcement of the first pulsar in February 1968, when the possibility of remote civilisations had already been ruled out, the Cambridge group withheld the coordinates of the other pulsars until April. Many astronomers felt they had been unfairly denied the opportunity to begin observing these strange new objects. A contributing factor may have been the Cambridge group's earlier experience with the 2C sky survey (see Chapter 4). The announcement of a sensational discovery, only to see it discredited, had taught a painful lesson.

3 B. Y. Mills in *Astronomical Insights* (Ed. P. Pockley), p. 31. The Molonglo group proposed the contraction PSR (from pulsar) to designate the new objects. Thus, PSR 0833–45 for the Vela pulsar gives its location at right ascension (first four digits) and declination (last two digits). This notation was later adopted by international convention, in preference to the original practice of naming a pulsar after the observatory making the discovery.

4 V. Radhakrishnan, 'Polarisation and its message in astronomy' *Q. J. R. Astron. Soc.* **30** (1989) 181–94. In their polar cap model of pulsars, Radhakrishnan and Komesaroff drew on their work on Jupiter which was under investigation at this time. Jupiter's magnetic axis is tilted to its rotation axis, its radiation is strongly polarised and there is also the same sweep in the plane of polarisation—see Chapter 9.

5 For the first Vela glitch see V. Radhakrishnan and R. N. Manchester, *Nature* **222** (1969) 228–9; for the fifth glitch see P. M. McCulloch *et al.*, *Nature* **302** (1983) 319–21. Vela was 'caught in the act' with the eighth glitch, observed as it occurred on Christmas Day 1988 with a 14 m dish at the University of Tasmania's observatory near Hobart.

6 See P. T. Wallace *et al.*, 'Detection of optical pulses from the Vela pulsar' *Nature* **266** (1977) 692–4. The twelve authors of this paper were from the Royal Greenwich Observatory, the Anglo-Australian Observatory and Radiophysics. The precise radio position leading to the optical identification was made in a series of observations at Tidbinbilla, Fleurs and Molonglo. The discovery provides another good example of the increasingly collaborative nature of astronomy.

7 R. N. Manchester *et al.*, 'The second Molonglo pulsar survey—discovery of 155 pulsars' *Mon. Not. R. Astron. Soc.* **185** (1978) 409–21.

8 See e.g. D. H. Clark, J. H. Parkinson and J. L. Caswell, 'Is Cir X-1 a runaway binary?' *Nature* **254** (1975) 674–6.

9 For both the Parkes discoveries referred to see P. M. McCulloch *et al., Nature* **303** (1983) 307–8 and J. G. Ables *et al., Nature* **342** (1989) 158–61. Another group at Parkes has recently questioned the existence of this binary pulsar in 47 Tuc—see R. N. Manchester *et al., Nature* **345** (1990) 598–600. It is interesting to note that one early triumph of general relativity was Einstein's calculation in 1915 of an effect known as the advance in the periastron of the planet Mercury, known by astronomers to be 43 arcsec per century. The binary system in 47 Tuc, if it does exist, is expected to show an advance in periastron of 0.6 degrees per day, an effect almost 2 million times larger than that observed for Mercury! General relativity was tested on the first binary pulsar discovered in 1974 and, at least within observational errors, came through with flying colours.

10 See A. G. Lyne *et al.*, 'An eclipsing millisecond pulsar in the globular cluster Terzan 5' *Nature* **347** (1990) 650–2.

11 V. Radhakrishnan, Abstracts, Astron. Soc. Aust. meeting 1988.

12 See for example J. V. Hindman, 'A high resolution study of the distribution and motions of neutral hydrogen in the small cloud of Magellan' *Aust. J. Phys.* **20** (1967) 147–71; R. X. McGee and J. A. Milton, '21 cm hydrogen-line survey of the Large Magellanic Cloud, Part II' *Aust. J. Phys.* **19** (1966) 343–74.

13 See for example D. S. Mathewson, 'The Clouds of Magellan' *Scient. Am.* **254**(4) (1985) 84–92. The Magellanic Stream was discovered with the 1024 channel digital correlator mentioned in Chapter 11. Mathewson adds: 'It was its *first* night of operation at Parkes. The correlator had the usual teething problems and the rather attractive vacation student from Monash that I had working with me did nothing to keep the engineers' minds on their task. However, it eventually worked and hydrogen signals at exactly the predicted velocity were detected at twenty points where I suspected the Stream should be. It was a night I shall never forget. Of course it involved lots of mapping by both the 18 m and 64 m dishes before the discovery was announced in September 1973' (letter to author 8/11/90).

14 For measurements of *interstellar* hydrogen within the Virgo galaxies see B. J. Robinson and J. A. Koehler, *Nature* **208** (1965) 993. For *intergalactic* hydrogen in the Virgo cluster see B. J. Robinson, K. J. van Damme and J. A. Koehler, *Nature* **199** (1963) 1176–7. It is interesting to note that the intergalactic hydrogen was detected not by its emission, but by an absorption dip at 21 cm in the spectrum of the source M87 (Virgo A) (see later). Observations of a similar absorption dip at 21 cm in the spectrum of the newly discovered quasar 3C 273, which lies in the same direction as the Virgo cluster, provided confirmation of the discovery. One further point: the absorption dip for 3C 273 proved that this source must be at least as far, and probably beyond, the Virgo cluster. This observation was a fatal blow to the theory at this time that quasars are local galactic objects (see Chapter 9).

15 B. J. Robinson, 'Hydroxyl radicals in space' *Scient. Am.* **213**(1) (1965) 26–33.

16 The Radiophysics group had been aware of the deuterium line since 1948, following a letter to Pawsey in which Grote Reber suggested not only a search for this line but, more importantly, the 21 cm hydrogen line (see Chapter 4, note 18). The unsuccessful search at Dover Heights was made with the hole-in-the-ground telescope by Gordon Stanley and visiting American Robert Price. Interstellar deuterium was eventually detected, not as isolated atoms, but contained in various molecules. It has an abundance ratio to hydrogen of one in 5000, far less than the terrestrial ratio of one in 600. Part of the reason for its low abundance is that exposure to cosmic rays in interstellar space destroys the atom, a process known as spallation.

17 For the discovery of interstellar OH by the MIT group see S. Weinreb, A. H. Barrett, M. L. Meeks and J. C. Henry, *Nature* **200** (1963) 829. Confirmation of the discovery at Parkes was made by J. G. Bolton, K. J. van Damme, F. F. Gardner and B. J. Robinson, *Nature* **201** (1964) 279.

18 See F. F. Gardner, B. J. Robinson, J. G. Bolton and K. J. van Damme, *Phys. Rev. Lett.* **13** (1964) 3–5.

19 Interview with author, May 1985. For the discovery of the OH maser see H. Weaver, D. R. W. Williams, N. H. Dieter and W. T. Lum, *Nature* **208** (1965) 29–31.

20 L. Staveley-Smith, D. A. Allen, J. M. Chapman, R. P. Norris and J. B. Whiteoak, *Nature* **337** (1989) 625–7.

21 G. L. Verschuur, *The Invisible Universe: The Story of Radio Astronomy.* It should be noted that astronomers in the late 1930s had detected the first interstellar molecules (CH, CH^+ and CN) at visible wavelengths. However, these simple diatomic molecules did not create the same interest as the larger molecules discovered later by radio astronomers.

22 For the detection of interstellar thioformaldehyde see M. W. Sinclair, N. Fourikis, J. C. Ribes, B. J. Robinson, R. D. Brown and P. D. Godfrey, *Aust. J. Phys.* **26** (1973) 85–91; for methanimine see P. D. Godfrey, R. D. Brown, B. J. Robinson and M. W. Sinclair, *Astrophys. Lett.* **13** (1973) 119–21; for methylamine see N. Fourikis, K. Takagi and M. Morimoto, *Astrophys. J. Lett.* **191** (1974) L139–41.

23 Not included here are the so-called hydrogen recombination lines, another important area of interstellar spectroscopy studied at Parkes. As distinct from the familiar 21 cm line, there is also a series of radio lines associated with atomic hydrogen which, at sufficiently high temperatures, will break up into its constituent electron and proton. When the electron recombines with a proton it 'cascades' down through a series of quantum energy levels, emitting a series of photons at radio wavelengths. Extensive surveys at Parkes of these recombination lines, especially those at wavelengths of 2, 3.3 and 6 cm, have been used to study the hot gas clouds known as HII regions, and also to distinguish HII regions from supernova remnants.

24 See for example J. L. Caswell *et al., Aust. J. Phys.* **42** (1989) 331–44.

25 For the detection of extragalactic formaldehyde see F. F. Gardner and J. B. Whiteoak, *Nature* **247** (1974) 526–7; for the detection of the H_2O maser in the Large Magellanic Cloud see J. B. Whiteoak *et al., Mon. Not. R. Astron. Soc.* **205** (1983) 275–9.

26 See B. Zuckerman *et al., Astrophys. J.* **196** (1975) L99–102.

27 See note 22.

28 R. D. Brown in *In the Beginning: The Origin of Planets and Life* (Ed. J. P. Wild), p. 9. Foremost among supporters of the 'life cloud' theory have been Fred Hoyle and Chandra Wickramasinghe (see e.g. Chapter 2, note 18).

29 See G. Cocconi and P. Morrison, *Nature* **184** (1959) 844–6. For a report on Project Ozma see F. D. Drake, *Phys. Today* **14**(4) (1961) 40–6.

30 T. W. Cole and R. D. Ekers, 'A survey for sharply pulsed emissions' *Proc. Astron. Soc. Aust.* **3** (1979) 328–30.

Bibliography

Alpher, R. A., and Herman, R. 'Reflections on early work on 'big bang' cosmology' *Phys. Today* **41**(8) (1988) 24–34.

Ball, D. *A Suitable Piece of Real Estate: American Installations in Australia* (Hale & Iremonger, Sydney, 1980).

Barnard, M. *One Single Weapon* (Radiophysics, unpublished manuscript).

Barrett, A. H. 'Discovery of giant molecular clouds', In *Serendipitous Discoveries in Radio Astronomy* (Eds K. Kellermann and B. Sheets), pp. 280–90 (NRAO, Green Bank, 1983).

Bartusiak, M. *Thursday's Universe* (Times Books, New York, 1986).

Bogle, G. S. 'Masers — new amplifiers for microwaves' *Aust. Scientist* **1** (1961) 128–32.

Bok, B. J. 'Trends in astronomical research for Australia' *Aust. J. Sci.* **22** (1960) 331–3.

Bok, B. J. 'Proposed British Commonwealth Observatory' *Aust. J. Sci.* **26** (1963) 6–7.

Bok, B. J., and Bok, P. F. *The Milky Way* (Harvard Univ. Press, 1981).

Bolton, J. G. 'The Australian 210-foot reflector and its research program' *Proc. IRE (Aust.)* **24** (1963) 106–12.

Bolton, J. G. 'New spectrum for astronomy' *Discovery* **27**(5) (1966) 24–9.

Bolton, J. G. 'Prospects of astronomy in Australia' *Nature* **246** (1973) 282–4.

Bolton, J. G. 'The changing universe' *Aust. Physicist* **13** (1976) 129–33.

Bolton, J. G. 'Radio astronomy at Dover Heights' *Proc. Astron. Soc. Aust.* **4** (1982) 349–58.

Bolton, J. G. 'Radiophysics in exile: The 40th anniversary of extragalactic radio astronomy' *Proc. Astron. Soc. Aust.* **8** (1990) 381–3.

Boswell, R. W. 'Weapons Research Establishment, Salisbury' *Aust. J. Sci.* **23** (1960) 137.

Bowen, E. G. '210-foot radio telescope at Parkes' *Times Science Review* No. 57 (1965) 10–11.

Bowen, E. G. 'The pre-history of the Parkes 64-m telescope' *Proc. Astron. Soc. Aust.* **4** (1981) 267–73.

Bowen, E. G. 'The origins of radio astronomy in Australia', In *The Early Years of Radio Astronomy* (Ed. W. T. Sullivan), pp. 85–111 (Cambridge Univ. Press, 1984).

Bowen, E. G. *Radar Days* (Adam Hilger, Bristol, 1987).

Bowen, E. G. 'From wartime radar to postwar radio astronomy in Australia' *J. Elect. Electron. Eng. (Aust.)* **8** (1988) 1–11.

Bowen, E. G., and Minnett, H. C. 'The Australian 210-foot radio telescope' *Proc. IRE (Aust.)* **24** (1963) 98–105.

Brandt, J. C., and Maran, S. P. *New Horizons in Astronomy* (Freeman, San Francisco, 1972).

Brittain, J. E. 'The magnetron and the beginning of the microwave age' *Phys. Today* **38**(7) (1985) 60–4.

Brown, R. D. 'Organic matter in interstellar space', In *In the Beginning: The Origin of Planets and Life* (Ed. J. P. Wild), pp. 1–14 (Aust. Acad. Science, Canberra, 1974).

Burbidge, G. R., and Burbidge, E. M. *Quasi-stellar Objects* (Freeman, San Francisco, 1966).

Calder, N. *Radio Astronomy* (Phoenix House, London, 1958).

Cannon, R. D. 'Prospects of astronomy in Australia' *Nature* **246** (1973) 285–7.

Christiansen, W. N. 'The first decade of solar radio astronomy in Australia', In *The Early Years of Radio Astronomy* (Ed. W. T. Sullivan), pp. 113–31 (Cambridge Univ. Press, 1984).

Christiansen, W. N. 'History and propaganda in astronomy' *Proc. Astron. Soc. Aust.* **8** (1989) 96–101.

Christiansen, W. N. 'An omission in radio astronomy histories' *Q. J. R. Astron. Soc.* **30** (1989) 357–8.

Christiansen, W. N., and Mills, B. Y. 'J. L. Pawsey, Obituary' *Aust. Physicist* **1** (1964) 137–41.

Clark, R. W. *Tizard* (Methuen, London, 1965).

Clarke, R. W. 'Locating radio sources with the moon' *Scient. Am.* **214**(6) (1966) 30–41.

Cockburn, S., and Ellyard, D. *Oliphant: The Life and Times of Sir Mark Oliphant* (Axiom Books, Adelaide, 1981).

Cortright, E. (Ed.) *Apollo Expeditions to the Moon* (NASA, Washington DC, 1975).

Crowther, J. G. *The Cavendish Laboratory* (Macmillan, London, 1974).

Davies, G. 'Manned spaceflight anniversary: 25 years of achievement and disappointment' *Aust. Physicist* **23** (1986) 92–3.

Davies, R. D., and Palmer, H. P. *Radio Studies of the Universe* (Routledge & Kegan Paul, London, 1959).

Davies, S. 'R. v. d. R. Woolley in Australia' *Hist. Rec. Aust. Sci.* **6** (1984) 59–69.

Drake, F. D. 'Project Ozma' *Phys. Today* **14**(4) (1961) 40–6.

Eddington, A. *The Expanding Universe* (Cambridge Univ. Press, 1933).

Edge, D. O., and Mulkay, M. J. *Astronomy Transformed: The Emergence of Radio Astronomy in Britain* (Wiley, New York, 1976).

Evans, W. F. *History of the Radiophysics Advisory Board 1939–45* (microfiche) (CSIRO, Melbourne, 1970).

Evans, W. F. *History of the Radio Research Board 1926–45* (CSIRO, Melbourne, 1973).

Fletcher, N. H. 'The history of radio astronomy—An Australian perspective' (unpublished manuscript, 1988).

Gamow, G. *The Creation of the Universe* (Viking, New York, 1952).

Gascoigne, S. C. B. 'Astrophysics at Mt Stromlo: The Woolley era' *Proc. Astron. Soc. Aust.* **5** (1984) 597–605.

Gascoigne, S. C. B. 'Australian astronomy since the Second World War', In *Australian Science in the Making* (Ed. R. W. Home), pp. 345–73 (Cambridge Univ. Press, 1988).

Gascoigne, S. C. B., Proust, K. M., and Robins, M. O. *The Creation of the Anglo-Australian Observatory* (Cambridge Univ. Press, 1990).

Graham-Smith, F. *Radio Astronomy* (Penguin, London, 1966).

Graham-Smith, F. 'Early work on radio stars at Cambridge', In *The Early Years of Radio Astronomy* (Ed. W. T. Sullivan), pp. 237–48 (Cambridge Univ. Press, 1984).

Graham-Smith, F., and Lovell, B. 'On the discovery of extragalactic radio sources' *J. Hist. Astron.* **14** (1983) 155–65.

Greenberg, D. S. 'Big dish: How haste and secrecy helped Navy waste $63 million in race to build huge telescope' *Science* **144** (1964) 1111–12.

Greenstein, J. L. 'Quasi-stellar radio sources' *Scient. Am.* **209**(6) (1963) 54–62.

Greenstein, J. L. 'Optical and radio astronomers in the early years', In *Serendipitous Discoveries in Radio Astronomy* (Eds K. Kellermann and B. Sheets), pp. 79–88 (NRAO, Green Bank, 1983).

Greenstein, J. L. 'An astronomical life' *Annu. Rev. Astron. Astrophys.* **22** (1984) 1–35.

Hanbury Brown, R. 'Paraboloids, galaxies and stars: memories of Jodrell Bank', In *The Early Years of Radio Astronomy* (Ed. W. T. Sullivan), pp. 213–35 (Cambridge Univ. Press, 1984).

Hanbury Brown, R., and Lovell, B. *The Exploration of Space by Radio* (Chapman & Hall, London, 1962).

Hazard, C. 'The coming of age of QSOs', In *Active Galactic Nuclei* (Ed. J. E. Dyson), pp. 1–19 (Manchester Univ. Press, 1985).

Head, A. K. 'A new form for a giant radio telescope' *Nature* **179** (1957) 692–3.

Henbest, N. 'Jodrell under Merlin's spell' *New Scient.* **96** (1982) 442–8.

Henbest, N. (Ed.) *Observing the Universe* (Blackwell–New Scientist, London, 1984).

Henbest, N. 'Astronomy is on a new track in Australia' *New Scient.* **115** (1987) 45–51.

Hewish, A. 'Pulsars' *Scient. Am.* **219**(4) (1968) 25–35.

Hewish, A. 'Pulsars and high density physics' *Science* **188** (1975) 1079–83.

Hey, J. S. *The Evolution of Radio Astronomy* (Elek Science, London, 1973).

Hey, J. S. *The Radio Universe* (Pergamon, Oxford, 1983).

Higgs, A. J. 'Quasars and the Parkes Radio Telescope' *Hemisphere* **10**(6) (1966) 2–8.

Higgs, A. J. *Radio Astronomy* (Radiophysics, unpublished, 1968).

Home, R. W. 'Origins of the Australian physics community' *Historical Studies* **20** (1983) 383–400.

Home, R. W. 'Science on service, 1939–45', In *Australian Science in the Making* (Ed. R. W. Home), pp. 220–51 (Cambridge Univ. Press, 1988).

Hoyle, F. *Nature of the Universe* (Blackwell, Oxford, 1950).

Hoyle, F. *Frontiers of Astronomy* (Heinemann, London, 1955).

Hoyle, F. *Galaxies, Nuclei and Quasars* (Harper & Row, New York, 1965).

Hyland, A. R., and Faulkner, D. J. 'From the sun to the universe — The Woolley and Bok directorships at Mt Stromlo' *Proc. Astron. Soc. Aust.* **8** (1989) 216–28.

Jansky, C. M. 'The beginnings of radio astronomy' *Amer. Scient.* **45**(1) (1957) 5–12.

Jeffery, M. H. 'Construction and operation of the 210-ft radio telescope at Parkes, Australia' *Ann. New York Acad. Sci.* **116** (1964) 62–88.

Jones, G. 'Sir Barnes Wallis: Inventor without a monument' *New Scient.* **84** (1979) 434–5.

Kellermann, K., and Sheets, B. (Eds) *Serendipitous Discoveries in Radio Astronomy* (NRAO, Green Bank, 1983).

Kerr, F. J. 'Australia's 210-foot radio telescope project' *Sky & Telescope* **18** (1959) 666–8.

Kerr, F. J. '210-ft radio telescope's first results' *Sky & Telescope* **24** (1962) 254–60.

Kerr, F. J. 'Reminiscences of life in the Radiophysics Laboratory' (unpublished manuscript, 1983).

Kerr, F. J. 'Early days in radio and radar astronomy in Australia', In *The Early Years of Radio Astronomy* (Ed. W. T. Sullivan), pp. 133–45 (Cambridge Univ. Press, 1984).

Kraus, J. *Big Ear* (Cygnus–Quasar Books, Ohio, 1976).

Krisciunas, K. *Astronomical Centers of the World* (Cambridge Univ. Press, 1988).

Lovell, B. 'Memoir of J. L. Pawsey' *Biog. Mem. R. Soc. London* **10** (1964) 229–43.

Lovell, B. *Our Present Knowledge of the Universe* (Melbourne Univ. Press, 1967).

Lovell, B. *The Story of Jodrell Bank* (Oxford Univ. Press, 1968).

Lovell, B. *Out of the Zenith* (Harper & Row, New York, 1973).

Lovell, B. 'The effects of defence science on the advance of astronomy' *J. Hist. Astron.* **8** (1977) 151–73.

Lovell, B. 'The origins and early history of Jodrell Bank', In *The Early Years of Radio Astronomy* (Ed. W. T. Sullivan), pp. 193–211 (Cambridge Univ. Press, 1984).

Lovell, B. *Astronomer by Chance* (Basic Books, New York, 1990).

Lovell, B., and Clegg, J. A. *Radio Astronomy* (Chapman & Hall, London, 1952).

Lovell, B., and Lovell, J. *Discovering the Universe* (Ernest Benn, London, 1963).

McClain, E. F. 'The 600-foot radio telescope' *Scient. Am.* **202**(1) (1960) 45–51.

McGee, R. X. 'Some historical aspects of the Division of Radiophysics', Information Sheet No. 1C (Radiophysics, 1982).

Macvey, J. W. *Whispers from Space* (Paladin, London, 1976).

Marsh, P. *The Space Business* (Penguin, London, 1985).

Mathewson, D. S. 'The Clouds of Magellan' *Scient. Am.* **252**(4) (1985) 84–92.

Mellor, D. P. *The Role of Science and Industry*, Series 4, Vol. 5 of *Australia in the War of 1939–45* (Australian War Memorial, Canberra, 1958).

Messel, H. 'The new radio astronomy centre of the University of Sydney' *Nature* **188** (1960) 528–9.

Millar, D. D. (Ed.) *The Messel Era* (Pergamon, Sydney, 1987).

Mills, B. Y. 'Radio sources and the log*N*–log*S* controversy', In *The Early Years of Radio Astronomy* (Ed. W. T. Sullivan), pp. 147–65 (Cambridge Univ. Press, 1984).

Mills, B. Y. 'Australian contribution to the science of radio astronomy' *J. Elect. Electron. Eng. (Aust.)* **8** (1988) 12–33.

Minnett, H. C. 'The Australian 210-foot radio telescope' *Sky & Telescope* **24** (1962) 184–9.

Minnett, H. C. 'The radio telescope at Parkes' (Radiophysics, unpublished manuscript, 1989).

Minnett, H. C., Roberts, J. A., Robinson, B. J., and Wild, J. P. 'Proposal for an Australian radio telescope for millimetre wavelengths' (Radiophysics, 1973).

Minnett, H. C., and Yabsley, D. E. 'Performance of the Parkes 210 foot paraboloid at wavelengths from 6 to 75 cm' *Proc. IREE (Aust.)* **27** (1966) 304–12.

Morpurgo, J. E. *Barnes Wallis — A Biography* (Penguin, London, 1973).

Norris, R. P. 'Encounter with Halley's Comet' *Aust. Physicist* **26** (1986) 132–4.

Ordway, F. I. (Ed.) *Advances in Space Science and Technology*, Vol. 7 (Academic, New York, 1965).

Osterbrock, D. E. 'Rudolph Minkowski: Observational astrophysicist' *Phys. Today* **38**(4) (1985) 50–7.

Palmer, H. P. 'The enigmatic quasars' *New Scient.* **25** (1965) 16–19.

Pawsey, J. L. 'Radio astronomy in Australia' *R. Astron. Soc. Canada* **47** (1953) 137–52.

Pawsey, J. L. 'Radio astronomy' *Aust. J. Sci.* **18** (3A) (1956) 27–35.

Pawsey, J. L. 'Vistas in radio astronomy', In *AAS Year Book*, pp. 65–80 (Aust. Acad. Science, Canberra, 1957).

Pawsey, J. L. 'Australian radio astronomy: How it developed in this country' *Aust. Scientist* **1** (1961) 181–6.

Pawsey, J. L. 'Introduction to the Radio Astronomy Issue' *Proc. IRE (Aust.)* **24** (1963) 95–7.
Pawsey, J. L., and Bracewell, R. N. *Radio Astronomy* (Oxford Univ. Press, 1955).
Pfeiffer, J. *The Changing Universe: The Story of the New Astronomy* (Scientific Book Club, London, 1956).
Piddington, J. H. *Radio Astronomy* (Hutchison, London, 1961).
Piddington, J. H., and Oliphant, M. L. E. 'Biographical memoir: D. F. Martyn' *Rec. Aust. Acad. Sci.* **2** (1971) 47–60.
Pockley, P. (Ed.) *Astronomical Insights* (Australian Broadcasting Commission, Sydney, 1970).
Price, R. M. 'The first years at Parkes', In *Serendipitous Discoveries in Radio Astronomy* (Eds K. Kellermann and B. Sheets), pp. 300–6 (NRAO, Green Bank, 1983).
Pugsby, A. 'Barnes Neville Wallis 1887–1979' *Biog. Mem. R. Soc. London* **27** (1981) 603–28.

Radiophysics. *A Proposal for a Giant Radio Telescope* (Radiophysics, Sydney, 1955).
Radiophysics. *Research Activities of the CSIRO Division of Radiophysics*, Years 1952, 1961/62, 1963/64, 1968, 1973, 1982 (Radiophysics, Sydney).
Radiophysics. *The 210-foot Radio Telescope and Apollo 11* (Radiophysics, Sydney, 1970).
Ralph, W. 'Looking for the edge of the universe' *Ecos* **34** (1982) 11–15.
Reber, G. 'Radio astronomy between Jansky and Reber', In *Serendipitous Discoveries in Radio Astronomy* (Eds K. Kellermann and B. Sheets), pp. 71–8 (NRAO, Green Bank, 1983).
Reber, G. 'A play entitled the beginning of radio astronomy' *J. R. Astron. Soc. Canada* **82** (1988) 93–106.
Rivett, R. *David Rivett: Fighter for Australian Science* (Rivett, Melbourne, 1972).
Roberts, J. A. (Ed.) *Radio Astronomy*, CSIRO Research for Australia Series No. 5 (CSIRO, Canberra, 1985).
Robertson, H. P. 'The Universe' *Scient. Am.* **195**(3) (1956) 73–81.
Robertson, P. 'John Bolton and Australian astronomy' *Aust. Physicist* **21** (1984) 178–80.
Robertson, P. 'Grote Reber: Last of the lone mavericks' *Search* **17** (1986) 118–21.
Robinson, B. J. 'Discovery of intergalactic hydrogen in the Virgo cluster of galaxies' (Radiophysics, unpublished manuscript, 1963).
Robinson, B. J. 'Hydroxyl radicals in space' *Scient. Am.* **213**(1) (1965) 26–33.
Rowan-Robinson, M. 'The great quasar odyssey' *New Scient.* **96** (1982) 305–8.
Ryle, M. 'Radio galaxies' *Scient. Am.* **195**(3) (1956) 205–14.
Ryle, M., Kurti, N., and Boyd, R. L. *Search and Research* (Mullard Ltd, London, 1971).

Saward, D. *Bernard Lovell: A Biography* (Hale, London, 1984).
Schedvin, C. B. 'The culture of CSIRO' *Aust. Cultural Hist.* **2** (1982) 76–89.
Schedvin, C. B. *Shaping Science and Industry: A History of Australia's Council for Scientific and Industrial Research, 1926–49* (Allen & Unwin, Sydney, 1987).
Schmidt, M. 'Discovery of quasars', In *Serendipitous Discoveries in Radio Astronomy* (Eds K. Kellermann and B. Sheets), pp. 171–4 (NRAO, Green Bank, 1983).
Sciama, D. W. *Modern Cosmology* (Cambridge Univ. Press, 1971).
Silk, J. *The Big Bang: The Creation and Evolution of the Universe* (Freeman, San Francisco, 1980).
Southall, I. *Woomera* (Angus & Robertson, Sydney, 1962).
Southworth, G. C. 'The early history of radio astronomy' *Amer. Scient.* **45** (1957) 13–27.
Steinberg, J. L., and Lequex, J. *Radio Astronomy* (McGraw–Hill, New York, 1963).
Sullivan, W. T. (Ed.) *Classics in Radio Astronomy* (Dover, New York, 1982).
Sullivan, W. T. 'Karl Jansky and the beginnings of radio astronomy', In *Serendipitous Discoveries in Radio Astronomy* (Eds K. Kellermann and B. Sheets), pp. 39–56 (NRAO, Green Bank, 1983).
Sullivan, W. T. 'Karl Jansky and the discovery of extraterrestrial radio waves', In *The Early Years of Radio Astronomy* (Ed. W. T. Sullivan), pp. 3–42 (Cambridge Univ. Press, 1984).
Sullivan, W. T. 'Frank Kerr and radio waves: From wartime radar to interstellar atoms', In *The Outer Galaxy* (Eds L. Batz and F. J. Lockman), pp. 268–87 (Springer, Berlin, 1988).
Sullivan, W. T. 'Early years of Australian radio astronomy', In *Australian Science in the Making* (Ed. R. W. Home), pp. 308–44 (Cambridge Univ. Press, 1988).
Sullivan, W. T. *A History of Radio Astronomy* (Cambridge Univ. Press, in press).

Taylor, J. H., and Manchester, R. N. *Pulsars* (Freeman, San Francisco, 1977).
Tindall, R. T. (Ed.) *Parkes: One Hundred Years of Local Government* (Parkes Centenary Book Committee, 1983).

Toulmin, S., and Goodfield, J. *The Fabric of the Heavens* (Scientific Book Club, London, 1961).

van de Hulst, H. C. (Ed.) *Radio Astronomy* (Cambridge Univ. Press, 1957).

Verschuur, G. L. *The Invisible Universe: The Story of Radio Astronomy* (English Universities Press, 1974).

Walters, E. M. 'The partnership philosophy: Australian–American space tracking relations', Ph.D. thesis, Univ. Georgia, 1970.

Westerhout, G. 'The early history of radio astronomy' *Ann. New York Acad. Sci.* **198** (1972) 211–18.

Westfold, K. C. (Ed.) *Review of Observatories: Report of Subcommittee* (Aust. Govnt Printing Service, Canberra, 1978).

White, F. W. G. 'Biographical memoir of J. P. V. Madsen' *Rec. Aust. Acad. Sci.* **2** (1970) 51–66.

White, F. W. G. 'Biographical memoir: Casey of Berwick and Westminster, Baron' *Rec. Aust. Acad. Sci.* **3** (1977) 54–83.

White, F. W. G. 'Robert Gordon Menzies 1894–1978' *Biog. Mem. R. Soc. London* **25** (1979) 445–76.

White, F. W. G., and Huxley, L. G. H. 'Radio research, Australia 1927–39' *Rec. Aust. Acad. Sci.* **3** (1974) 7–29.

White, F. W. G., and Kimpton, D. *Surprise and Enterprise: Fifty Years of Science for Australia* (CSIRO, Canberra, 1976).

Whiteoak, J. B. 'The Australia Telescope' *Aust. Physicist* **25** (1988) 174–7.

Wild, J. P. 'The Australian National Radio Astronomy Observatory' *Nature* **192** (1961) 1019.

Wild, J. P. 'Instrumentation for radio astronomy' *Phys. Today* **19**(7) (1966) 28–40.

Wild, J. P. 'The exploration of the sun by radio' *Aust. Physicist* **5** (1968) 117–22.

Wild, J. P. 'The beginnings of radio astronomy in Australia' *Rec. Aust. Acad. Sci.* **2** (1972) 52–61.

Wild, J. P. 'The beginnings of radio astronomy in Australia' *Proc. Astron. Soc. Aust.* **7** (1987) 95–102.

Wilson, J. L. J. (Ed.) 'Radio astronomy' *Current Affairs Bull.* **16** (1955) 114–28.

Wilson, J. L. J. (Ed.) 'Radio telescope' *Current Affairs Bull.* **28** (1961) 146–60.

Woolley, R. v. d. R. 'Opportunities for astrophysical work in Australia', Report of ANZAAS Meeting, Perth 1947 (Ed. A. D. Ross), pp. 64–79 (Govnt Printer, Perth, 1948).

Wright, A. 'The Australia Telescope' *Hemisphere* **27** (1983) 272–6.

Wright, A., and Wright, H. *At the Edge of the Universe* (Ellis Horwood, Chichester, 1989).

Photographic credits

The majority of photographs were supplied by courtesy of the CSIRO Division of Radiophysics. I also acknowledge the following organisations and individuals who kindly granted permission to reproduce their photographs: American Institute of Physics Niels Bohr Library (New York) pp. 90(right), 119(2), 233, 239; Anglo-Australian Telescope Board pp. 240, 248, 275; Argent Studios (London) p. 153; Australian Academy of Science p. 128(left); E. G. Bowen p. 20(2); Cavendish Laboratory (Cambridge) pp. 95, 306(above); CSIRO Archives pp. 27(below), 61, 197; CSIRO Editorial Services p. 46; Daily Telegraph (Sydney) p. 121; Defence Science & Technology Organisation (Salisbury) p. 255; F. Graham-Smith (copyright B. Elsmore) p. 53; Institute of Astronomy (Cambridge) pp. 90(left), 313(left); B. Lovell p. 143; D. S. Mathewson p. 317(below); D. K. Milne pp. 171, 312; Mt Stromlo & Siding Spring Observatory p. 110(2); National Radio Astronomy Observatory (Green Bank) pp. 11(below), 15(2); Parliament House Collection (portrait by Julian Ashton) p. 165; J. A. Roberts pp. 226(below), 229; D. E. Robertson p. 306(below); Royal Signals & Radar Establishment (Worcestershire) p. 39; Science Foundation for Physics (Sydney) pp. 128(right), 193(2); W. T. Sullivan, III p. 62; Sydney Morning Herald pp. 59, 92, 232, 237; Vickers (London) p. 145; A. E. Wright p. 246.

Index

Page numbers in *italics* refer to illustrations. Numbers accompanied by the letter n indicate notes on the text.